Springer Series in Statistics

Advisors:
P. Bickel, P. Diggle, S. Fienberg, K. Krickeberg,
I. Olkin, N. Wermuth, S. Zeger

Springer

New York
Berlin
Heidelberg
Barcelona
Hong Kong
London
Milan
Paris
Singapore
Tokyo

Springer Series in Statistics

(continued after index)

Barry C. Arnold
Enrique Castillo
José María Sarabia

Conditional Specification of Statistical Models

With 54 Illustrations

 Springer

Barry C. Arnold
Department of Statistics
University of California–Riverside
Riverside, CA 92521
USA

Enrique Castillo
Department de Mathematica Aplicanda y
   Ciencas de las Computacion
University of Cantabria
39005 Santander
Spain
casti@ccaix3.unican.es

José María Sarabia
Departmento de Economia (Statistics)
Facultad de Ciencas Econimicas y EE
University of Cantabria
39005 Santander
Spain
sarabia@ccaix3.unican.es

Library of Congress Cataloging-in-Publication Data
Arnold, Barry C.
    Conditional specification of Statistical Models/
  Barry C. Arnold, Enrique Castillo, José María Sarabia.
      p.   cm. — (Springer series in statistics)
    Expanded, up-to-date version of 1992 ed.
    Includes bibliographical references and index.
    ISBN 0-387-98761-4 (hc.: alk. paper)
    1. Distribution (Probability theory)   I. Castillo, Enrique, 1946–   .
  II. Sarabia, José María.   III. Title.   IV. Series.
  QA273.6.A76   1999
  519.2′4—dc21   99-30378

Printed on acid-free paper.

Production managed by Frank McGuckin; manufacturing supervised by Jeffrey Taub.
Camera-ready copy prepared from the authors' LaTeX files using Springer's svsing.sty macro.
Printed and bound by Maple-Vail Book Manufacturing Group, York, PA.
Printed in the United States of America.

9 8 7 6 5 4 3 2 1

ISBN 0-387-98761-4 Springer-Verlag New York Berlin Heidelberg   SPIN 10710885

Barry: To my Dad.
Enrique: To my wife Mary Carmen.
José María: To my son José María.

To Janos Aczél, whose work on functional equations
made this book possible.

Preface

This book is an expanded up-to-date version of our Lecture Notes in Statistics monograph entitled *Conditionally Specified Distributions*. Chapters in that monograph have been edited, expanded, and brought up-to-date (1998).

The concept of conditional specification of distributions is not new but, except in normal families, it has not been well developed in the literature. Computational difficulties undoubtedly hindered or discouraged developments in this direction. However, such roadblocks are of diminished importance today. Questions of compatibility of conditional and marginal specifications of distributions are of fundamental importance in modeling scenarios. Such issues are carefully analyzed in this book. Building on a normal conditionals model, which dates back at least to Bhattacharyya (1943), a broad spectrum of conditionally specified models is developed. Models with conditionals in exponential families are particularly tractable and provide useful models in a broad variety of settings.

Chapter 1 covers basic results on characterization of compatibility of conditional distributions and uniqueness of the corresponding joint distribution in a variety of settings. In addition, important functional equation results are presented. These prove to be basic tools in subsequent development of families of distributions with conditionals in specified parametric families. Chapter 2 is focussed on the finite discrete case. In it, a variety of compatibility and near-compatibility results are described.

Especially in Bayesian prior elicitation contexts, inconsistent conditional specifications are to be expected. In such situations, interest will center on most nearly compatible distributions. That is, distributions whose conditionals differ minimally from those given by the informed expert providing prior information. Progress in this area is also discussed in Chapter 2.

Chapter 3 includes a careful development of the normal conditionals model. Chapter 4 treats conditionals in prescribed exponential familes. A spectrum of conditionally specified distributions not involving exponential families is studied in Chapter 5. Chapter 6 focusses on the utility of certain improper conditionally specified models which appear to have predictive utility despite their failure to qualify as genuine joint distributions. The situation here is somewhat analogous to cases in which Bayesians fearlessly use nonintegrable priors and density estimators use estimates which are sometimes negative to estimate positive quantities. Chapter 7 discusses distributional characterizations involving a mixture of regression assumptions and conditional specifications. In certain circumstances such a mixture of assumptions might well describe a researcher's views on the nature of the joint distribution and it is of interest to determine what classes of distributions are determined by such assumptions. Much of the material in Chapters 1-7 can be extended to higher dimensions, at a price of increased notational complexity. Details of such extensions are provided in Chapter 8. In addition, certain characterizations of classical multivariate distributions via conditional specification are described.

Inference procedures for conditionally specified models require creativity and/or computer intensive approaches. Nevertheless, relatively efficient straightforward approaches are possible in some cases. Estimation techniques, both classical and Bayesian, are discussed in Chapter 9. Conditional specification provides a broad spectrum of feasible multivariate models. Appropriate inferential techniques for these models require considerable further development. There are still many open questions.

Simulations for conditionally specified distributions provide a tailor-made scenario for use of the Gibbs sampler. Discussion of this and other relevant material on simulation is gathered in Appendix A.

In Chapter 10, we discuss the general problem of specifying a multivariate distribution using marginal and/or conditional densities. Compatibility and uniqueness issues here are necessarily somewhat more complicated than they are in two dimensions.

There are several alternative conditional specification routes that can be considered. The conditional density approach has been most commonly studied; but, for example, in reliability contexts, conditional survival function specification might be more appropriate. This and related models are surveyed in Chapter 11. In Chapter 12, models for bivariate extremes are discussed.

Conditionally specified distributions can often play roles as natural flexible conjugate prior families in many standard data analysis settings. In this context, the availability of the Gibbs sampler is particularly convenient in allowing easy simulations of posterior distributions. The development of such flexible informative multiparameter priors is documented in detail in

Chapter 13. The relationship between conditionally specified models and simultaneous equations models is discussed in Chapter 14. Chapter 15 is somewhat unstructured, containing as it does, a collection of conditional specification topics that are either only partially developed or just didn't seem to fit in elsewhere.

In some sections of this book, ample references are given. The absence of references in a particular section sometimes indicates that the material is appearing for the first time. At other times the material may be a close paraphrase of earlier papers of the authors and references are not explicitly given. At the end of each chapter we provide brief bibliographic commentaries to help the reader put the material in context and to provide access to references for further reading.

We are grateful to the University of Cantabria, the University of Castilla-La Mancha, and to the Spanish CICYT for financial assistance during the past seven years that we have worked in this area. Special thanks are given to Iberdrola and José Antonio Garrido for their finantial support.

Several years ago, after hearing one of the authors give an introductory talk on conditionally specified models, Mike Hidiroglou came up and said:

"Well, you seem to have found a nice sand-box to play in."

He was right. We invite you to join us.

Riverside, CA   Barry C. Arnold
Santander, Spain   Enrique Castillo
Santander, Spain   José María Sarabia

August 1999

Contents

1

Conditional Specification: Concepts and Theorems

1.1 Why Conditional Specified Models?

In efforts to specify bivariate probability models, the researcher is frequently hindered by an inability to visualize the implications of assuming that a given bivariate family of densities will contain a member which will adequately describe the given phenomenon. We contend that it is often easier to visualize conditional densities or features of conditional densities than marginal or joint densities. Thus to cite a classical example, it is not unreasonable to visualize that in some human population, the distribution of heights for a given weight will be unimodal with the mode of the conditional distribution varying monotonically with weight. Similarly, we may visualize a unimodal distribution of weights for a given height, this time with the mode varying monotonically with the height. It is not as easy to visualize features of the appropriate joint distribution; unless we blindly follow Galton's assertion that a unimodal bivariate distribution with elliptical contours is clearly appropriate. Actually, for some of Galton's data sets, the elliptical nature of the contours is far from self-evident. The point is that, even for the height-weight data, an assumption of normality for both sets of conditional distributions might be the most we could justify. As we shall see, such normal conditionals distributions comprise a flexible family which subsumes and extends the classical bivariate normal model.

In this book, we will study the concept of conditional specification of joint densities. We consider questions of compatibility and near compatibility of given families of conditional distributions. We also consider cases in which conditional densities are only assumed to be known to belong to specified parametric families (as in the height-weight example above). The models thus derived are called conditionally specified models. We discuss aspects of their distributions and address the issues of parametric estimation and simulation for such models.

Before embarking on this study, we will make a few brief comments on alternative methods of specifying joint distributions. In addition, we will trace a few historical precedents for the study of conditionally specified distributions.

1.2   How May One Specify a Bivariate Distribution?

Let (X, Y) be a two-dimensional random variable. Clearly its probabilistic behavior for most purposes is adequately specified by knowledge of its joint distribution function

$$F_{X,Y}(x, y) = P(X \leq x, Y \leq y); \quad x, y \in \mathbb{R}. \tag{1.1}$$

Naturally for $F_{X,Y}(x, y)$ to be a legitimate distribution, it must be monotone in x and in y, must satisfy $F(-\infty, y) = 0, F(x, -\infty) = 0, F(\infty, \infty) = 1$, and must assign nonnegative mass to every rectangle in $\mathbb{R}^2$. At a very basic level the distribution of (X, Y) will be determined by identifying the probability space on which X and Y are defined, say $(\Omega, \mathcal{F}, P)$, and explicitly defining the mappings $X : \Omega \to \mathbb{R}$ and $Y : \Omega \to \mathbb{R}$. Of course, we usually don't get down to basics like that very often. More likely we will specify $F_{X,Y}(x, y)$ by defining it in terms of some joint density $f_{X,Y}(x, y)$ with respect to some measure on $\mathbb{R}^2$. The density $f_{X,Y}(x, y)$ is then required to be nonnegative and integrate to 1. An alternative is to specify a large number of moments and mixed moments of X and Y. With luck, this will completely determine $F_{X,Y}(x, y)$. More exotic characterization methods exist but they are often specific to the given form of $F_{X,Y}(x, y)$, and cannot be viewed as general techniques for characterizing bivariate distributions (the key references if one wishes to pursue such matters are Kagan, Linnik, and Rao (1973), Galambos and Kotz (1978)), Ramachandran and Lau (1991), and Rao and Shanbhag (1994).

A variety of transforms can be used to characterize $F_{X,Y}(x, y)$. The joint characteristic function

$$\phi_{X,Y}(t_1, t_2) = E(\exp[i(t_1 X + t_2 Y)]), \quad t_1, t_2 \in \mathbb{R}, \tag{1.2}$$

will uniquely determine $F_{X,Y}(x, y)$. The joint Laplace transform

$$\Psi_{X,Y}(t_1, t_2) = E(\exp[-(t_1 X + t_2 Y)]), \quad t_1, t_2 > 0, \tag{1.3}$$

will characterize $F_{X,Y}(x,y)$'s corresponding to nonnegative random variables X and Y. The joint moment generating function, the joint mean residual life function, and the joint hazard function, when they are well defined, will all uniquely determine $F_{X,Y}(x,y)$.

Knowledge of the marginal distributions $F_X(x)$ and $F_Y(y)$ has long been known to be inadequate to determine $F_{X,Y}(x,y)$. A vast array of joint distributions with given marginals has developed over the years (surveys may be found in Mardia (1970) and Ord (1972) for example, see also Hutchinson and Lai (1990)). An extensive up-to-date catalog of discrete multivariate distributions may be found in Johnson, Kotz, and Balakrishnan (1997).

If we incorporate conditional specification instead of, or together with, marginal specification the picture brightens. It is sometimes possible to characterize distributions this way.

First it is clearly enough to know one marginal distribution and the family of corresponding conditional distributions, i.e., knowledge of $F_X(x)$ and

$$F_{X|Y}(x|y) = P(X \le x|Y = y), \tag{1.4}$$

for every y, will completely determine the joint distribution of (X,Y). Actually in some circumstances we can get away with a little less. Knowledge of (1.4) and knowledge that $X \stackrel{d}{=} Y$ will often characterize $F_{X,Y}(x,y)$ (see Section 15.5).

In reliability contexts other modeling approaches are sometimes used. The "dynamic construction" prescribes the joint distribution of (X,Y) by specifying:

(i) the distribution of $\min(X,Y)$;

(ii) the probability $p(t)$, that $\min(X,Y) = X$ given that $\min(X,Y) = t, t > 0$; and

(iii) the conditional distribution of Y given $\min(X,Y) = X = t$ and the conditional distribution of X given $\min(X,Y) = Y = t, t > 0$.

For details see Shaked and Shanthikumar (1987).

What if we are given both families of conditional distributions, $F_{X|Y}(x|y)$ for every possible value y of Y and $F_{Y|X}(y|x)$ for every possible value x of X? Provided the families of conditional distributions are compatible, in a sense to be discussed later in this chapter, and provided a related Markov process is indecomposable, then indeed these families of conditional distributions will uniquely determine the joint distribution of (X,Y). A survey of results related to such characterizations is to be found in Arnold and Press (1989b). Sections 1.5-1.7 draw heavily on this source.

Perhaps the earliest work in this area was that of Patil (1965). He considered the discrete case and under a mild regularity condition showed that the conditional distributions of X given Y and <u>one</u> conditional distribution of Y given $X = x_0$ will uniquely determine the joint distribution of

(X, Y). As a corollary he concludes that, in the discrete case under a mild regularity condition, both sets of conditionals will determine the joint distribution. The next contribution to this area was that of Amemiya (1975), augmented by Nerlove and Press (1986) (based in part on a working paper dated 1976). They discussed conditions sufficient for compatibility in the discrete case. Sufficient conditions in a more general setting were presented by Gourieroux and Montfort (1979) who also showed by a counterexample that some regularity conditions were needed to guarantee the uniqueness of a joint distribution corresponding to given compatible conditionals. Abrahams and Thomas (1984) essentially stated the compatibility condition (Theorem 1.2 in the present work) correctly, but overlooked the possible lack of uniqueness indicated in Gourieroux and Montfort's work. A synthesis of their results was presented in Arnold and Press (1989b). They also dealt with more abstract settings and considered multivariate extensions of the results.

Rather than specify completely the conditional distributions of X given Y and of Y given X, we may wish to specify only that these conditional distributions are members of some well-defined parametric families of distributions. This is the conditional specification paradigm which is a major theme of this book. A brief review of early contributions to this area is provided in the next section.

1.3   Early Work on Conditionally Specified Models

A conditionally specified bivariate distribution is associated with two parametric families of distributions $\mathcal{F}_1 = \{F_1(x; \theta) : \underline{\theta} \in \Theta\}$ and $\mathcal{F}_2 = \{F_2(y : \tau) : \underline{\tau} \in T\}$. The joint distribution of (X, Y) is required to have the property that for each possible value y of Y, the conditional distribution of X given $Y = y$ is a member of $\mathcal{F}_1$ with parameter $\underline{\theta}$ possibly dependent on y. In addition, each conditional distribution of Y given $X = x$ must be a member of $\mathcal{F}_2$ for some choice of $\underline{\tau}$ which may depend on x. One of the earliest contributions to the study of such models was the work of Patil (1965). He showed that if every distribution of X given $Y = y$ was a power series distribution and if every distribution of Y given $X = x$ was a power series distribution, then the joint distribution was a power series distribution. Besag (1974), in the context of spatial processes, discussed conditional specification. Several of the models he introduced reduce in the bivariate case to models involving conditionals in exponential families to be discussed in Chapter 4. A major breakthrough, in which the important role of functional equations was brought into focus, was provided by Castillo and Galambos (1987a). They completely characterized the class of distributions with normal conditionals. Abrahams and Thomas (1984), Besag (1974), and Bhattacharyya (1943) had actually described normal conditionals models with nonlinear regressions but had not attempted to determine explicitly the class of all distributions with normal conditionals. Incidently, Brucker (1979) did show that if we require normal conditionals <u>with</u> linear regression and constant conditional variance, then we are led to the classical bivariate normal distribution, independently verifying one of Bhattacharyya's observations.

Mimicking the work of Castillo and Galambos, Arnold (1987) described and studied the class of all distributions with Pareto conditionals. Subsequently, Arnold and Strauss (1988a) dealt with exponential conditionals. This was followed by their paper (Arnold and Strauss (1991)) which discussed conditionals in prescribed exponential families. Chapter 4 is based on this work which unified and extended several earlier papers. Much of the material in Chapter 5 is based on a series of papers by Castillo and Sarabia using functional equations to treat a variety of conditional specification models not involving exponential families.

1.4 The Conditional Specification Paradigm

Our goal is to discuss general conditions under which candidate families of conditional distributions for X given Y and for Y given X are compatible. When we say they are compatible, we mean that there will exist at least one joint distribution for (X, Y) with the given families as its conditional distributions.

In cases in which compatibility is confirmed the question of the possible uniqueness of the compatible distribution must be addressed. These concepts lead naturally to what will be a central focus of this book, the study of distributions with conditionals in prescribed parametric families. A pivotal role in the development of such models will be played by a classical theorem dealing with the solutions of a particular kind of functional equation (Theorem 1.3 below). It dates back to 1904, though special cases of it may have been resolved even earlier.

In discussions of the compatibility of families of conditional distributions, the concepts are most easily described and visualized when densities exist. In particular, the discussion is most transparent when X, Y are discrete and each has only a finite set of possible values. Our discussion begins in such settings.

1.5 Compatible Conditionals: Finite Discrete Case

Consider X and Y to be discrete random variables with possible values $x_1, x_2, \ldots, x_I$ and $y_1, y_2, \ldots, y_J$, respectively. A putative conditional model for the joint distribution of (X, Y) can be associated with two $I \times J$ matrices

A and B with elements a_{ij} and b_{ij}. It is assumed that the a_{ij}'s and b_{ij}'s are nonnegative. Our question is: What further conditions must A and B satisfy in order that there might exist a random vector (X, Y) with the property that $\forall i, j$

$$a_{ij} = P(X = x_i | Y = y_j) \tag{1.5}$$

and

$$b_{ij} = P(Y = y_j | X = x_i)? \tag{1.6}$$

In addition we may address the question of the possible uniqueness of the distribution of (X, Y) satisfying (1.5) and (1.6). Two obvious constraints that we must require are

$$\sum_{i=1}^{I} a_{ij} = 1, \quad \forall j, \tag{1.7}$$

and

$$\sum_{j=1}^{J} b_{ij} = 1, \quad \forall i. \tag{1.8}$$

Another obvious constraint is that A and B must have a common incidence set.

Definition 1.1 (Incidence set of a matrix). Given a matrix A the set $\{(i, j) : a_{ij} > 0\}$ is called the incidence set of A and is denoted by N^A. □

If a bivariate random variable (X, Y) is to exist with conditionals (1.5) and (1.6) then its corresponding marginal distributions will be severely constrained. Let us introduce the following notation for these marginal distributions, if they exist,

$$\tau_i = P(X = x_i), \quad i = 1, 2, \ldots, I, \tag{1.9}$$

$$\eta_j = P(Y = y_j), \quad j = 1, 2, \ldots, J. \tag{1.10}$$

Since $P(X = x_i, Y = y_j)$ can be written in two ways by conditioning on either X or Y, $A, B, \underline{\tau}$, and $\underline{\eta}$ must satisfy

$$\tau_i b_{ij} = \eta_j a_{ij}, \quad \forall i, j \in N^A. \tag{1.11}$$

These observations lead immediately to the following theorem:

Theorem 1.1   *A and B, satisfying (1.7) and (1.8), are compatible iff:*

(i) $N^A = N^B = N$ say; and

(ii) there exist vectors $\underline{u}$ and $\underline{v}$ of appropriate dimensions for which

$$c_{ij} = a_{ij}/b_{ij} = u_i v_j, \quad \forall (i, j) \in N. \tag{1.12}$$

Proof. If A and B are compatible, then from (1.11) we have $c_{ij} = \tau_i/\eta_j$ so (1.12) holds. Conversely, if (1.12) holds, an appropriate choice for $\underline{\tau}$ is provided by $\tau_i = u_i/\sum_{i=1}^{I} u_i$ and an appropriate choice for $\underline{\eta}$ is provided by $\eta_j = v_j^{-1}/\sum_{j=1}^{J} v_j^{-1}$. $\qquad\square$

Note that the condition (1.12) is intimately related to the concept of quasi-independence as encountered in the study of contingency tables. It is convenient to define an $I \times J$ matrix C with elements $c_{ij} = a_{ij}/b_{ij}$ when $(i,j) \in N$ and with $c_{ij} = 0$ when $(i,j) \notin N$. To get a feeling for the implications of the theorem, the reader is invited to verify that the following pair of candidate conditionals distribution arrays A, B are compatible:

$$A = \begin{pmatrix} 1/6 & 0 & 3/14 \\ 0 & 1/4 & 4/14 \\ 5/6 & 3/4 & 7/14 \end{pmatrix}, \tag{1.13}$$

$$B = \begin{pmatrix} 1/4 & 0 & 3/4 \\ 0 & 1/3 & 2/3 \\ 5/18 & 6/18 & 7/18 \end{pmatrix}. \tag{1.14}$$

Acceptable choices for $\underline{u}$ and $\underline{v}$ are provided by $\underline{u} = (8, 12, 36)$ and $\underline{v} = (\frac{1}{12}, \frac{1}{16}, \frac{1}{28})$. If B is replaced by $\tilde{B}$, as follows,

$$\tilde{B} = \begin{pmatrix} 3/4 & 0 & 1/4 \\ 0 & 1/3 & 2/3 \\ 5/18 & 6/18 & 7/18 \end{pmatrix}, \tag{1.15}$$

then we may verify that A and $\tilde{B}$, (1.13) and (1.15), are <u>not</u> compatible. Note that in order to show incompatibility, i.e., that (1.12) fails, it is enough to identify a "rectangle" of four nonzero entries in $C = (c_{ij})$, say $c_{i_1,j_1}, c_{i_1,j_2}, c_{i_2,j_1}, c_{i_2,j_2}$, for which

$$c_{i_1,j_1} c_{i_2,j_2} \neq c_{i_1,j_2} c_{i_2,j_1}.$$

If all entries in A and B are positive so that $c_{ij} > 0, \forall i = 1, 2, \ldots, I$, and $j = 1, 2, \ldots, J$, then the condition for compatibility is simply expressible as

$$c_{ij} c_{..} = c_{i.} c_{.j}, \quad \forall (i,j),$$

where $c_{i.} = \sum_j c_{ij}, c_{.j} = \sum_i c_{ij}$ and $c_{..} = \sum_i \sum_j c_{ij}$.

In Section 2.2 we will catalog a variety of alternative compatibility specifications in the finite discrete case. In addition, in that section we will discuss "approximately" or "almost" compatible cases. For the moment, we will stick with the compatibility criterion provided in Theorem 1.1 and move on to consider cases where X and Y no longer have a finite list of possible values.

1.6   Compatibility in More General Settings

When we relax the condition that (X, Y) be discrete random variables with only a finite number of possible values, most of our observations of Section 1.4 will carry over with only notational changes. The only complicating factor is that, whereas in the finite case we could sum over rows and columns without concern, now we may need to postulate suitable summability and/or integrability conditions to justify such operations. We assume that (X, Y) is a random vector which is absolutely continuous with respect to some product measure $\mu_1 \times \mu_2$ on $S(X) \times S(Y)$ where $S(X)$ (respectively $S(Y)$) denotes the support of $X(Y)$. Note that this allows one variable to be discrete and the other continuous, an important special case. The joint, marginal, and conditional densities of X and Y will be denoted by $f_{X,Y}(x, y), f_X(x), f_Y(y), f_{X|Y}(x|y)$, and $f_{Y|X}(y|x)$. The support sets $S(X)$ and $S(Y)$ can be finite, countable, or uncountable.

We will denote the families of candidate conditional densities (with respect to μ_1 and μ_2) by

$$a(x, y) = f_{X|Y}(x|y), \quad x \in S(X), \quad y \in S(Y), \tag{1.16}$$

and

$$b(x, y) = f_{Y|X}(y|x), \quad x \in S(X), \quad y \in S(Y). \tag{1.17}$$

It is convenient to introduce the notation

$$N_a = \{(x, y) : a(x, y) > 0\}, \tag{1.18}$$

$$N_b = \{(x, y) : b(x, y) > 0\}. \tag{1.19}$$

The discussion in Section 1.4, allows us to immediately enunciate the appropriate compatibility theorem.

Theorem 1.2   *A joint density $f(x, y)$, with $a(x, y)$ and $b(x, y)$ as its conditional densities, will exist iff (i) $N_a = N_b = N$, and (ii) there exist functions u and v such that for all $x, y \in N$*

$$a(x, y)/b(x, y) = u(x)v(y), \tag{1.20}$$

where $\int_{S(X)} u(x) \; d\mu_1(x) < \infty.$ □

Proof.   In order for $a(x, y)$ and $b(x, y)$ to be compatible, suitable marginal densities $f(x)$ and $g(y)$ must exist. Clearly (1.20) must hold with $f(x) \propto u(x)$ and $g(y) \propto 1/v(y)$. The condition $\int u(x) \; d\mu_1(x) < \infty$ is equivalent (via Tonelli's theorem) to the condition $\int [1/v(y)] \; d\mu_2(y) < \infty$ and only one needs to be checked in practice. These integrability conditions reflect the fact that the marginal densities must be integrable and indeed must integrate to 1.   □

An example of the application of Theorem 1.2 follows. Consider the following candidate family of conditional densities (with respect to Lebesgue measure):

$$f_{X|Y}(x|y) = a(x, y) = (y + 2)e^{-(y+2)x} I(x > 0),$$

and

$$f_{Y|X}(y|x) = b(x, y) = (x + 3)e^{-(x+3)y} I(y > 0).$$

Here $S(X) = S(Y) = (0, \infty)$ and to confirm compatibility we must verify that (1.20) holds. In this example we have

$$\frac{a(x, y)}{b(x, y)} = \frac{(y + 2)e^{-(y+2)x}}{(x + 3)e^{-(x+3)y}}$$

$$= \left(\frac{e^{-2x}}{x+3}\right)\left(\frac{y+2}{e^{-3y}}\right)$$

and clearly (1.20) holds with

$$u(x) = e^{-2x}/(x + 3)$$

and

$$v(y) = (y + 2)e^{3y}.$$

For this choice of $u(x)$ we have $\int_0^\infty u(x)\, dx < \infty$ and compatibility of the two families of conditional densities is confirmed. The marginal density of X is proportional to $u(x)$. Such densities will be discussed further in Chapter 4.

The crucial nature of the integrability condition may be seen in the following simple example in which certain uniform conditional distributions are candidates. The putative conditional densities are given by

$$f_{X|Y}(x|y) = a(x, y) = y\, I(0 < x < y^{-1})I(y > 0),$$

and

$$f_{Y|X}(y|x) = b(x, y) = x\, I(0 < y < x^{-1})I(x > 0).$$

Here $S(X) = S(Y) = (0, \infty)$ and $N = N_a = N_b = \{(x, y) : x > 0, y > 0, xy < 1\}$. The ratio

$$\frac{a(x, y)}{b(x, y)} = \frac{y}{x}$$

factors nicely. Unfortunately, the function $u(x) = x^{-1}$ is not integrable so that the two families are not compatible. No joint distribution exists with such conditional distributions. This example will be discussed again in Chapter 6.

Returning to the general Theorem 1.2, if we define $c(x, y) = a(x, y)/b(x, y)$ our goal is to factor this expression into a function of x and a function of y.

We need to show that for any (x_1, x_2, y_1, y_2) with $(x_1, y_1) \in N, (x_1, y_2) \in N, (x_2, y_1) \in N$, and $(x_2, y_2) \in N$ we have

$$c(x_1, y_1)c(x_2, y_2) = c(x_1, y_2)c(x_2, y_1).$$

Perhaps the simplest situation involves the case in which N is a Cartesian product $S(X) \times S(Y)$ and in which serendipitously $c(x, y)$ is integrable over $S(X) \times S(Y)$ (a condition which may well not obtain even for perfectly compatible $a(x, y)$ and $b(x, y)$). In this case we would only check to see if

$$c(x, y) = \frac{\int_{S(X)} c(x, y) \ d\mu_1(x) \int_{S(Y)} c(x, y) \ d\mu_2(y)}{\int_{S(X)} \int_{S(Y)} c(x, y) \ d\mu_1(x) \ d\mu_2(y)}$$

holds for every $x \in S(X)$ and $y \in S(Y)$. Integrability of $c(x, y)$ over N is equivalent to integrability of both $u(x)$ and $v(y)$. The former $u(x)$ must be integrable for compatibility, the latter $v(y)$ may well fail to be integrable and factorization cannot in such cases be verified by simple integration.

If a and b are compatible, the following straightforward algorithm can be used to obtain the corresponding joint density of (X, Y):

Algorithm 1.1 (Obtaining a compatible joint distribution of (X, Y) given the conditionals $X|Y$ and $Y|X$).

Input. *Two conditional probability density functions $a(x, y) = f_{X|Y}(x|y)$ and $b(x, y) = f_{Y|X}(y|x)$, and an Error.*

Output. *The corresponding compatible Y-marginal density function $f_Y(y)$ and the joint probability density function $f_{X,Y}(x, y)$ or a close alternative.*

Step 1. *Make $Error1 = 1$ and choose an arbitrary Y-marginal probability density function $f_0(y)$.*

Step 2. *Make $f_Y(y) = f_0(y)$.*

Step 3. *Calculate the joint density $f_{X,Y}(x, y)$ using*

$$f_{X,Y}(x, y) = f_Y(y)a(x, y).$$

Step 4. *Calculate the X-marginal density $f_X(x)$ using*

$$f_X(x) = \frac{f_{X,Y}(x, y)}{\int_{S(Y)} f_{X,Y}(x, y) \ dy}.$$

Step 5. *Calculate the updated Y-marginal probability density function $f_0(y)$ using*

$$f_0(y) = \int_{S(X)} b(y, x)f_X(x) \ dx.$$

Step 6. *Calculate the error by*

$$Error1 = \int_{S(Y)} |f_Y(y) - f_0(y)| \, dy.$$

Step 7. *If $Error1 > Error$, go to Step 2; otherwise return the marginal probability density $f_0(y)$ and the joint probability density function $f_{X,Y}(x, y)$ and exit.*

$\square$

1.7   Uniqueness

Once we have determined that $a(x, y)$ and $b(x, y)$ are compatible, we may reasonably ask whether the associated joint density $f_{X,Y}(x, y)$ is unique. A negative answer is sometimes appropriate. Perhaps the most transparent case in which nonuniqueness occurs is the following. Focus on discrete random variables (X, Y) with $S_X = S_Y = \{0, 1, 2, \ldots\}$. Now consider the family of conditional densities.

$$a(x, y) = b(x, y) = \begin{cases} 1, & x = y, \\ 0, & \text{otherwise.} \end{cases} \tag{1.21}$$

Thus we are postulating $X = Y$ with probability 1. Evidently (1.21) describes compatible conditionals but there is no uniqueness in sight. Any marginal distribution for X may be paired with such a family of conditional distributions. The corresponding family of joint distributions for (X, Y) with conditionals given by (1.21) is

$$P(X = i, Y = j) = \begin{cases} p_i, & i = j = 0, 1, 2, \ldots, \\ 0, & i \neq j, \end{cases} \tag{1.22}$$

in which $\{p_0, p_1, \ldots\}$ is any sequence of positive numbers summing to 1.

Necessary and sufficient conditions for uniqueness may be described by recasting the problem in a Markov chain setting. Suppose (X, Y) is absolutely continuous with respect to $\mu_1 \times \mu_2$ with supports $S(X)$ and $S(Y)$. Suppose that $a(x, y)$ and $b(x, y)$ (defined by (1.16) and (1.17)) are compatible with a marginal density $\tau(x)$ for X. When is $\tau(x)$ unique? Define a stochastic kernel ba by

$$ba(x|z) = \int_{S(Y)} a(x, y)b(z, y) \, d\mu_2(y). \tag{1.23}$$

Now consider a Markov chain with state space $S(X)$ and transition kernel ba. For such a chain, τ is a stationary distribution. It will be unique iff the chain is indecomposable. Note that the chain associated with a and b defined by (1.21) is far from being indecomposable! One situation in which indecomposability (and hence uniqueness) is readily verified corresponds to the case in which $N_a = N_b = S(X) \times S(Y)$ (in such a case $ba(x|z) > 0$ for every x and z).

1.8   Conditionals in Prescribed Families

With the question of compatibility now completely resolved, it is time to turn to a related but clearly distinct question regarding conditional specification of bivariate distributions. We may conveniently recall the motivating example described in Chapter 1. In that example we asked (i) is it possible to have a bivariate distribution with $X|Y = y \sim$ normal $(\mu(y), \sigma^2(y))$, $\forall y$ and $Y|X = x \sim$ normal $(\nu(x), \tau^2(x))$, $\forall x$, and (ii) if such models exist, can we characterize the complete class of such distributions? In this scenario, the conditional densities of X given Y are required only to belong to some parametric family and the conditionals of Y given X are required only to belong to some (possibly different) parametric family. Such conditionally specified distributions, subject of course to compatibility conditions, will be the major focus of this book.

Consider a k-parameter family of densities on $\mathbb{R}$ with respect to μ_1 denoted by $\{f_1(x; \underline{\theta}) : \underline{\theta} \in \Theta\}$ where $\Theta \subseteq \mathbb{R}^k$. Consider a possible different ℓ-parameter family of densities on $\mathbb{R}$ with respect to μ_2 denoted by $\{f_2(y; \underline{\tau}) : \tau \in T\}$ where $T \subseteq \mathbb{R}^\ell$. We are interested in all possible bivariate distributions which have all conditionals of X given Y in the family f_1 and all conditionals of Y given X in the family f_2. Thus we demand that

$$f_{X|Y}(x|y) = f_1(x; \underline{\theta}(y)), \quad \forall x \in S(X), \ y \in S(Y), \tag{1.24}$$

and

$$f_{Y|X}(y|x) = f_2(y; \underline{\tau}(x)), \quad \forall x \in S(X), \ y \in S(Y). \tag{1.25}$$

If (1.24) and (1.25) are to hold, then there must exist marginal distributions for X and Y denoted by $f_X(x)$ and $f_Y(y)$ such that

$$f_Y(y) f_1(x; \underline{\theta}(y)) = f_X(x) f_2(y; \underline{\tau}(x)), \quad \forall x \in S(X), \ y \in S(Y). \tag{1.26}$$

Whether such a functional equation can be solved for $\underline{\theta}(y)$ and $\underline{\tau}(x)$ depends crucially on the nature of the known functions f_1 and f_2. In many cases no solution is possible. Even in those cases in which nontrivial solutions are found we must be careful to check whether the resulting solutions correspond to valid (i.e., nonnegative and integrable) joint densities.

A remarkable number of families of conditional densities are expressible in the form

$$f(x; \underline{\theta}) = \phi \left(\sum_{i=1}^{k} T_i(x) \theta_i \right), \quad x \in S(X), \tag{1.27}$$

where ϕ is invertible. For example, multiparameter exponential families are of this form, as are Cauchy distributions. If we are dealing with conditionals from a family such as (1.27) then, quite often, the problem can be solved. An important tool in obtaining the solution is the following theorem presented in a form due to Aczél (1966) (see also Castillo and Ruiz-Cobo (1992)). The basic idea of the theorem can be traced back to Stephanos (1904), Levi-Civita (1913), and Suto (1914).

Theorem 1.3 *All solutions of the equation*

$$\sum_{k=1}^{n} f_k(x) g_k(y) = 0, \quad x \in S(X), \ y \in S(Y), \tag{1.28}$$

can be written in the form

$$\begin{bmatrix} f_1(x) \\ f_2(x) \\ \cdots \\ f_n(x) \end{bmatrix} = \begin{bmatrix} a_{11} & a_{12} & \cdots & a_{1r} \\ a_{21} & a_{22} & \cdots & a_{2r} \\ \cdots & \cdots & \cdots & \cdots \\ a_{n1} & a_{n2} & \cdots & a_{nr} \end{bmatrix} \begin{bmatrix} \phi_1(x) \\ \phi_2(x) \\ \cdots \\ \phi_r(x) \end{bmatrix},$$

$$\begin{bmatrix} g_1(y) \\ g_2(y) \\ \cdots \\ g_n(y) \end{bmatrix} = \begin{bmatrix} b_{1r+1} & b_{1r+2} & \cdots & b_{1n} \\ b_{2r+1} & b_{2r+2} & \cdots & b_{2n} \\ \cdots & \cdots & \cdots & \cdots \\ b_{nr+1} & b_{nr+2} & \cdots & b_{nn} \end{bmatrix} \begin{bmatrix} \Psi_{r+1}(y) \\ \Psi_{r+2}(y) \\ \cdots \\ \Psi_n(y) \end{bmatrix}, \tag{1.29}$$

where r is an integer between 0 and n, and $\phi_1(x), \phi_2(x), \ldots, \phi_r(x)$ on the one hand and $\Psi_{r+1}(x), \Psi_{r+2}(x), \ldots, \Psi_n(x)$ on the other are arbitrary systems of mutually linearly independent functions and the constants a_{ij} and b_{ij} satisfy

$$\begin{bmatrix} a_{11} & a_{21} & \cdots & a_{n1} \\ a_{12} & a_{22} & \cdots & a_{n2} \\ \cdots & \cdots & \cdots & \cdots \\ a_{1r} & a_{2r} & \cdots & a_{nr} \end{bmatrix} \begin{bmatrix} b_{1r+1} & b_{1r+2} & \cdots & b_{1n} \\ b_{2r+1} & b_{2r+2} & \cdots & b_{2n} \\ \cdots & \cdots & \cdots & \cdots \\ b_{nr+1} & b_{nr+2} & \cdots & b_{nn} \end{bmatrix} = 0. \tag{1.30}$$

$\square$

Two simple and straightforward consequences of the Aczél theorem are of sufficient utility to merit separate listing.

Theorem 1.4 *All solutions of the equation*

$$\sum_{i=1}^{r} f_i(x) \phi_i(y) = \sum_{j=1}^{s} g_j(y) \Psi_j(x), \quad x \in S(X), \ y \in S(Y), \tag{1.31}$$

where $\{\phi_i\}_{i=1}^{r}$ and $\{\Psi_j\}_{j=1}^{s}$ are given systems of mutually linearly independent functions, are of the form

$$\underline{f}(x) = C \underline{\Psi}(x) \tag{1.32}$$

and

$$g(y) = D\underline{\phi}(y) \tag{1.33}$$

where $D = C'$.   □

Proof.   Take $A = \begin{pmatrix} I \\ D \end{pmatrix}$ and $B = \begin{pmatrix} D \\ -I \end{pmatrix}$ in Theorem 1.3.   □

Theorem 1.5   *If for some invertible function γ we have*

$$
\begin{aligned}
g(x,y) &= \gamma(\textstyle\sum_{i=1}^{r} f_i(x)\phi_i(y)) \\
&= \gamma(\textstyle\sum_{j=1}^{s} g_j(y)\Psi_j(x)), \quad x \in S(X), \ y \in S(Y),
\end{aligned} \tag{1.34}
$$

where $\{\phi_i\}_{i=1}^{r}$ and $\{\Psi_j\}_{j=1}^{s}$ are given systems of mutually linearly independent functions, then

$$g(x,y) = \gamma(\underline{\phi}'(y)C\underline{\Psi}(x)). \tag{1.35}$$

□

Theorem 1.5 will allow us to completely characterize distributions with conditionals in given exponential families (Chapters 3 and 4). It will also yield solutions for Cauchy conditionals, Pareto conditionals, and other non-exponential family situations (Chapter 5). It must be remarked that, if we use Theorem 1.5 to identify a joint density, there may be further restriction necessary on the elements of C in (1.35) to ensure that the density is non-negative and integrable. Such conditions are sometimes vexingly difficult to pin down. The final fly in the ointment involves normalization. The joint density must integrate to 1. Frequently, as we shall see, Theorem 1.5 easily provides us with conditionally specified densities up to an awkward normalizing constant. It is possible to simulate data from such distributions and to estimate parameters without specifically knowing the normalizing constant, so this does not seriously hamper our use of such conditionally specified models.

Note that Theorem 1.5 will only be useful in situations in which the support of the joint density is a Cartesian product, i.e., in which the support of $f_{X|Y}(x|y)$ does not depend on y and that of $f_{Y|X}(y|x)$ does not depend on x. Several of the examples to be discussed in Chapter 5 do not fit this paradigm. They can be resolved but a generally applicable theorem is lacking and they are handled on a case by case basis. Examples treated include the uniform conditionals case and the translated exponential conditionals case.

1.9   An Example

Theorems 1.3, 1.4, and 1.5 are quite general and will be some of our basic investigative tools, but it goes without saying that in some simple cases they are more powerful than necessary. Simpler arguments may well be adequate. We can illustrate this phenomenon in the exponential conditionals case. In this situation we assume that for each $y > 0$, X given $Y = y$ is exponential with intensity $\mu(y)$ and for each $x > 0$, Y given $X = x$ is exponential with intensity $\nu(x)$. It follows that, writing the joint density of (X, Y) as a product of marginal and conditional densities in both possible manners, we have

$$f_X(x)\nu(x)e^{-\nu(x)y} = f_Y(y)\mu(y)e^{-\mu(y)x}, \quad \forall x > 0, \ y > 0. \qquad (1.36)$$

We can of course apply Theorem 1.5 here. First rewrite (1.36) in the form

$$\exp\{h_1(x) \cdot 1 - \nu(x)y\} = \exp\{h_2(y) \cdot 1 - \mu(y)x\}, \qquad (1.37)$$

where $h_1(x) = \log[f_X(x)\nu(x)]$ and $h_2(y) = \log[f_Y(y)\mu(y)]$. Clearly (1.37) is of the form (1.34) so that Theorem 1.5 can be applied. The appropriate identification between (1.37) and (1.34) is provided by the following relations:

$$
\begin{array}{rclcrcl}
\gamma(u) & = & e^{-u}, & \quad & r, s & = & 2, \\
f_1(x) & = & h_1(x), & \quad & \phi_1(y) & = & 1, \\
f_2(x) & = & \nu(x), & \quad & \phi_2(y) & = & -y, \\
g_1(y) & = & h_2(y), & \quad & \Psi_1(x) & = & 1, \\
g_2(y) & = & \mu(y), & \quad & \Psi_2(x) & = & -x.
\end{array}
\qquad (1.38)
$$

It then follows from Theorem 1.5 that the joint density must be of the form

$$f_{X,Y}(x, y) = \exp(c_{11} - c_{21}x - c_{12}y + c_{22}xy)I(x > 0)I(y > 0), \qquad (1.39)$$

subject to appropriate constraints on the c_{ij}'s. We will (Section 4.4) meet this density again with a slightly different parametrization. Specifically, it will be written

$$f_{X,Y}(x, y) = \exp(m_{00} - m_{10}x - m_{01}y + m_{11}xy)I(x > 0)I(y > 0).$$

The same result could, in this case, have been obtained by more elementary means. Merely take logarithms on both sides of (1.36) and apply a differencing argument with respect to x and y (mimicking a differentiation argument without assuming differentiability). From this it follows that

$$\mu(y) = \alpha y + \beta_1$$

and

$$\nu(x) = \alpha x + \beta_2.$$

We may then plug this back into (1.36) and conclude that

$$f_X(x) \propto (\alpha x + \beta_2)^{-1} e^{-\beta_1 x}.$$

From this it follows that

$$f_{X,Y}(x,y) \propto \exp\left[-(\beta_1 x + \beta_2 y + \alpha xy)\right] I(x > 0)I(y > 0)$$

which is equivalent to (1.39). In order for (1.39) to be integrable over the positive quadrant we must have $m_{10} > 0$ and $m_{01} > 0$. For the same reason, we must have $m_{11} \leq 0$ (the case $m_{11} = 0$ corresponds to independent marginals). Once $m_{10}, m_{01},$ and m_{11} are fixed, m_{00} is completely determined by the requirement that the joint density integrates to 1. Specifically, we have

$$m_{00} = \log\left[\frac{1}{m_{10}m_{01}} \int_0^\infty e^{-u}\left(1 + \frac{m_{11}u}{m_{10}m_{01}}\right)^{-1} du\right]. \qquad (1.40)$$

This can be evaluated numerically or, via a change of variable, can be expressed in terms of the tabulated exponential integral function. It is a classic instance of the kinds of awkward normalizing constants to be encountered when distributions are conditionally specified.

1.10   Bibliographic Notes

The material in Sections 1.5, 1.6, and 1.7 draws heavily on Arnold and Press (1989a). Versions of the general theorems in Section 1.8 were used in several papers by Castillo and Sarabia (1990a, 1990b, 1991). Section 1.9 is based in part on Arnold and Strauss (1988a).

Exercises

1.1 Using Theorem 1.3, obtain the most general solution of the following functional equations, where $f_i(x)$ and $g_i(x), i = 1, 2, 3,$ are unknown functions:

(a) $f_1(x)y + f_2(x)y^2 + f_3(x)y^3 = xg_1(y) + x^2g_2(y) + x^3g_3(y).$

(b) $f_1(x)\log y + f_2(x)\log(1+y) + f_3(x) = g_1(y)\log x + g_2(y)\log(1+x) + g_3(y).$

(c) $f_1(x) + xf_1(x)g_2(y) = g_1(y) + yg_1(y)f_2(x).$

1.2 Find the most general surface of the form $z = z(x,y)$ such that its cross sections with planes $x = 0$ and $y = 0$ are second-order polynomials, that is,

$$z(x,y) = a(y)x^2 + b(y)x + c(y),$$

and

$$z(x, y) = d(x)y^2 + e(x)y + f(x).$$

Obtain the most general expression for $z = z(x, y)$ and plot some examples.

(Castillo and Ruiz-Cobo (1992).)

1.3 Let (X, Y) be a bivariate random variable with joint pdf $f(x, y)$. Let $f(x|y)$ and $f(y|x)$ be the conditional densities $f(x|y)$ of X given $Y = y$, and Y given $X = x$, respectively.

Prove that:

(a) If $f(x_0|y) > 0$ for all y, then

$$f(x, y) \propto \frac{f(x|y)f(y|x_0)}{f(x_0|y)}.$$

(b) A sufficient condition for $f(x|y)$ and $f(y|x)$ to be compatible with a joint density for (X, Y) is that

$$\frac{f(x_2|y)f(y|x_1)}{f(x_1|y)f(y|x_2)}$$

does not depend on y for any choice of (x_1, x_2) such that $x_1 \neq x_2$.

1.4 The equation

$$[a_1(x)y + b_1(x)]^{c_1(x)} = [a_2(y)x + b_2(y)]^{c_2(y)},$$

where $a_1(x)$, $b_1(x)$, $c_1(x)$, $a_2(y)$, $b_2(y)$, and $c_2(y)$ are the unknown functions, $y \geq -b_1(x)/a_1(x)$ and $x \geq -b_2(y)/a_2(y)$, and the functions $a_1(x)$, $a_2(y)$, $c_1(x)$, and $c_2(y)$ are positive, is known as the *Castillo–Galambos functional equation*. Solve this functional equation, assuming that

$$\varlimsup_{x \to \infty} \left| \frac{b_1(x)}{a_1(x)} \right| < \infty,$$

$$\varlimsup_{y \to \infty} \left| \frac{b_2(y)}{a_2(y)} \right| < \infty.$$

(Castillo and Galambos (1987b).)

1.5 Solve the functional equation,

$$\alpha_1(y) + \beta_1(y)x^{\gamma_1(y)} = \alpha_2(x) + \beta_2(x)y^{\gamma_2(x)},$$

where $\alpha_i(x)$, $\beta_i(x)$, and $\gamma_i(x)$ are unknown functions.

2

Exact and Near Compatibility in Distributions with Finite Support Sets

2.1 Introduction

In the finite discrete case, a variety of compatibility conditions can be derived. Such conditions provide a spectrum of alternative ways of measuring discrepancy between incompatible conditionals. In addition, they suggests alternative ways in which most nearly compatible distributions can be defined in incompatible cases. A related concept of ϵ-compatibility arises naturally in the discussion of incompatible cases.

2.2 Review and Extensions of Compatibility Results

We are interested in discrete random variables X and Y with possible values x_1, x_2, ..., x_I and $y_1, y_2, \ldots, y_J$, respectively. A candidate conditional model for the joint distribution of (X, Y) consists of $I \times J$ matrices A and B with nonnegative elements in which A has columns which sum to 1 and B has rows which sum to 1.

   A and B form a compatible conditional specification for the distribution of (X, Y) if there exists some $I \times J$ matrix P with nonnegative entries p_{ij} and with $\sum_{i=1}^{I} \sum_{j=1}^{J} p_{ij} = 1$ such that for every i, j

$$a_{ij} = p_{ij}/p_{\cdot j}$$

and

$$b_{ij} = p_{ij}/p_{i.},$$

where $p_{i.} = \sum_{j=1}^{J} p_{ij}$ and $p_{.j} = \sum_{i=1}^{I} p_{ij}$.

If such a matrix P exists then, if we assume that

$$p_{ij} = P(X = x_i, Y = y_j), \quad i = 1, 2, \ldots, I, \quad j = 1, 2, \ldots, J,$$

we will have

$$a_{ij} = P(X = x_i | Y = y_j), \quad i = 1, 2, \ldots, I, \quad j = 1, 2, \ldots, J,$$

and

$$b_{ij} = P(Y = y_j | X = x_i), \quad i = 1, 2, \ldots, I, \quad j = 1, 2, \ldots, J.$$

Definition 2.1 (Compatible conditional probability matrices). Two conditional probability matrices A and B are compatible if there exists a joint distribution (i.e., P as described above) which has the columns and rows, respectively, of A and B as its conditional distributions. □

As in Chapter 1, denote the incidence set of a matrix A by N^A (i.e., $N^A = \{(i, j) : a_{ij} > 0\}$).

As remarked earlier, compatibility of A and B can only occur if $N^A = N^B$. We will always assume that each row and each column of A (and B) contains at least one positive element (otherwise we would redefine the list of possible values to leave out the zero rows and/or columns). Our first characterization of compatibility was given in Theorem 1.1: A and B are compatible iff they have identical incidence sets and if there exist vectors $\underline{u}$ and $\underline{v}$ for which

$$c_{ij} = a_{ij}/b_{ij} = u_i v_j, \quad \forall (i, j) \in N^A. \tag{2.1}$$

Equivalently, and perhaps more transparently, A and B are compatible if they have identical incidence sets and if there exist stochastic vectors

$$\underline{\tau} = (\tau_1, \tau_2, \ldots, \tau_I) \quad and \quad \underline{\eta} = (\eta_1, \eta_2, \ldots, \eta_J)$$

such that

$$\eta_j a_{ij} = \tau_i b_{ij}, \quad \forall i, j. \tag{2.2}$$

In the case of compatibility, $\underline{\tau}$ and $\underline{\eta}$ can be readily interpreted as the resulting marginal distributions of X and Y, respectively.

In discussing alternative compatibility criteria, it is convenient to first consider the case where all elements of A and B are positive before considering more general cases.

2.2.1   *Compatibility of Positive Conditional Probability Matrices*

If all the elements of A and B are positive, conditions (2.1) and (2.2) can be related to the concepts of cross-product ratios and uniform marginal representations.

Definition 2.2 (Cross-product ratio of a 2×2 matrix). The cross-product ratio of a 2×2 matrix $D = (d_{ij})$ with positive elements is defined to be

$$d_{11}d_{22}/d_{12}d_{21}. \qquad (2.3)$$

$\square$

Definition 2.3 (Cross-product ratios of a $I \times J$ matrix). The set of cross-product ratios of a nonnegative $I \times J$ matrix A consists of the cross-product ratios associated with all positive 2×2 submatrices of the form

$$\begin{pmatrix} a_{i_1 j_1} & a_{i_1 j_2} \\ a_{i_2 j_1} & a_{i_2 j_2} \end{pmatrix}, \qquad (2.4)$$

where $1 \leq i_1 < i_2 \leq I$ and $1 \leq j_1 < j_2 \leq J$. $\square$

It is evident that if (2.2) holds (with $a_{ij} > 0, \forall i, j$) then every cross-product ratio of A will be equal to the corresponding cross-product ratio of B. Conversely, equating of the cross-product ratios guarantees the existence of vectors $\underline{\tau}$ and $\underline{\eta}$ such that (2.2) holds.

Cross-product ratios essentially reflect and describe the dependence structure of a joint distribution or of a contingency table, or more generally, of a matrix with nonnegative elements. Mosteller (1968) introduced the concept of a uniform marginal representation of a contingency table to separate marginal information from dependence structure information in such tables. The concept is meaningful for any matrix with nonnegative elements.

Definition 2.4 (Uniform marginal representation of a matrix). Given an $I \times J$ matrix with nonnegative elements (with at least one positive element in each row and column), we iteratively normalize rows and columns to have sums $1/I$ and $1/J$, respectively, until the procedure converges. The limiting matrix is called the uniform marginal representation (UMR) of the original matrix. $\square$

It is possible to interpret the UMR of a matrix P with nonnegative elements which sum to 1, as that matrix Q with nonnegative elements which sum to 1 and with uniform marginals, which is closest to P in terms of minimal Kullback-Leibler information distance (see, e.g., Arnold and Gokhale (1994)). This distance measure will be useful in our further discussion of compatibility.

Definition 2.5 (Kullback-Leibler information pseudo-distance).
Given two matrices P and Q, the Kullback-Leibler information distance
between them is defined as

$$I(Q, P) = \sum_i \sum_j q_{ij} \log(q_{ij}/p_{ij}).$$

(2.5)

□

The following algorithm gives the UMR matrix associated with a given
nonnegative matrix A.

Algorithm 2.1 (Obtaining the UMR matrix).

Input. *A matrix $A_{I \times J}$ and an Error.*

Output. *The corresponding UMR matrix.*

Step 1. *Make $Error1 = 1$.*

Step 2. *If $Error1 > Error$, make $B = A$ and go to Step 3; otherwise
return matrix A and exit.*

Step 3. *Calculate i-marginals by $s_i = \sum_{j=1}^{J} a_{i,j}$, $i = 1, \ldots, I$.*

Step 4. *Normalize rows: $a_{ij} = a_{ij}/(Is_i)$, $i = 1, \ldots, I$, $j = 1, \ldots, J$.*

Step 5. *Calculate j-marginals by $t_j = \sum_{i=1}^{I} a_{i,j}$, $j = 1, \ldots, J$.*

Step 6. *Normalize columns: $a_{ij} = a_{ij}/(Jt_j)$, $i = 1, \ldots, I$, $j = 1, \ldots, J$.*

Step 7. *Calculate $Error1 = \sum_{i,j} |a_{ij} - b_{ij}|$ and go to Step 2.*   □

It is evident from the defining algorithm for the UMR representation
that the UMR representation of A will have the same cross-product ratios
as A. Consequently, two matrices have the same UMRs if and only if they
have identical cross-product ratios. These observations are summarized in
the following theorem:

Theorem 2.1 (Compatibility of conditional probability matrices).
*(Arnold and Gokhale (1994).) Suppose that A and B contain only positive
elements, then the following statements are equivalent:*

(i) A and B are compatible.

*(ii) For every 2×2 subtable of A and the corresponding subtable of B,
the cross-product ratios are equal.*

(iii) A and B have identical uniform marginal representations.   □

It is evident that if two conditional matrices are compatibles their UMR matrices and the UMR of the compatible joint probability distribution P coincide.

Example 2.1 (A compatible case).   As an illustration consider two candidate conditional matrices A and B as follows:

$$A = \begin{pmatrix} 1/7 & 1/4 & 3/7 & 1/7 \\ 2/7 & 1/2 & 1/7 & 2/7 \\ 4/7 & 1/4 & 3/7 & 4/7 \end{pmatrix} \qquad (2.6)$$

and

$$B = \begin{pmatrix} 1/6 & 1/6 & 1/2 & 1/6 \\ 2/7 & 2/7 & 1/7 & 2/7 \\ 1/3 & 1/12 & 1/4 & 1/3 \end{pmatrix}. \qquad (2.7)$$

To determine compatibility we could compute the cross-product ratios of all 2×2 subtables of A and B. For example, the cross-product ratios corresponding to the upper left 2×2 subtables of A and B are $\left(\dfrac{1}{7} \cdot \dfrac{1}{2}\right) / \left(\dfrac{2}{7} \cdot \dfrac{1}{4}\right)$ and $\left(\dfrac{1}{6} \cdot \dfrac{2}{7}\right) / \left(\dfrac{2}{7} \cdot \dfrac{1}{6}\right)$, in both cases equal to 1. The other 17 cross-product ratios of A and B are also equal. Instead of considering such cross-product ratios, we can check for consistency by computing the UMRs of A and B. By successive row and column renormalizations a common UMR for A and B may be found; namely

$$\text{UMR}(A) = \text{UMR}(B) = \begin{pmatrix} 0.05478 & 0.08188 & 0.14190 & 0.05478 \\ 0.08489 & 0.12690 & 0.03665 & 0.08489 \\ 0.11030 & 0.04123 & 0.07145 & 0.11030 \end{pmatrix}. \qquad (2.8)$$

Thus, compatibility of A and B given in (2.6) and (2.7) is assured.   □

It is not essential that A and B contain only positive elements for Theorem 2.1 to hold. However, some restrictions on the common incidence set of A and B are necessary.

Example 2.2 (Counterexample).   Consider

$$A = \begin{pmatrix} 1/2 & 1/2 & 0 \\ 0 & 1/2 & 1/2 \\ 1/2 & 0 & 1/2 \end{pmatrix} \qquad (2.9)$$

and

$$B = \begin{pmatrix} 1/3 & 2/3 & 0 \\ 0 & 1/3 & 2/3 \\ 1/3 & 0 & 2/3 \end{pmatrix}. \qquad (2.10)$$

It may be verified that A and B have equal cross-product ratios (there are no positive 2×2 submatrices) but they do not have identical uniform marginal representations, thus A and B are not compatible.   □

It is evident that the array of cross-product ratios of a positive $I \times J$ matrix contains a considerable amount of redundant information. The basic cross-product ratio information of a matrix A with positive elements can be summarized in the form of the cross-product ratio matrix.

Definition 2.6 (Cross-product ratio matrix). The $F_{(I-1) \times (J-1)}$ matrix with elements

$$f_{ij} = \frac{a_{ij} a_{IJ}}{a_{iJ} a_{Ij}}, \quad i = 1, \ldots, I-1, \quad j = 1, \ldots, J-1, \qquad (2.11)$$

is called the cross-product ratio matrix corresponding to A.   □

An alternative specification of compatibility of two conditional distributions A and B with all elements positive is then possible, in terms of equality of their corresponding cross-product ratio matrices. Indeed, this may be the simplest criterion to check.

Example 2.3 (Checking compatibility by means of cross-product ratio matrices). If

$$A = \begin{pmatrix} 1/5 & 2/7 & 3/8 \\ 3/5 & 2/7 & 1/8 \\ 1/5 & 3/7 & 1/2 \end{pmatrix} \qquad (2.12)$$

and

$$B = \begin{pmatrix} 1/6 & 1/3 & 1/2 \\ 1/2 & 1/3 & 1/6 \\ 1/8 & 3/8 & 1/2 \end{pmatrix}, \qquad (2.13)$$

compatibility is assured since both A and B share the common cross-product ratio matrix

$$F = \begin{pmatrix} 4/3 & 8/9 \\ 12 & 8/3 \end{pmatrix}. \qquad (2.14)$$

□

2.2.2   Compatibility of General Conditional Probability Matrices

To determine whether A and B (no longer assumed to have all elements positive) are compatible it is helpful to go back to the original definition. They will be compatible if there exists a joint distribution P which has A

and B as its corresponding conditionals, i.e., such that, for every $(i,j) \in N(= N^A = N^B)$,

$$p_{ij} = a_{ij} \sum_i p_{ij} \tag{2.15}$$

and

$$p_{ij} = b_{ij} \sum_j p_{ij}. \tag{2.16}$$

Thus we are seeking solutions to linear equations. Additional constraints that must be invoked are

$$p_{ij} \geq 0, \quad \forall i,j, \tag{2.17}$$

and

$$\sum_{i=1}^{I} \sum_{j=1}^{J} p_{ij} = 1. \tag{2.18}$$

Equation (2.18) is just one more linear equation to add to the list. The constraint (2.17) is a lot more troublesome. A solution to (2.15), (2.16), and (2.18) will be of no use to us unless it is nonnegative.

There are two other ways in which we can describe our search for a compatible P in terms of linear equations subject to inequality constraints. It seems redundant to list three ways at this juncture, but as we shall see, all three ways will prove to be useful in noncompatibility cases when we, in some sense, try to identify an "almost" compatible P. The second method is based on the observation that if we could find compatible marginals we could, using A and B, readily obtain P. The third method is based on the fact that we really only need to find one compatible marginal, say that corresponding to the random variable X. It combined with B will give us P.

The three methods to determine compatibility may be summarized as follows:

Method I. Seek one probability matrix P satisfying

$$\begin{aligned}
p_{ij} - a_{ij} \sum_{i=1}^{I} p_{ij} &= 0, \quad \forall i,j, \\
p_{ij} - b_{ij} \sum_{j=1}^{J} p_{ij} &= 0, \quad \forall i,j, \\
\sum_{i=1}^{I} \sum_{j=1}^{J} p_{ij} &= 1,
\end{aligned} \tag{2.19}$$

and

$$p_{ij} \geq 0, \quad \forall i,j. \tag{2.20}$$

Method II. Seek two probability vectors $\underline{\tau}$ and $\underline{\eta}$ satisfying

$$\eta_j a_{ij} - \tau_i b_{ij} = 0, \quad \forall i, j,$$
$$\sum_{i=1}^{I} \tau_i = 1,$$
$$\sum_{j=1}^{J} \eta_j = 1, \tag{2.21}$$

and

$$\tau_i \geq 0, \quad \eta_j \geq 0, \quad \forall i, j. \tag{2.22}$$

Method III. Seek one probability vector $\underline{\tau}$ satisfying

$$a_{ij} \sum_{k=1}^{I} \tau_k b_{kj} - \tau_i b_{ij} = 0, \quad \forall i, j,$$
$$\sum_{i=1}^{I} \tau_i = 1,$$
$$\tau_i \geq 0, \quad \forall i. \tag{2.23}$$

In Method I, P is directly sought. In Method II we seek the marginals which combined with A and B will give us P. In Method III, we seek the X-marginal $\underline{\tau}$ which, combined with B, will give us P.

All three methods involve linear equations to be solved subject to non-negativity constraints.

Method I involves $2|N| + 1$ equations in $|N|$ unknowns (here $|N|$ is the cardinality of the incidence set $N^A = N^B$).

Method II involves $|N| + 2$ equations in $I + J$ unknowns while Method III involves $|N| + 1$ equations in I unknowns. System (2.23) will probably be the one we will try to solve in practice since it involves less equations and less unknowns. Systems (2.19)–(2.20) and (2.21)–(2.22) will be of interest to us in Section 2.8 in the context of ϵ-compatibility.

To check the existence of solutions and to identify all solutions to systems of equations under nonnegativity constraints, such as those above, we can use the following theorems (Castillo, Cobo, Jubete, and Pruneda (1998)).

First we need some definitions.

Definition 2.7 (Cone generated by a matrix). Let A be a real matrix of dimension $m \times n$. The polyhedral convex cone generated by A, denoted by $\pi(A)$, is the set of all vectors in $\mathbb{R}^m$ which can be expressed as linear combinations of the columns of A with nonnegative coefficients.   □

Definition 2.8 (Dual or polar cone). Let π be a cone in $\mathbb{R}^m$. The dual or polar cone of π is the set

$$\Omega(\pi) = \{\underline{u} \in \mathbb{R}^m : \underline{v}'\underline{u} \leq 0, \quad \forall \underline{v} \in \pi\}.$$

In the case in which π is the cone generated by A, the dual cone admits a simple description involving the matrix A. Thus

$$\Omega(\pi(A)) = \{\underline{u} \in \mathbb{R}^m : A^T \underline{u} \le \underline{0}\}.$$

$\square$

It can be verified that $\Omega(\pi(A))$ itself is a polyhedral convex cone. It too can be viewed as a cone generated by some finite set of vectors in $\mathbb{R}^m$. In any cone π, there may be some vectors $\underline{u}$ such that $-\underline{u} \in \pi$ and some vectors $\underline{u}$ such that $-\underline{u} \notin \pi$. Consequently, a cone can be represented as the sum of a linear space and a cone. The polar cone $\Omega(\pi(A))$ is thus of the form

$$\Omega(\pi(A)) = L(V) + \pi(W), \tag{2.24}$$

where $L(V)$ is the linear space generated by the columns of an $(m \times k_1)$-dimensional matrix V and $\pi(W)$ is the cone generated by the columns of an $(m \times k_2)$-dimensional matrix W. The matrices V and W that appear in (2.24) are called the generators of the dual cone $\Omega(\pi(A))$. Details regarding the construction of the specific matrices V and W appearing in (2.24) may be found in Castillo et al. (1998).

Theorem 2.2 (Existence of a nonnegative solution in a linear system). *The system,*

$$C\underline{x} = \underline{a} \;\; subject \; to \;\; \underline{x} \ge \underline{0}, \tag{2.25}$$

where C is a $m \times n$ constant matrix, $\underline{x}$ is a column matrix of n unknowns, and $\underline{a}$ is a m column matrix of real numbers, has a solution iff

$$\begin{aligned} V^T \underline{a} &= \underline{0}, \\ W^T \underline{a} &\le \underline{0}, \end{aligned} \tag{2.26}$$

where V and W are the generators of $\Omega(\pi(C))$, the dual or polar cone of the cone generated by C (i.e., those matrices appearing in the representation (2.24)). $\square$

Thus, analyzing the compatibility of the system of equations $C\underline{x} = \underline{a}$ reduces to finding the polar cone $\Omega(\pi(C))$ and checking whether or not $V^T \underline{a} = \underline{0}$ and $W^T \underline{a} \le \underline{0}$.

We will illustrate the use of this theorem for Method III above. In that situation, the role of $\underline{x}$ is played by $\underline{\tau}$, $\underline{a} = (0, \dots, 0, 1)$ and the coefficients of the matrix C can be identified from (2.23).

Due to the simple structure of $\underline{a}$ in our application, we need only check the last component of the dual cone generators, that is, $v_{in} = 0, \forall i$ and $w_{jn} \le 0, \forall j$.

Theorem 2.3 (Solution of a system of linear equations and inequalities) *Consider a system of linear equations and inequalities. Without loss of generality, we can assume that it can be written in the following form:*

$$
\begin{aligned}
B\underline{x} &= \underline{0}, \\
C\underline{x} &\leq \underline{0}, \\
x_n &= 1.
\end{aligned}
\tag{2.27}
$$

where B and C are matrices associated with the equalities and inequalities of the system. The set $\mathbf{X}$ of solutions of (2.27) is

$$
\mathbf{X} \equiv \{\underline{x} \in \Omega(L(B) + \pi(C)) | x_n = 1\}.
\tag{2.28}
$$

This implies that to solve (2.27) we can find first the polar cone $\Omega(L(B) + \pi(C))$ associated with the first two sets of constraints, and then impose the condition $x_n = 1$. □

Example 2.4 (Compatibility using Method III). Consider the matrices A and B:

$$
A = \begin{pmatrix} 1/4 & 1/2 \\ 3/4 & 1/2 \end{pmatrix},
\tag{2.29}
$$

$$
B = \begin{pmatrix} 1/3 & 2/3 \\ 3/5 & 2/5 \end{pmatrix}.
\tag{2.30}
$$

If we use Method III, then the system (2.23) becomes

$$
\begin{pmatrix}
1/4 & -3/20 \\
1/3 & -1/5 \\
-1/4 & 3/20 \\
-1/3 & 1/5 \\
1 & 1
\end{pmatrix}
\underline{\tau} =
\begin{pmatrix} 0 \\ 0 \\ 0 \\ 0 \\ 1 \end{pmatrix}
; \quad \underline{\tau} \geq \underline{0}.
\tag{2.31}
$$

According to Theorem 2.2 we need to obtain the generators of the dual cone $\Omega(\pi(C)) = L(V) + \pi(W)$, which are given in Table 2.1. The reader interested in the derivation of these generators is referred to Castillo, Cobo, Jubete, and Pruneda (1998). In general it is not a trivial exercise.

Since the generators V and W do satisfy conditions (2.26), the conditional probability matrices A and B are compatible. □

Example 2.5 (Incompatibility using Method III). Consider the matrices A and B:

$$
A = \begin{pmatrix} 1/4 & 1/2 \\ 3/4 & 1/2 \end{pmatrix},
\tag{2.32}
$$

$$
B = \begin{pmatrix} 1/6 & 5/6 \\ 3/5 & 2/5 \end{pmatrix}.
\tag{2.33}
$$

TABLE 2.1. Generators of the dual cone $\Omega(\pi(C))$ in Example 2.4.

w_1	w_2	v_1	v_2	v_3
$-5/2$	$5/2$	1	$4/3$	$-4/3$
0	0	0	0	1
0	0	1	0	0
0	0	0	1	0
$-3/8$	$-5/8$	0	0	0

TABLE 2.2. Generators of the dual cone $\Omega(\pi(C))$ in Example 2.5.

w_1	w_2	v_1
$16/3$	$100/9$	$148/9$
-4	$-10/3$	$-22/3$
0	0	1

Using again Method III, the system (2.23) becomes

$$\begin{pmatrix} 1/8 & -3/20 \\ 5/12 & -1/5 \\ -1/8 & 3/20 \\ -5/12 & 1/5 \\ 1 & 1 \end{pmatrix} \underline{\tau} = \begin{pmatrix} 0 \\ 0 \\ 0 \\ 0 \\ 1 \end{pmatrix}, \quad \underline{\tau} \geq \underline{0}. \tag{2.34}$$

At this stage, if desired, we can remove redundant rows and linearly dependent rows before proceeding to computing the generators of the dual cone. For example, we can remove Equations 3 and 4, because they are exactly the same (sign changed) as Equations 1 and 2, respectively, to obtain the new system:

$$\begin{pmatrix} 1/8 & -3/20 \\ 5/12 & -1/5 \\ 1 & 1 \end{pmatrix} \underline{\tau} = \begin{pmatrix} 0 \\ 0 \\ 1 \end{pmatrix}, \quad \underline{\tau} \geq \underline{0}. \tag{2.35}$$

Following Theorem 2.2 we obtain the generators of the dual $\Omega(\pi(C^*)) = L(V) + \pi(W)$, which are given in Table 2.2.

These matrices V and W do not satisfy (2.26), thus the conditional probability matrices A and B are incompatible.   □

Example 2.6 (A difficult case). Consider the matrices A and B:

$$A = \begin{pmatrix} 1/2 & 1/2 & 0 \\ 0 & 1/2 & 1/2 \\ 1/2 & 0 & 1/2 \end{pmatrix}, \tag{2.36}$$

TABLE 2.3. Generators of the dual cone $\Omega(\pi(C))$ in Example 2.6.

w_1	w_2	w_3	v_1
6	12	12	30
-6	-6	-6	-18
-3	-6	-3	-12
0	0	0	1

$$B = \begin{pmatrix} 1/3 & 2/3 & 0 \\ 0 & 1/3 & 2/3 \\ 1/3 & 0 & 2/3 \end{pmatrix}. \tag{2.37}$$

We cannot resolve this case using cross-product ratio matrices (there are no positive 2×2 submatrices). However, Method III and Theorem 2.2 may be used without difficulty.

In this case, the system (2.23) becomes

$$\begin{pmatrix} 1/6 & 0 & -1/6 \\ 1/3 & -1/6 & 0 \\ 0 & 0 & 0 \\ 0 & 0 & 0 \\ -1/3 & 1/6 & 0 \\ 0 & 1/3 & -1/3 \\ -1/6 & 0 & 1/6 \\ 0 & 0 & 0 \\ 0 & -1/3 & 1/3 \\ 1 & 1 & 1 \end{pmatrix} \underline{\tau} = \begin{pmatrix} 0 \\ 0 \\ 0 \\ 0 \\ 0 \\ 0 \\ 0 \\ 0 \\ 0 \\ 1 \end{pmatrix}, \quad \underline{\tau} \geq \underline{0}, \tag{2.38}$$

which by removing redundant equations can be written as:

$$\begin{pmatrix} 1/6 & 0 & -1/6 \\ 1/3 & -1/6 & 0 \\ 0 & 1/3 & -1/3 \\ 1 & 1 & 1 \end{pmatrix} \underline{\tau} = \begin{pmatrix} 0 \\ 0 \\ 0 \\ 1 \end{pmatrix}, \quad \underline{\tau} \geq \underline{0}. \tag{2.39}$$

The generators of the dual $\Omega(\pi(C)) = L(V) + \pi(W)$, where C is the matrix associated with system (2.39), are given in Table 2.3.

Since V and W do not satisfy (2.26), the conditional probability matrices A and B are incompatible. □

2.3   Near Compatibility

Suppose that we are given two families of conditional distributions of X given Y and of Y given X (i.e., A and B above) which are not compatible.

How can we measure such incompatibility? And, how can we find a distribution P that is, in some sense, minimally incompatible with the given conditional specifications? Such results will be of potential interest in the context of elicitation of joint prior distributions in Bayesian analysis. In the case of a two-dimensional parameter θ, the informed expert might give conditional probabilities for θ_1 given particular choices of values for θ_2 and conditional probabilities for θ_2 given values of θ_1. If our expert is fallible, it is quite possible that the collection of conditional probabilities thus elicited might be incompatible. A suitable choice of prior to use in subsequent analysis might then be that joint distribution $f(\theta_1, \theta_2)$ that is least at variance with the given elicited conditional probabilities. More generally, we might envision obtaining partial or complete conditional specification from more than one expert. Such information would most likely lack consistency and, again a minimally discrepant distribution might be sought. The Kullback-Leibler information function provides a convenient discrepancy measure in such settings. As we shall see, not only does it provide a discrepancy measure but, using it, a straightforward algorithm can be described which will yield the most nearly compatible distribution. We will mention alternatives to the Kullback-Leibler measure but, in many ways, it seems the most attractive choice.

2.4 Minimal Incompatibility in Terms of Kullback-Leibler Pseudo-Distance

Suppose that A and B, two families of conditional distributions, are not compatible and perhaps do not even have identical incidence sets. We seek a probability matrix P with nonnegative entries summing to 1, which has conditionals as close as possible to those given by A and B. Thus, we are seeking $P = (p_{ij})_{I \times J}$ with $\sum_{i=1}^{I} \sum_{j=1}^{J} p_{ij} = 1$ and with

$$p_{ij}/p_{.j} \approx a_{ij}, \quad \forall i, j,$$

and

$$p_{ij}/p_{i.} \approx b_{ij}, \quad \forall i, j.$$

To measure discrepancy between distributions we will use the Kullback-Leibler information function as a measure of (pseudo) distance. Using it, it is reasonable to search for a matrix P that will minimize the following objective function

$$\sum_{i=1}^{I} \sum_{j=1}^{J} b_{ij} \log \left(\frac{b_{ij} p_{i.}}{p_{ij}} \right) + \sum_{i=1}^{I} \sum_{j=1}^{J} a_{ij} \log \left(\frac{a_{ij} p_{.j}}{p_{ij}} \right). \tag{2.40}$$

Define

$$D = A + B,$$

with elements $d_{ij} = a_{ij} + b_{ij}$. In order to ensure that a unique minimizing choice of P exists for the objective function (2.40), it is necessary to make some assumptions about the incidence set of the matrix D. For example, D must not be block diagonal. A reasonable requirement that we will assume is that some power of D, perhaps D itself, have all elements strictly positive.

If we differentiate (2.40) and use a Lagrange multiplier for the constraint $\sum_{i=1}^{I} \sum_{j=1}^{J} p_{ij} = 1$, we may verify:

Theorem 2.4 *Denote by P^* the choice of P which minimizes (2.40). Then P^* must satisfy the following system of equations:*

$$\frac{p_{ij}^*}{p_{i\cdot}^*} + \frac{p_{ij}^*}{p_{\cdot j}^*} = d_{ij}, \quad i = 1, 2, \ldots, I, \quad j = 1, 2, \ldots, J. \quad (2.41)$$

$\square$

It is possible to solve (2.41) using a simple iterative algorithm:

$$p_{i,j}^{(n+1)} = \frac{d_{ij}/[1/p_{i\cdot}^{(n)} + 1/p_{\cdot j}^{(n)}]}{\sum_{i=1}^{I} \sum_{j=1}^{J} d_{ij}/[1/p_{i\cdot}^{(n)} + 1/p_{\cdot j}^{(n)}]}, \quad (2.42)$$

beginning initially with $p_{ij}^{(0)} = 1/IJ$, $i = 1, 2, \ldots, I$, $j = 1, 2, \ldots, J$. The algorithm (2.42) does appear to usually converge. Indeed it converges quite rapidly in all examples tried. However, a proof of this convergence is lacking. Fortunately there is another way to seek the optimal choice P^* to minimize (2.40).

Let P^* be the solution to (2.41). Introduce the notation

$$p_{1|2}^*(i|j) = p_{ij}^*/p_{\cdot j}^*,$$

$$p_{2|1}^*(j|i) = p_{ij}^*/p_{i\cdot}^*.$$

Now for each i, j we have

$$p_{1|2}^*(i|j) + p_{2|1}^*(j|i) = d_{ij}.$$

Thus if $d_{ij} \neq 0$, there exist numbers q_{ij1}^* and q_{ij2}^* such that

$$p_{1|2}^*(i|j) = d_{ij} q_{ij1}^*, \quad (2.43)$$

$$p_{2|1}^*(j|i) = d_{ij} q_{ij2}^*, \quad (2.44)$$

where

$$q_{ij1}^* + q_{ij2}^* = 1, \quad i = 1, 2, \ldots, I, \quad j = 1, 2, \ldots, J, \quad (2.45)$$

and indeed

$$q_{ijk}^* \in [0, 1], \quad \forall i, j, k.$$

In addition, the following relations hold

$$1 = \sum_{i=1}^{I} p^*_{1|2}(i|j) = \sum_{i=1}^{I} d_{ij} q^*_{ij1}, \quad j = 1, 2, \ldots, J, \qquad (2.46)$$

and

$$1 = \sum_{j=1}^{J} p^*_{2|1}(j|i) = \sum_{j=1}^{J} d_{ij} q^*_{ij2}, \quad i = 1, 2, \ldots, I. \qquad (2.47)$$

We seek an $I \times J \times 2$ array, Q^* of nonnegative q^*_{ijk}'s satisfying (2.45), (2.46), and (2.47).

Visualizing our array as having rows, columns and "stacks," we see that (2.47) can be achieved by row normalizations, (2.46) by column normalizations, and (2.45) by stack normalizations. If we begin with the initial array $q^*_{ijk} \equiv 0.5$ and iteratively apply such row, column, and stack normalizations we are guaranteed convergence to a unique Q^* since we are simply using a variation of the Darroch-Ratcliff (1972) iterative scaling algorithm.

Algorithm 2.2 (Obtaining the matrix Q^*).

Input. *Matrices A and B, of dimension $I \times J$, such that the columns of A and the rows of B sum to 1, and $DesiredErr$, the desired error.*

Output. *Matrix Q^*.*

Step 1. *Calculate $D = A + B$.*

Step 2. *Initialize $q^*_{ijk} = 0.5$, $\forall i, j, k$, and make $TrueErr = DesiredErr$.*

Step 3. *If $TrueErr > DesiredErr$, then make $r^*_{ijk} = q^*_{ijk}$, $\forall i, j, k$, and go to Step 4. Otherwise, go to Step 8.*

Step 4. *Normalize rows: $q^*_{ij2} = \dfrac{q^*_{ij2}}{\sum\limits_{j=1}^{J} d_{ij} q^*_{ij2}}$, $\forall i, j$.*

Step 5. *Normalize columns: $q^*_{ij1} = \dfrac{q^*_{ij1}}{\sum\limits_{i=1}^{I} d_{ij} q^*_{ij1}}$; $\forall i, j$.*

Step 6. *Normalize stacks: $q^*_{ijk} = q_{ijk}/(q^*_{ij1} + q^*_{ij2})$, $\forall i, j, k$.*

Step 7. *Calculate $TrueErr = \max\limits_{i,j,k} |q^*_{ijk} - r^*_{i,j,k}|$.*

Step 8. *Return matrix Q^*.* □

Arnold and Gokhale (1998c) report that in cases that they have considered, the determination of P^* via the Darroch-Ratcliff determination of Q^* agrees with the determination of P^* based on the usually quicker iterative scheme (2.42). This supports, but of course does not prove, the claim that the simple scheme (2.42) may be used routinely in practice.

Example 2.7 (Obtaining a nearly compatible probability matrix).
Consider the following candidate arrays A and B:

$$A = \begin{pmatrix} 0.3 & 0.3 & 0.0 & 0.2 \\ 0.2 & 0.1 & 0.2 & 0.3 \\ 0.5 & 0.2 & 0.4 & 0.4 \\ 0.0 & 0.4 & 0.4 & 0.1 \end{pmatrix}, \tag{2.48}$$

$$B = \begin{pmatrix} 0.2 & 0.3 & 0.4 & 0.1 \\ 0.5 & 0.0 & 0.2 & 0.3 \\ 0.6 & 0.1 & 0.2 & 0.1 \\ 0.0 & 0.4 & 0.6 & 0.0 \end{pmatrix}. \tag{2.49}$$

It is not difficult to verify that they are incompatible, they do not even share a common coincidence matrix!

If we write the $I \times J \times 2$ array Q^* as an $I \times 2J$ matrix whose first J columns give the q^*_{ij1}'s and whose second J columns give the q^*_{ij2}'s, iterative normalization of rows, columns, and stacks yields in the limit:

$$Q^* = \begin{pmatrix} 0.346 & 0.505 & 0.352 & 0.610 & 0.654 & 0.495 & 0.648 & 0.390 \\ 0.346 & 0.506 & 0.353 & 0.610 & 0.654 & 0.494 & 0.647 & 0.390 \\ 0.532 & 0.687 & 0.539 & 0.771 & 0.468 & 0.313 & 0.461 & 0.229 \\ 0.388 & 0.550 & 0.395 & 0.652 & 0.613 & 0.450 & 0.606 & 0.348 \end{pmatrix}.$$

Referring to (2.43) and (2.44), we can then write the corresponding conditional distributions which are closest to A and B in the sense of minimizing (2.40). They are

$$P^*_{1|2}(i|j) = \begin{pmatrix} 0.1728 & 0.3032 & 0.1410 & 0.1830 \\ 0.2422 & 0.0506 & 0.1411 & 0.3662 \\ 0.5850 & 0.2061 & 0.3235 & 0.3854 \\ 0.0000 & 0.4402 & 0.3945 & 0.0652 \end{pmatrix} \tag{2.50}$$

and

$$P^*_{2|1}(j|i) = \begin{pmatrix} 0.3272 & 0.2968 & 0.2590 & 0.1170 \\ 0.4578 & 0.0494 & 0.2589 & 0.2338 \\ 0.5150 & 0.0939 & 0.2765 & 0.1146 \\ 0.0000 & 0.3598 & 0.6055 & 0.0348 \end{pmatrix}. \tag{2.51}$$

The matrices displayed in (2.50) and (2.51) satisfy condition (ii) of Theorem 1.1 and thus are compatible with some probability distribution P^*. To fully identify P^*, it remains only to identify one of its marginal distributions. In general, this can be accomplished by solving a system of $I \times J$

linear equations. However, if column j of $P^*_{1|2}$ contains no zeros (as is the case with columns $2, 3$, and 4 in our example), we can write

$$p^*_{i.} = \frac{p^*_{1|2}(i|j)/p^*_{2|1}(j|i)}{\sum_{i=1}^{4} p^*_{1|2}(i|j)/p^*_{2|1}(j|i)}.$$

(2.52)

Evaluating (2.52), say for $j = 2$, using the figures displayed in arrays (2.50) and (2.51) we obtain

$$\begin{aligned} p^*_{1.} &= 0.18701, \quad p^*_{2.} = 0.18741, \\ p^*_{3.} &= 0.40174, \quad p^*_{4.} = 0.22384. \end{aligned}$$

(2.53)

For completeness, if we wish we may also use an analogous approach to obtain the other marginal distribution:

$$\begin{aligned} p^*_{.1} &= 0.35371, \quad p^*_{.2} = 0.18302, \\ p^*_{.3} &= 0.34347, \quad p^*_{.4} = 0.11981. \end{aligned}$$

(2.54)

Combining the values in (2.53) (the marginal distribution of X) with the values in (2.51) (the conditional distributions of Y given X) we finally obtain the following optimal distribution (i.e., that which is least discordant with A and B):

$$P^* = \begin{pmatrix} 0.0612 & 0.0555 & 0.0484 & 0.0219 \\ 0.0857 & 0.0093 & 0.0485 & 0.0439 \\ 0.2069 & 0.0377 & 0.1111 & 0.0461 \\ 0.0000 & 0.0805 & 0.1355 & 0.0078 \end{pmatrix}.$$

(2.55)

□

The algorithm used to determine P^* works for almost any pair of candidate conditional distributions A, B (we only need something like $(A+B)^k > 0$ for some k). It works even when cross-product ratios are noninformative, as in the pathological example given in expressions (2.9) and (2.10). In addition, the algorithm can be used to determine whether two matrices A and B are compatible. We simply find the corresponding P^* and compare its conditional distributions (which were found in the development of P^*) with A (and/or B). If they agree, A and B are compatible; if not, A and B are incompatible.

Example 2.8 (Checking compatibility). Recall A and B, as given in (2.9) and (2.10):

$$A = \begin{pmatrix} 1/2 & 1/2 & 0 \\ 0 & 1/2 & 1/2 \\ 1/2 & 0 & 1/2 \end{pmatrix}$$

(2.56)

and

$$B = \begin{pmatrix} 1/3 & 2/3 & 0 \\ 0 & 1/3 & 2/3 \\ 1/3 & 0 & 2/3 \end{pmatrix}. \tag{2.57}$$

It is perhaps not obvious whether these are incompatible or not. We apply the above algorithm to obtain

$$Q^* = \begin{pmatrix} 0.5673 & 0.4519 & 0.3415 & 0.4327 & 0.5481 & 0.6585 \\ 0.6758 & 0.5673 & 0.4519 & 0.3242 & 0.4327 & 0.5481 \\ 0.6327 & 0.5200 & 0.4052 & 0.3673 & 0.4800 & 0.5948 \end{pmatrix}, \tag{2.58}$$

$$P^*_{1|2}(i|j) = \begin{pmatrix} 0.4728 & 0.5272 & 0.0000 \\ 0.0000 & 0.4728 & 0.5272 \\ 0.5272 & 0.0000 & 0.4728 \end{pmatrix}, \tag{2.59}$$

$$P^*_{2|1}(j|i) = \begin{pmatrix} 0.3606 & 0.6394 & 0.000 \\ 0.000 & 0.3606 & 0.6394 \\ 0.3061 & 0.000 & 0.6939 \end{pmatrix}, \tag{2.60}$$

with corresponding marginals

$$p^*_{1.} = 0.2562, \quad p^*_{2.} = 0.4073, \quad p^*_{3.} = 0.3365,$$
$$p^*_{.1} = 0.1954, \quad p^*_{.2} = 0.3107, \quad p^*_{.3} = 0.4940.$$

Thus, we finally get as our most nearly compatible distribution

$$\tilde{P} = \begin{pmatrix} 0.0924 & 0.1638 & 0.0000 \\ 0.0000 & 0.1468 & 0.2605 \\ 0.1030 & 0.0000 & 0.2335 \end{pmatrix}. \tag{2.61}$$

Since (2.59) differs from (2.56) and (2.60) differs from (2.57), we confirm the incompatibility of (2.56) and (2.57). □

The value of the criterion (2.40) can be used as a measure of incompatibility. If two pairs (A, B) and (C, D) of incompatible matrices of the same order are given, the pair with the lower value of (2.40) can be ranked as less incompatible. Indeed, if matrices (A, B) are compatible the minimal value of (2.40) is zero and the converse also holds.

2.5   More Than One Expert

Instead of just one pair of conditional distributions, we might have several, undoubtedly lacking in consistency, from a variety of forecasters. More generally, we might have a number, say $n^{(1)}_{ij}$, of determinations of a_{ij} (possibly a different number of determinations for each pair (i, j)) and $n^{(2)}_{ij}$ determinations of b_{ij}. We might then seek a joint distribution, P, that is minimally discrepant from all of these conditional probabilities. If our determinations are judged to be equally reliable than a reasonable approach would begin by determining averages such as:

$$\tilde{a}_{ij} = \frac{1}{n_{ij}^{(1)}} \sum_{k=1}^{n_{ij}^{(1)}} a_{ij(k)} \tag{2.62}$$

and

$$\tilde{b}_{ij} = \frac{1}{n_{ij}^{(2)}} \sum_{k=1}^{n_{ij}^{(2)}} b_{ij(k)}. \tag{2.63}$$

Assume that the $n_{ij}^{(1)}$'s and $n_{ij}^{(2)}$'s are positive (i.e., at least one determination has been provided for each of the conditional probabilities). Also assume that each row and column of $\tilde{A}$ and $\tilde{B}$ has at least one nonzero entry. Next, column normalize $\tilde{A}$ to make its columns sum to 1, to obtain $\tilde{\tilde{A}}$. Analogously, row normalize $\tilde{B}$ to obtain $\tilde{\tilde{B}}$. Then seek a distribution P^* that is minimally discrepant from $\tilde{\tilde{A}}$ and $\tilde{\tilde{B}}$, using the techniques of Section 2.4, i.e., using criterion (2.40).

If the determination of the a_{ij}'s and b_{ij}'s are judged to be of unequal reliability, then weights may be appropriately introduced to the definitions of $\tilde{a}_{ij}$ and $\tilde{b}_{ij}$ in (2.62) and (2.63) above.

Other criteria may of course be used. The advantage of the one just described is that it is simple and can be implemented using the same methods as were used in the case of one expert (just replacing (A, B) by $(\tilde{\tilde{A}}, \tilde{\tilde{B}})$).

A modified version of the objective function (2.40) allowing multiple determinations of the a_{ij}'s and b_{ij}'s can be set up as follows:

$$\sum_{i=1}^{I} \sum_{j=1}^{J} \sum_{k=1}^{n_{ij}^{(1)}} a_{ij(k)} \log \left(\frac{a_{ij(k)} p_{.j}}{p_{ij}} \right) + \sum_{i=1}^{I} \sum_{j=1}^{J} \sum_{k=1}^{n_{ij}^{(2)}} b_{ij(k)} \log \left(\frac{b_{ij(k)} p_{i.}}{p_{ij}} \right). \tag{2.64}$$

In the case in which all the $n_{ij}^{(1)}$'s and all the $n_{ij}^{(2)}$'s are equal say to n (the case of n experts each providing complete conditional distributions (A_k, B_k), $k = 1, 2, \ldots, n$), use of criterion (2.64) will be completely equivalent to use of criterion (2.40) applied to the average matrices $\tilde{A}$ and $\tilde{B}$.

2.6   Related Discrepancy Measures

There are, of course, a variety of distance and pseudo-distance measures that can be applied to determine how far the conditional distributions of a matrix P are from two given conditional probability matrices A and B.

The Kullback-Liebler criterion (2.40) was a reasonable candidate and turned out to be remarkably tractable. It is worth noting that, if we switched the roles of a_{ij}, b_{ij} and $p_{1|2}(i|j)$ and $p_{2|1}(j|i)$ in (2.40) then, since the Kullback-Liebler measure is not symmetric, we obtain a related but not equivalent objective function:

$$\sum_{i=1}^{I}\sum_{j=1}^{J}\frac{p_{ij}}{p_{i.}}\log\left(\frac{p_{ij}}{p_{i.}b_{ij}}\right)+\sum_{i=1}^{I}\sum_{j=1}^{J}\frac{p_{ij}}{p_{.j}}\log\left(\frac{p_{ij}}{p_{.j}a_{ij}}\right). \qquad (2.65)$$

Criterion (2.65) is much more troublesome to work with than is criterion (2.40). Absent any strong argument for using (2.65), use of (2.40) seems preferable.

However, some tractable alternatives are available. Recall that our goal is to choose P so that p_{ij} is well approximated by $a_{ij}p_{.j}, \forall i, j$, and so that also p_{ij} is well approximated by $b_{ij}p_{i.}, \forall i, j$. This suggests consideration of a quadratic measure of discrepancy such as

$$Q = \sum_{i=1}^{I}\sum_{j=1}^{J}[(p_{ij}-b_{ij}p_{i.})^2 + (p_{ij}-a_{ij}p_{.j})^2]. \qquad (2.66)$$

which can also be written as

$$Q = \sum_{i=1}^{I}\sum_{j=1}^{J}\left[2\left(p_{ij}-(b_{ij}p_{i.}+a_{ij}p_{.j})/2\right)^2 + (a_{ij}p_{.j}-b_{ij}p_{i.})^2/2\right]$$
$$(2.67)$$
$$\geq \sum_{i=1}^{I}\sum_{j=1}^{J}\left[(a_{ij}p_{.j}-b_{ij}p_{i.})^2/2\right].$$

Thus, the minimizing value of P corresponding to this objective function Q can be found by solving the following system of linear equations:

$$p_{ij} = (p_{.j}a_{ij}+p_{i.}b_{ij})/2, \quad i=1,2,\ldots,I, \quad j=1,2,\ldots,J, \qquad (2.68)$$

subject to $\sum_{i=1}^{I}\sum_{j=1}^{J}p_{ij}=1$.

An efficient manner of solving system (2.68) involves solving first for row and column sums of (2.68) (namely the $p_{i.}$'s and $p_{.j}$'s).

If we sum (2.68) over j for each fixed i, and over i for each fixed j, we are led to the following system of $I+J$ linear equations:

$$p_{i.} = \frac{1}{2}\left[\sum_{j=1}^{J}p_{.j}a_{ij}+p_{i.}\right], \quad i=1,2,\ldots,I, \qquad (2.69)$$

$$p_{.j} = \frac{1}{2}\left[p_{.j}+\sum_{i=1}^{I}p_{i.}b_{ij}\right], \quad j=1,2,\ldots,J. \qquad (2.70)$$

If we define an $(I + J)$-dimensional stochastic vector

$$\underline{q} = \frac{1}{2}(p_{1.}, p_{2.}, \ldots, p_{I.}, p_{.1}, p_{.2}, \ldots, p_{.J})$$

we can write the system (2.69)-(2.70) in the form

$$\underline{q} = \underline{q}R, \tag{2.71}$$

where

$$R = \frac{1}{2}\begin{pmatrix} I & B \\ A' & I \end{pmatrix}. \tag{2.72}$$

The matrix R is stochastic and the solution to (2.71) will be a long-run stationary distribution of a Markov chain with transitions governed by R. The fact that the diagonal elements of R are positive, guarantees that the chain is aperiodic. It will have a unique long-run distribution (stationary distribution) provided that R is irreducible. Iterative application of (2.71) will converge to a solution to (2.71) in the irreducible case.

Having found the $p_{i.}$'s and $p_{.j}$'s by solving (2.71), it is a straightforward matter to obtain the p_{ij}'s directly from (2.68).

In fact (2.68) itself can be interpreted in a Markovian context. If we rewrite (p_{ij}) in stacked form as an $(I \times J)$-dimensional row vector, then we may verify that (p_{ij}) will be a left eigenvector corresponding to the eigenvalue 1 normalized to sum to 1, of a somewhat complicated stochastic matrix. So, in most cases (irreducibility considerations), the following iterative scheme will converge to a solution to (2.68):

$$p_{ij}^{(n+1)} = \frac{1}{2}[p_{.j}^{(n)}a_{ij} + p_{i.}^{(n)}b_{ij}]. \tag{2.73}$$

Although the iterative scheme (2.71) is of smaller dimension than scheme (2.73) (since it involves solving a system with $I + J - 2$ instead of $IJ - 1$ unknowns), the simplicity of algorithm (2.73) is appealing unless I and/or J is large.

Example 2.9 (Iterative method). Application of algorithm (2.73) (criterion (2.66)) to the matrices (2.48) and (2.49) leads to

$$P^{**} = \begin{pmatrix} 0.0754 & 0.0562 & 0.0382 & 0.0213 \\ 0.0859 & 0.0092 & 0.0518 & 0.0466 \\ 0.2141 & 0.0384 & 0.1049 & 0.0435 \\ 0.0000 & 0.0796 & 0.1292 & 0.0059 \end{pmatrix}, \tag{2.74}$$

which is slightly different from the matrix

$$P^{*} = \begin{pmatrix} 0.0612 & 0.0555 & 0.0484 & 0.0219 \\ 0.0857 & 0.0093 & 0.0485 & 0.0439 \\ 0.2069 & 0.0377 & 0.1111 & 0.0461 \\ 0.0000 & 0.0805 & 0.1355 & 0.0078 \end{pmatrix}, \tag{2.75}$$

which was obtained in (2.55) by minimizing the KL criterion. □

Example 2.10 (Iterative method applied to matrices with zeros).
Applying the iterative method to the matrices (2.56) and (2.57) we obtain
the matrix

$$P^{**} = \begin{pmatrix} 0.0917 & 0.1583 & 0.0000 \\ 0.0000 & 0.1417 & 0.2583 \\ 0.1083 & 0.0000 & 0.2417 \end{pmatrix}. \tag{2.76}$$

□

**Example 2.11 (Obtaining the compatible probability matrix by
the iterative method).** Consider again the matrices in (2.6) and (2.7):

$$A = \begin{pmatrix} 1/7 & 1/4 & 3/7 & 1/7 \\ 2/7 & 1/2 & 1/7 & 2/7 \\ 4/7 & 1/4 & 3/7 & 4/7 \end{pmatrix} \tag{2.77}$$

and

$$B = \begin{pmatrix} 1/6 & 1/6 & 1/2 & 1/6 \\ 2/7 & 2/7 & 1/7 & 2/7 \\ 1/3 & 1/12 & 1/4 & 1/3 \end{pmatrix}. \tag{2.78}$$

We can use the iterative method (2.73) above to obtain the probability
matrix

$$P^{**} = \begin{pmatrix} 1/25 & 1/25 & 3/25 & 1/25 \\ 2/25 & 2/25 & 1/25 & 2/25 \\ 4/25 & 1/25 & 3/25 & 4/25 \end{pmatrix} \tag{2.79}$$

which is compatible with the above A and B conditionals.   □

Other criteria might be of interest. Replacing squares in (2.66) by absolute values leads to

$$Q' = \sum_{i=1}^{I} \sum_{j=1}^{J} [|p_{ij} - b_{ij} p_{i.}| + |p_{ij} - a_{ij} p_{.j}|]. \tag{2.80}$$

Linear programming techniques could be useful in managing this objective
function.

It might be considered more natural to seek P so that $p_{ij}/p_{.j}$ is close to
$a_{ij}, \forall i, j$ and $p_{ij}/p_{i.}$ is close to $b_{ij}, \forall i, j$. The price to be paid if we choose
this route is that the more natural objective functions will be more difficult
to minimize, e.g.,

$$Q'' = \sum_{i=1}^{I} \sum_{j=1}^{J} \left[\left(a_{ij} - \frac{p_{ij}}{p_{.j}} \right)^2 + \left(b_{ij} - \frac{p_{ij}}{p_{i.}} \right)^2 \right], \tag{2.81}$$

or

$$Q''' = \sum_{i=1}^{I} \sum_{j=1}^{J} \left[\left| a_{ij} - \frac{p_{ij}}{p_{.j}} \right| + \left| b_{ij} - \frac{p_{ij}}{p_{i.}} \right| \right]. \tag{2.82}$$

2.7   Markovian Measures of Discrepancy

If we have compatible conditional specifications, then a Gibbs sampler-type argument can be used to identify the corresponding marginal distributions (cf. the discussion in Section 1.7). In the finite discrete case it is particularly simple to describe. The long-run distribution of the I state chain with transition matrix BA' will coincide with $(p_{1.}, p_{2.}, \ldots, p_{I.})$, the X marginal distribution of P. Analogously, the long-run distribution of the J state chain $A'B$ will coincide with $(p_{.1}, p_{.2}, \ldots, p_{.J})$, the Y marginal of P.

If we denote the long-run distributions corresponding to BA' and $A'B$ by $\underline{\pi} = (\pi_1, \pi_2, \ldots, \pi_I)$ and $\underline{\eta} = (\eta_1, \eta_2, \ldots, \eta_J)$, respectively, then, provided A and B are compatible, we will have

$$a_{ij}\eta_j = b_{ij}\pi_i, \qquad i = 1, 2, \ldots, I, \quad J = 1, 2, \ldots, J. \qquad (2.83)$$

If A and B are incompatible the left and right sides of (2.83) will not all be equal.

This suggests use of the following index of incompatibility

$$D = \sum_{i=1}^{I} \sum_{j=1}^{J} (a_{ij}\eta_j - b_{ij}\tau_i)^2, \qquad (2.84)$$

where $\underline{\eta}$ and $\underline{\tau}$ are solutions of the system

$$\underline{\pi} BA' = \underline{\pi} \qquad (2.85)$$

and

$$\underline{\eta} A'B = \underline{\eta}. \qquad (2.86)$$

Note that in fact only one of the two systems (2.85) and (2.86) needs to be solved since the solutions are related by

$$\underline{\eta} = \underline{\pi} B. \qquad (2.87)$$

It must be emphasized that solutions to (2.85) and (2.86) which satisfy (2.87) will almost always exist, whether or not A and B are compatible. It is only in the compatible case however that (2.83) holds.

Liu (1996) briefly discusses the difference between the arrays $(a_{ij}\eta_j)$ and $(b_{ij}\tau_i)$ in the incompatible case in the context of Gibbs sampler simulations.

Example 2.12 (Markovian measure of incompatibility; incompatible case). Consider again the matrices in (2.48) and (2.49). Then we have

$$
BA' = \begin{pmatrix} 0.17 & 0.18 & 0.36 & 0.29 \\ 0.21 & 0.23 & 0.45 & 0.11 \\ 0.23 & 0.20 & 0.44 & 0.13 \\ 0.12 & 0.16 & 0.32 & 0.40 \end{pmatrix}, \quad A'B = \begin{pmatrix} 0.46 & 0.14 & 0.26 & 0.14 \\ 0.23 & 0.27 & 0.42 & 0.08 \\ 0.34 & 0.20 & 0.36 & 0.10 \\ 0.43 & 0.14 & 0.28 & 0.15 \end{pmatrix},
$$
$$(2.88)$$

and then

$$\underline{\pi} = (\,0.191056 \quad 0.193394 \quad 0.40089 \quad 0.214659\,) \tag{2.89}$$

and

$$\underline{\eta} = (\,0.375442 \quad 0.18327 \quad 0.324075 \quad 0.117213\,), \tag{2.90}$$

which lead to a value of $D = 0.019$ for the Markovian measure of incompatibility. $\qquad\square$

The actual value assumed by D does not appear to have a ready interpretation. It can be used to order pairs (A, B) in terms of discrepancy but does not seem to provide an intrinsic absolute measure of discrepancy. Of course, in the case of compatibility, $D = 0$.

Example 2.13 (Markovian measure of incompatibility; compatible case). Consider again the matrices in (2.6) and (2.7). Then we have

$$BA' = \begin{pmatrix} 17/56 & 1/4 & 25/56 \\ 3/14 & 16/49 & 45/98 \\ 25/112 & 15/56 & 57/112 \end{pmatrix} \tag{2.91}$$

and

$$A'B = \begin{pmatrix} 29/98 & 15/98 & 25/98 & 29/98 \\ 15/56 & 23/112 & 29/112 & 15/56 \\ 25/98 & 29/196 & 67/196 & 25/98 \\ 29/98 & 15/98 & 25/98 & 29/98 \end{pmatrix}, \tag{2.92}$$

so that

$$\underline{\pi} = (\,0.24 \quad 0.28 \quad 0.48\,) \tag{2.93}$$

and

$$\underline{\eta} = (\,0.28 \quad 0.16 \quad 0.28 \quad 0.28\,), \tag{2.94}$$

which lead to a value of $D = 0$ for our Markovian measure of incompatibility. $\qquad\square$

A variant measure of incompatibility is suggested by the definition of D (i.e., (2.84)). We know that, in the compatible case and only in that case, there will exist vectors $\underline{\eta}$ and $\underline{\tau}$ such that (2.83) holds. These vectors $\underline{\eta}$ and $\underline{\tau}$ admit interpretation as marginals of (X, Y) or as long-run distributions of related Markov chains, and we do not have to insist on these interpretations in the formulation of our objective function. We can instead set up the objective function

$$\tilde{D}(\underline{u}, \underline{v}) = \sum_{i=1}^{I} \sum_{j=1}^{J} (a_{ij}v_j - b_{ij}u_i)^2 \tag{2.95}$$

and seek stochastic vectors $\underline{u}^*$ and $\underline{v}^*$ to minimize $\tilde{D}(\underline{u}, \underline{v})$. The achieved value $D^* = D(\underline{u}^*, \underline{v}^*)$ will be our index of inconsistency. If A, B are compatible then $D^* = 0$ and $\underline{u}^* = \underline{\pi}$ and $\underline{v}^* = \underline{\eta}$.

2.8 ϵ-Compatibility

In Section 2.2.2 we described three linear equation formulations of the search for a matrix P compatible with given conditional matrices A and B. If, instead of precise compatibility, we are willing to accept approximate compatibility we will need to modify our objectives only slightly. First we postulate the existence of a weight matrix W which quantifies the relative importance of accuracy in determining the various elements in P. Thus if w_{ij} is small, then quite precise determination of p_{ij} is deemed to be desirable. If, for some i, j, w_{ij} is large, then we are not so worried about precise determination of that particular element p_{ij}. We will then say that we can find an ϵ-compatible matrix P corresponding to A and B if we can approximately solve systems in Methods I, II, and III of Section 2.2.2 with an error tolerance of ϵw_{ij} in all i, j. Thus our revised options in the search for an (ϵ, W)-compatible matrix P (for given A, B) are as follows:

Option 1. Seek a probability matrix P to satisfy

$$\left| p_{ij} - a_{ij} \sum_{i=1}^{I} p_{ij} \right| \leq \epsilon w_{ij}, \quad \forall i, j,$$
$$\left| p_{ij} - b_{ij} \sum_{j=1}^{J} p_{ij} \right| \leq \epsilon w_{ij}, \quad \forall i, j, \qquad (2.96)$$
$$\sum_{i=1}^{I} \sum_{j=1}^{J} p_{ij} \qquad = 1,$$

and

$$p_{ij} \geq 0, \quad \forall i, j. \qquad (2.97)$$

Option 2. Seek two probability matrices $\underline{\tau}$ and $\underline{\eta}$ such that

$$|\eta_j a_{ij} - \tau_i b_{ij}| \leq \epsilon w_{ij}, \quad \forall i, j,$$
$$\sum_{i=1}^{I} \tau_i = 1, \qquad (2.98)$$
$$\sum_{j=1}^{J} \eta_j = 1,$$

and

$$\tau_i \geq 0, \eta_j \geq 0, \forall i, j. \qquad (2.99)$$

Option 3. Seek one probability vector $\underline{\tau}$ satisfying

$$|a_{ij} \sum_{i=1}^{I} \tau_i b_{ij} - \tau_i b_{ij}| \leq \epsilon w_{ij}, \quad \forall i, j,$$

$$\sum_{i=1}^{I} \tau_i = 1, \tag{2.100}$$

and

$$\tau_i \geq 0, \quad \forall i. \tag{2.101}$$

Of course each of the constraints involving absolute values in (2.96)-(2.100) can be replaced by two linear inequality constraints. So all our constraints and equations are linear.

The above options immediately motivate three different concepts of ϵ-compatibility.

Definition 2.9 (ϵ-compatible matrices). Two conditional probability matrices A and B are said to be ϵ_1-compatible iff system (2.96) ((2.98) or (2.100)) has a solution for $\epsilon \geq \epsilon_1$ and not for $\epsilon < \epsilon_1$, i.e. iff ϵ_1 is the minimum value of ϵ that allows system (2.96) ((2.98) or (2.100)) to have a solution. □

Note that two matrices A and B are compatible iff they are 0-compatible, a special case of ϵ-compatibility.

To analyze and discuss the near-compatibility problem derived from each of these concepts we can use, for example, any of the following three methods:

Method 1. (Check for existence of a solution). Determine the set of ϵ values that allows the system to have a solution.

Method 2. (Solving the system of inequalities). Find the most general solution of the system considering ϵ as one more variable. This can be done using the methods described in Theorem 2.3.

Method 3. (Solving a linear programming problem). Use linear programming methods to minimize the function $f(\epsilon, \underline{p}) = \epsilon$, where $\underline{p}$ denotes the vector obtained by stacking the columns of matrix P, or $f(\underline{\tau}, \underline{\eta}, \epsilon) = \epsilon$ or $f(\underline{\tau}, \epsilon) = \epsilon$ subject to the appropriate constraints (2.96)–(2.97), (2.98)–(2.99), or (2.100)–(2.101).

We of course might apply these methods and options to either compatible or incompatible pairs A, B.

A full list of examples would then involve $3 \times 3 \times 2 = 18$ detailed computations. We content ourselves with a stratified sample of such examples. Throughout, for simplicity, we assume $w_{ij} = 1, \forall i, j$. However, other values are possible.

Example 2.14 (A compatible example). Consider the following matrices A and B:

$$A = \begin{pmatrix} 1/4 & 1/2 \\ 3/4 & 1/2 \end{pmatrix}, \tag{2.102}$$

$$B = \begin{pmatrix} 1/3 & 2/3 \\ 3/5 & 2/5 \end{pmatrix}. \tag{2.103}$$

Consider Option 1. The system (2.96) can be written as

$$
\begin{pmatrix}
3/4 & 0 & -1/4 & 0 \\
-3/4 & 0 & 1/4 & 0 \\
0 & 1/2 & 0 & -1/2 \\
0 & -1/2 & 0 & 1/2 \\
-3/4 & 0 & 1/4 & 0 \\
3/4 & 0 & -1/4 & 0 \\
0 & -1/2 & 0 & 1/2 \\
0 & 1/2 & 0 & -1/2 \\
2/3 & -1/3 & 0 & 0 \\
-2/3 & 1/3 & 0 & 0 \\
-2/3 & 1/3 & 0 & 0 \\
2/3 & -1/3 & 0 & 0 \\
0 & 0 & 2/5 & -3/5 \\
0 & 0 & -2/5 & 3/5 \\
0 & 0 & -2/5 & 3/5 \\
0 & 0 & 2/5 & -3/5 \\
-1 & 0 & 0 & 0 \\
0 & -1 & 0 & 0 \\
0 & 0 & -1 & 0 \\
0 & 0 & 0 & -1
\end{pmatrix}
\underline{p} \le
\begin{pmatrix}
\epsilon \\ \epsilon \\ \epsilon \\ \epsilon \\ \epsilon \\ \epsilon \\ \epsilon \\ \epsilon \\ \epsilon \\ \epsilon \\ \epsilon \\ \epsilon \\ \epsilon \\ \epsilon \\ \epsilon \\ \epsilon \\ 0 \\ 0 \\ 0 \\ 0 \\ 0
\end{pmatrix}.
$$

$$(1 \quad 1 \quad 1 \quad 1) \qquad \underline{p} = \quad 1. \tag{2.104}$$

We briefly describe the above three methods.

Method 1. The system (2.104) is compatible for any value of ϵ. Thus, matrices A and B are compatible.

Method 2. Solving system (2.104), considering ϵ as one more variable, leads to a general solution that involves 25 extreme points but is too long to be included here. However, from it we can determine that the minimum value of ϵ is $\epsilon = 0$ and the corresponding unique solution is

$$P = \begin{pmatrix} 1/8 & 1/4 \\ 3/8 & 1/4 \end{pmatrix}. \tag{2.105}$$

Method 3. Minimizing the function

$$f(\epsilon, p_{11}, p_{12}, p_{21}, p_{22}) = \epsilon,$$

subject to the constraints in (2.104), leads to a zero (minimum) value of the objective function that is attained at the P given in (2.105).   □

Example 2.15 (An incompatible example). Consider the matrices A and B:

$$A = \begin{pmatrix} 1/4 & 1/2 \\ 3/4 & 1/2 \end{pmatrix}, \tag{2.106}$$

$$B = \begin{pmatrix} 1/6 & 5/6 \\ 3/5 & 2/5 \end{pmatrix}. \tag{2.107}$$

Consider Option 2. The system (2.98) can be written as

$$\begin{pmatrix} -1/6 & 0 & 1/4 & 0 \\ 1/6 & 0 & -1/4 & 0 \\ -5/6 & 0 & 0 & 1/2 \\ 5/6 & 0 & 0 & -1/2 \\ 0 & -3/5 & 3/4 & 0 \\ 0 & 3/5 & -3/4 & 0 \\ 0 & -2/5 & 0 & 1/2 \\ 0 & 2/5 & 0 & -1/2 \\ -1 & 0 & 0 & 0 \\ 0 & -1 & 0 & 0 \\ 0 & 0 & -1 & 0 \\ 0 & 0 & 0 & -1 \end{pmatrix} \begin{pmatrix} \underline{\tau} \\ \underline{\eta} \end{pmatrix} \leq \begin{pmatrix} \epsilon \\ \epsilon \\ \epsilon \\ \epsilon \\ \epsilon \\ \epsilon \\ \epsilon \\ \epsilon \\ 0 \\ 0 \\ 0 \\ 0 \end{pmatrix}, \tag{2.108}$$

$$\begin{pmatrix} 1 & 1 & 0 & 0 \\ 0 & 0 & 1 & 1 \end{pmatrix} \qquad \begin{pmatrix} \underline{\tau} \\ \underline{\eta} \end{pmatrix} = \begin{pmatrix} 1 \\ 1 \end{pmatrix}.$$

We briefly describe the above three methods.

Method 1. It can be shown, using Theorem 2.2, that system (2.108) is compatible for $\epsilon \geq 9/214$. Thus, matrices A and B are incompatible but $(9/214)$-compatible.

Method 2. The solution of system (2.108), considering ϵ as one more variable, is

$$\begin{pmatrix} \eta_1 \\ \eta_1 \\ \tau_1 \\ \tau_2 \\ \epsilon \end{pmatrix} = \pi_1 \begin{pmatrix} 0 \\ 0 \\ 0 \\ 0 \\ 1 \end{pmatrix} + H\underline{\lambda},$$

$$\pi_1 \geq 0,$$

$$\sum_{i=1}^{13} \lambda_i = 1,$$

where

$$\underline{\lambda} = \begin{pmatrix} \lambda_1 & \lambda_2 & \lambda_3 & \lambda_4 & \lambda_5 & \lambda_6 & \lambda_7 & \lambda_8 & \lambda_9 & \lambda_{10} & \lambda_{11} & \lambda_{12} & \lambda_{13} \end{pmatrix}^T .$$

and

$$H = \begin{pmatrix} \frac{12}{37} & \frac{6}{41} & \frac{42}{107} & 0 & 0 & 1 & \frac{1}{2} & 0 & 1 & \frac{1}{4} & \frac{9}{14} & 0 & 1 \\ \frac{25}{37} & \frac{35}{41} & \frac{65}{107} & 1 & 1 & 0 & \frac{1}{2} & 1 & 0 & \frac{3}{4} & \frac{5}{14} & 1 & 0 \\ \frac{26}{37} & \frac{22}{41} & \frac{46}{107} & \frac{3}{5} & \frac{2}{5} & \frac{1}{6} & 0 & 0 & 0 & 1 & 1 & 1 & 1 \\ \frac{11}{37} & \frac{19}{41} & \frac{61}{107} & \frac{2}{5} & \frac{3}{5} & \frac{5}{6} & 1 & 1 & 1 & 0 & 0 & 0 & 0 \\ \frac{9}{74} & \frac{9}{82} & \frac{9}{214} & \frac{1}{5} & \frac{3}{10} & \frac{5}{12} & \frac{3}{10} & \frac{3}{5} & \frac{1}{2} & \frac{3}{10} & \frac{15}{28} & \frac{2}{5} & \frac{5}{6} \end{pmatrix},$$

From this it is clear that the minimum value of ϵ leading to a solution is $\epsilon = 9/214$, and then

$$\underline{\tau} = \begin{pmatrix} 42/107 \\ 65/107 \end{pmatrix}, \quad \underline{\eta} = \begin{pmatrix} 46/107 \\ 61/107 \end{pmatrix}. \tag{2.109}$$

With these values of $\boldsymbol{\eta}$ and $\boldsymbol{\tau}$ we obtain two probability matrices:

$$P^1 = \begin{pmatrix} 23/214 & 61/214 \\ 69/214 & 61/214 \end{pmatrix}, \quad (p_{ij}^1 = a_{ij}\eta_j), \tag{2.110}$$

$$P^2 = \begin{pmatrix} 14/214 & 70/214 \\ 78/214 & 52/214 \end{pmatrix}, \quad (p_{ij}^2 = b_{ij}\tau_i), \tag{2.111}$$

such that the maximum absolute error is $9/214$, since

$$P^1 - P^2 = \begin{pmatrix} 9/214 & -9/214 \\ -9/214 & 9/214 \end{pmatrix}.$$

Method 3. Minimizing the function

$$f(\epsilon, \underline{\eta}, \underline{\tau}) = \epsilon$$

subject to the constraints in (2.108) leads to a minimum value of the objective function of $9/214$, which is attained at the $(\boldsymbol{\eta}, \underline{\tau})$ given in (2.109). $\square$

Example 2.16 (Incompatible, once more). Again consider A and B as in (2.106) and (2.107). However now we will use Option 3. The system of inequalities (2.100) can be written as:

$$\begin{pmatrix} 1/8 & -3/20 \\ 5/12 & -1/5 \\ -1/8 & 3/20 \\ -5/12 & 1/5 \\ -1/8 & 3/20 \\ -5/12 & 1/5 \\ 1/8 & -3/20 \\ 5/12 & -1/5 \end{pmatrix} \underline{\tau} \leq \begin{pmatrix} \epsilon \\ \epsilon \\ \epsilon \\ \epsilon \\ \epsilon \\ \epsilon \\ \epsilon \\ \epsilon \end{pmatrix},$$

$$\begin{pmatrix} 1 & 1 \end{pmatrix} \underline{\tau} = 1,$$

$$\underline{p} \geq 0. \tag{2.112}$$

TABLE 2.4. Generators of the dual cone $\Omega(\pi(C))$ in Example 2.16.

w_1	w_2	w_3	w_4	w_5	w_6	w_7	w_8	w_9
0	−148	0	0	0	−20	−1	0	0
0	0	−1	0	0	0	0	−5	−66
0	0	0	0	−8	0	−1	0	−148
−12	−66	−1	0	0	0	0	0	0
−5	−9	0	−1	−1	−3	0	−1	9

Method 1. Removing redundant constraints and using a set of four non-negative slackness variables $\underline{y} = (y_1, y_2, y_3, y_4)$, (2.112) can be written as

$$
\begin{pmatrix}
1/8 & -3/20 & 1 & 0 & 0 & 0 \\
5/12 & -1/5 & 0 & 1 & 0 & 0 \\
-1/8 & 3/20 & 0 & 0 & 1 & 0 \\
-5/12 & 1/5 & 0 & 0 & 0 & 1 \\
1 & 1 & 0 & 0 & 0 & 0
\end{pmatrix}
\begin{pmatrix} \underline{\tau} \\ \underline{y} \end{pmatrix}
=
\begin{pmatrix} \epsilon \\ \epsilon \\ \epsilon \\ \epsilon \\ 1 \end{pmatrix},
\qquad (2.113)
$$
$$
\underline{\tau}, \underline{y} \geq 0.
$$

According to Theorem 2.2, we need to calculate the dual of the cone generated by the columns of the matrix C of coefficients in (2.113). A minimal set of generators of this dual is given in Table 2.4.

For the system (2.113) to have a solution we must have $W\underline{a} \leq \underline{0}$. This implies

$$(-66 - 148)\epsilon + 9 \leq 0 \quad \Leftrightarrow \quad \epsilon \geq 9/214. \qquad (2.114)$$

Thus the value $9/214$ is a measure of compatibility for this case, which has a clear interpretation (the maximum deviation in the alternative evaluations of p_{ij}). Thus if we define matrices

$$P_1 = (b_{ij}\tau_i) \qquad (2.115)$$

and

$$P_2 = \left(a_{ij} \sum_{i=1}^{I} b_{ij}\tau_i \right), \qquad (2.116)$$

then by judicious choice of $\underline{\tau}$ we can make the maximal deviation between elements of P_1 and P_2 as small as $9/214$.

Method 2. Solving the system of inequalities (2.112) considering ϵ as one more variable. This leads to the solution

$$\begin{pmatrix} \epsilon \\ \tau_1 \\ \tau_2 \end{pmatrix} = \begin{pmatrix} 1 \\ 0 \\ 0 \end{pmatrix} \pi_1 + \begin{pmatrix} 9/214 & 9/82 & 1/5 & 5/12 \\ 42/107 & 6/41 & 0 & 1 \\ 65/107 & 35/41 & 1 & 0 \end{pmatrix} \begin{pmatrix} \lambda_1 \\ \lambda_2 \\ \lambda_3 \\ \lambda_4 \end{pmatrix},$$

$$\sum_{i=1}^{4} \lambda_i = 1 \qquad\qquad (2.117)$$

$$\lambda_i \geq 0, \quad i = 1, 2, 3, 4,$$

$$\tau_1 \geq 0.$$

It is obvious from (2.117) that the minimum value of ϵ is obtained for $\pi_1 = 0, \lambda_1 = 1, \lambda_i = 0, i \neq 1$, leading to 9/214, as found before. The corresponding values of the X-marginal are $\tau_1 = 42/107$ and $\tau_2 = 65/107$.
From (2.115) and (2.116) we get

$$P_1 = \frac{\lambda_1}{107} \begin{pmatrix} 7 & 35 \\ 39 & 26 \end{pmatrix} + \frac{\lambda_2}{41} \begin{pmatrix} 1 & 5 \\ 21 & 14 \end{pmatrix} + \frac{\lambda_3}{3} \begin{pmatrix} 1 & 2 \\ 0 & 0 \end{pmatrix} + \frac{\lambda_4}{6} \begin{pmatrix} 1 & 5 \\ 0 & 0 \end{pmatrix},$$

$$P_2 = \frac{\lambda_1}{214} \begin{pmatrix} 23 & 61 \\ 69 & 61 \end{pmatrix} + \frac{\lambda_2}{82} \begin{pmatrix} 11 & 19 \\ 33 & 19 \end{pmatrix} + \frac{\lambda_3}{20} \begin{pmatrix} 3 & 4 \\ 9 & 4 \end{pmatrix} + \frac{\lambda_4}{24} \begin{pmatrix} 1 & 10 \\ 3 & 10 \end{pmatrix},$$

$$(2.118)$$

which for the optimum case $\epsilon = 9/214$ leads to

$$P_1 = \frac{1}{107} \begin{pmatrix} 7 & 35 \\ 39 & 26 \end{pmatrix},$$

$$P_2 = \frac{1}{214} \begin{pmatrix} 23 & 61 \\ 69 & 61 \end{pmatrix}.$$

Method 3. Solving the linear programming problem, we get the same solution for $\underline{\tau}$ and ϵ above. □

We can use the concepts of ϵ-compatibility to give us yet more definitions of a most nearly compatible matrix P for a given pair A, B. If we use Option 1, and if A, B are ϵ-compatible, then the matrix P^* which satisfies (2.96) with $\epsilon = \epsilon_1$ will be judged to be most nearly compatible. If we use Option 2 and if A, B are ϵ_1-compatible, then a reasonable choice for a most nearly compatible matrix P^* will be

$$P^* = \frac{1}{2}(\eta_j^* a_{ij} + \pi_i^* b_{ij}),$$

where $\underline{\pi}^*, \underline{\eta}^*$ satisfy (2.98) with $\epsilon = \epsilon_1$. Finally, if we use Option 3 and if A, B are ϵ_1-compatible, then a plausible choice for a most nearly compatible matrix P^* will be

$$P^* = (\pi_i^* b_{ij}),$$

where $\underline{\pi}^*$ satisfies (2.100) with $\epsilon = \epsilon_1$.

2.9   Extensions to More General Settings

Some of the material in this chapter extends readily to cases where (X, Y) has a joint density that is absolutely continuous with respect to a convenient product measure $\mu_1 \times \mu_2$ on $S(X) \times S(Y)$ (where $S(X)$ denotes the support of X). Integrals will replace sums in the discussion and, provided we check the integrability of solutions, few technical difficulties will be encountered. Thus compatibility, as described in Theorem 1.2, can be defined in terms of a cross-product ratio function

$$\frac{f(x_1, y_1)f(x_2, y_2)}{f(x_1, y_2)f(x_2, y_1)} \tag{2.119}$$

to obtain an extended version of the equivalence of conditions (i) and (ii) in Theorem 2.1. The concepts of near compatibility discussed in the latter section do indeed continue to be meaningful in more abstract settings (with integrals instead of sums), however we will encounter difficulties in implementing the iterative algorithms if the support sets of X and Y are infinite. Concepts of ϵ-compatibility can also be developed in more general settings but they will be technically difficult to deal with.

2.10   Bibliographic Notes

Material on compatibility using cross-product ratios and uniform marginal representations may be found in Arnold and Gokhale (1994, 1998b). The discussion of compatibility via solution of linear equations subject to constraints is based on Arnold, Castillo, and Sarabia (1999a). The concept of ϵ-compatibility is introduced in Arnold, Castillo and Sarabia (1999a). Markovian discrepancy measures are discussed in Arnold and Gokhale (1998a). Sections 2.4, 2.5, and 2.6 are based on Arnold and Gokhale (1998b).

Exercises

2.1 Consider the conditional probability matrices in Example 2.15. Instead of uniform weights, assume the following weight matrix:

$$W = \begin{pmatrix} 1 & 2 \\ 2 & 1 \end{pmatrix}.$$

(a) Determine the nearest compatible conditional probability matrices to A and B.

(b) Compare the results with those in Example 2.15.

(c) Determine which of the error conditions are active and which are not.

(d) Decrease the values of the elements in the above weight matrix W as much as possible, without altering the nearest compatible matrices.

2.2 Consider again matrices A and B in Example 2.15. Assume that $\epsilon = 0.1$ and that a weight matrix of the form

$$W = \begin{pmatrix} a & b \\ 2a & b/2 \end{pmatrix},$$

where a and b are constants, has been selected.

(a) Determine the conditions to be satisfied by a and b for the near compatibility problem to have a solution.

(b) Determine the extra conditions for having a maximum set of active constraints.

2.3 Given the two conditional probability matrices

$$A = \begin{pmatrix} 0.2 & 0.3 & 0.1 \\ 0.1 & 0.4 & 0.4 \\ 0.7 & 0.3 & 0.5 \end{pmatrix} \quad \text{and} \quad B = \begin{pmatrix} 0.2 & 0.1 & 0.7 \\ 0.3 & 0.4 & 0.3 \\ 0.1 & 0.4 & 0.5 \end{pmatrix}.$$

determine whether or not they are compatible by:

(a) The cross-product ratio method.

(b) The uniform marginal representation method.

(c) One of Methods I, II, or III, together with Theorem 2.2.

(d) The iterative algorithm in (2.42) to obtain Q^*, A^*, B^*, and the associated P^*.

(e) The iterative method in (2.73).

(f) The Markovian measure of discrepancy method.

(g) The ϵ-compatibility approach with one of the options 1, 2, or 3.

(h) Linear programming techniques.

2.4 Given the following conditional probability matrices:

$$A = \begin{pmatrix} 0.2 & 0.3 & 0.1 \\ 0.1 & 0.4 - a & 0.4 \\ 0.7 & 0.3 + a & 0.5 \end{pmatrix} \quad \text{and} \quad B = \begin{pmatrix} 0.2 + b & 0.1 & 0.7 - b \\ 0.3 & 0.4 & 0.3 \\ 0.1 & 0.4 & 0.5 \end{pmatrix},$$

determine whether or not there are values a and b such that they are compatible.

If the above answer is positive, determine such values.

2.5 In Option 2, Section 2.8, force the P^1 and P^2 probability matrices in (2.110) and (2.111) to have the same Y-marginals. Compare the resulting solution with that of Option 3.

2.6 Consider the conditional probability matrices

$$A = \begin{pmatrix} 0 & 1/3 & 0 \\ 1 & 1/3 & 1/2 \\ 0 & 1/3 & 1/2 \end{pmatrix}, \quad B = \begin{pmatrix} 0 & 1 & 0 \\ 1/4 & 1/2 & 1/4 \\ 0 & 1/5 & 4/5 \end{pmatrix}.$$

Verify that A and B are not compatible even though $\text{UMR}(A) = \text{UMR}(B)$.

2.7 Consider the following conditional probability matrices:

$$A_1 = \begin{pmatrix} 0 & 0 & 1/10 & 0 & 0 \\ 0 & 1/3 & 3/10 & 4/7 & 0 \\ 1 & 1/3 & 2/10 & 2/7 & 1 \\ 0 & 1/3 & 3/10 & 1/7 & 0 \\ 0 & 0 & 1/10 & 0 & 0 \end{pmatrix}, \quad B_1 = \begin{pmatrix} 0 & 0 & 1 & 0 & 0 \\ 0 & 1/5 & 1/5 & 3/5 & 0 \\ 1/9 & 2/9 & 3/9 & 2/9 & 1/9 \\ 0 & 2/5 & 1/5 & 2/5 & 0 \\ 0 & 0 & 1 & 0 & 0 \end{pmatrix},$$

$$A_2 = \begin{pmatrix} 0 & 1/3 & 0 \\ 1 & 0 & 1 \\ 0 & 2/3 & 0 \end{pmatrix}, \quad B_2 = \begin{pmatrix} 0 & 1 & 0 \\ 1/5 & 0 & 4/5 \\ 0 & 1 & 0 \end{pmatrix}.$$

(a) Verify that A_1, B_1 are incompatible while A_2, B_2 are compatible.

(b) Show that UMRs do not exist for any of the matrices A_1, B_1, A_2, or B_2.

3

Distributions with Normal Conditionals

3.1 Introduction

At first glance, it is surprising that there can exist bivariate distributions with all their conditional densities of the normal form, which are not the classical bell-shaped bivariate normal that we are so familiar with. But such distributions do exist and have actually been at least partially known and understood for more than 50 years. After reviewing the history of such distributions we focus on developing a convenient parametric representation of all such normal conditionals distributions. The role of the classical bivariate normal distribution as a special case is investigated in some detail. In fact we begin our discussion with this topic.

3.2 Variations on the Classical Bivariate Normal Theme

If (X, Y) has a classical bivariate normal distribution with density

$$
\begin{aligned}
f_{X,Y}(x, y) = {} & \frac{1}{2\pi\sigma_1\sigma_2\sqrt{1 - \rho^2}} \\
& \times \exp\left\{ -\frac{1}{2(1 - \rho^2)} \left[\left(\frac{x - \mu_1}{\sigma_1} \right)^2 - 2\rho\left(\frac{x - \mu_1}{\sigma_1} \right)\left(\frac{y - \mu_2}{\sigma_2} \right) + \left(\frac{y - \mu_2}{\sigma_2} \right)^2 \right] \right\},
\end{aligned}
\tag{3.1}
$$

then it is well known that both marginal densities are univariate normal and all conditional densities are univariate normal. In addition, the regression functions are linear and the conditional variances do not depend on the value of the conditioned variable. Moreover the contours of the joint density are ellipses. Individually none of the above properties is restrictive enough to characterize the bivariate normal. Collectively they do characterize the bivariate normal distribution but, in fact, far less than the complete list is sufficient to provide such a characterization. Marginal normality is of course not enough. It is easy to construct examples of distributions not of the form (3.1) which have normal marginals. Perhaps the simplest example is

$$f_{X,Y}(x,y) = \begin{cases} \dfrac{1}{\pi}e^{-(x^2+y^2)/2}, & xy > 0, \\ 0, & xy < 0. \end{cases} \tag{3.2}$$

which has standard normal marginals but has possible values only in two quadrants of the plane. Some reference to conditional distributions or conditional moments seems necessary to characterize the bivariate normal. An indication of the difficulties is provided by the following putative characterization. If the marginal density of X is normal and if the conditional density of Y given $X = x$ is normal for every x, can we conclude that (X, Y) is bivariate normal? Clearly the answer is no unless we assume in addition that $E(Y|X = x)$ is linear in x and that var$(Y|X = x)$ does not depend on x. With these two additional assumptions bivariate normality is guaranteed, otherwise we would have quite an arbitrary regression of Y on X and an equally arbitrary conditional variance function of Y given $X = x$; neither of which is permitted in the bivariate normal model.

In a ground-breaking paper Bhattacharyya (1943) provided the following interesting array of characterizations involving normal conditionals:

1. If for each fixed y, the conditional distribution of X given $Y = y$ is normal and the equiprobable contours of $f_{X,Y}(x,y)$ are similar concentric ellipses, then the bivariate density is normal.

2. If the regression of X on Y is linear, if the conditional distribution of X given $Y = y$ is normal and homoscedastic for each y, and if the marginal distribution of Y is normal, then $f_{X,Y}(x,y)$ must be bivariate normal.

3. If the conditional distributions of X given $Y = y$ for each y, and of Y given $X = x$ for each x are normal and one of these conditional distributions is homoscedastic, then (X, Y) is bivariate normal.

4. If the regressions, of X on Y and of Y on X, are both linear, and the conditional distribution of each variable for every fixed value of the other variable is normal, then $f_{X,Y}(x,y)$ is either normal or may (with suitable choice of origin and scale) be written in the form

$$f_{X,Y}(x, y) \propto \exp\{-(x^2 + a^2)(y^2 + b^2)\}. \qquad (3.3)$$

Example 4 in Bhattacharyya's list tells us there exist joint distributions with normal conditionals and even with linear regressions which are not bivariate normal. As we shall see, Bhattacharyya is not too far away from identifying all possible normal conditionals distributions. Much later, Brucker (1979) verified that normal conditionals (of X given Y and of Y given X) with linear regressions <u>and</u> constant conditional variances are enough to characterize the bivariate normal distribution. Variations on this theme were described by Fraser and Streit (1980) and Ahsanullah (1985). Stoyanov ((1987), pp. 85–86) also considers (3.3) and an analogous trivariate density.

It may be instructive to view some distributions with normal conditionals which are not bivariate normal. For our first example, as in Castillo and Galambos (1989), consider

$$f_{X,Y}(x, y) = C \exp\{-[x^2 + y^2 + 2xy(x + y + xy)]\}; \quad \forall x, y, \qquad (3.4)$$

where $C > 0$ is a constant such that $f_{X,Y}(x, y)$ integrates to 1. Even though both conditionals are normal, $f_{X,Y}(x, y)$ is not bivariate normal, because for this density the conditional variance of Y, given $X = x$ is $\sigma^2(Y|X = x) = 1/(2 + 4x + 4x^2)$ (nonconstant) and the conditional expectation of Y, given $X = x$, is $(-x^2)/(1 + 2x + 2x^2)$ (nonlinear). A perhaps more striking example is provided by

$$f_{X,Y}(x, y) = C \exp\left\{-\frac{9}{2}x - 2y - \frac{9}{2}x^2 - \frac{1}{2}y^2 - 4xy - 4x^2 y - xy^2 - x^2 y^2\right\}, \forall x, y. \qquad (3.5)$$

In this case both conditional densities are normal and the conditional expectation of Y given $X = x$ ($\mu(Y|X = x) = -2$) and the conditional expectations of X given $Y = y$ ($\mu(X|Y = y) = -1/2$) are constant. Clearly, however, this bivariate density is not normal, since the variances of the conditional distributions are not constant $[\sigma^2(Y|X = x) = 1/(1 + 2x + 2x^2), \sigma^2(X|Y = y) = 1/(9 + 8y + 2y^2)]$.

Looking at (3.4) and (3.5) it is easy to verify that the corresponding conditional densities are indeed normal and with a little algebra it is possible to write down the correct expressions for the conditional means and variances. What is perhaps not obvious is how one dreams up such examples. The picture will clear in the next section.

3.3   Normal Conditionals

As in Castillo and Galambos (1987a, 1989) we begin with a joint density $f_{X,Y}(x, y)$ assumed to have all conditionals in the univariate normal family.

Specifically we assume the existence of functions $\mu_2(x), \sigma_2(x), \sigma_1(y)$, and $\mu_1(y)$ such that

$$f_{X|Y}(x|y) = \frac{1}{\sqrt{2\pi}\sigma_1(y)} \exp\left[-\frac{1}{2}\left(\frac{x - \mu_1(y)}{\sigma_1(y)}\right)^2\right] \tag{3.6}$$

and

$$f_{Y|X}(y|x) = \frac{1}{\sqrt{2\pi}\sigma_2(x)} \exp\left[-\frac{1}{2}\left(\frac{y - \mu_2(x)}{\sigma_2(x)}\right)^2\right]. \tag{3.7}$$

Denote the corresponding marginal densities of X and Y by $f_1(x)$ and $f_2(y)$, respectively. Note that

$$f_1(x) > 0, \quad f_2(y) > 0, \quad \sigma_2(x) > 0, \quad \text{and } \sigma_1(y) > 0.$$

If we write the joint density as a product of a marginal and a conditional density in both ways we find

$$\frac{f_1(x)}{\sigma_2(x)} \exp\left[-\frac{1}{2}\left(\frac{y - \mu_2(x)}{\sigma_2(x)}\right)^2\right] = \frac{f_2(y)}{\sigma_1(y)} \exp\left[-\frac{1}{2}\left(\frac{x - \mu_1(y)}{\sigma_1(y)}\right)^2\right]. \tag{3.8}$$

If we define

$$u(x) = \log(f_1(x)/\sigma_2(x)) \tag{3.9}$$

and

$$v(y) = \log(f_2(y)/\sigma_1(y)), \tag{3.10}$$

then (3.8) assumes a form appropriate for direct application of Theorem 2.5 using the function $\gamma(t) = e^t$. Rather than quote Theorem 2.5 we shall use our basic result, Theorem 2.3, to characterize our distributions.

Taking logarithms of both sides of (3.8) we obtain

$$u(x) - \frac{1}{2}\left(\frac{y - \mu_2(x)}{\sigma_2(x)}\right)^2 = v(y) - \frac{1}{2}\left(\frac{x - \mu_1(y)}{\sigma_1(y)}\right)^2 \tag{3.11}$$

which can be rearranged to yield

$$\begin{aligned}[2u(x)\sigma_2^2(x) - \mu_2^2(x)]\sigma_1^2(y) + \sigma_2^2(x)[\mu_1^2(y) - 2v(y)\sigma_1^2(y)] - y^2\sigma_1^2(y) \\ + x^2\sigma_2^2(x) + 2\mu_2(x)y\sigma_1^2(y) - 2x\sigma_2^2(x)\mu_1(y) = 0.\end{aligned} \tag{3.12}$$

This is an equation of the form (2.24). The sets $\{\sigma_2^2(x), x\sigma_2^2(x), x^2\sigma_2^2(x)\}$ and $\{\sigma_1^2(y), y\sigma_1^2(y), y^2\sigma_1^2(y)\}$ are sets of linearly independent functions. They can play the roles of the ϕ_i's and ψ_j's in (2.25). Thus from (2.25) we obtain

$$\begin{pmatrix} 2u(x)\sigma_2^2(x) - \mu_2^2(x) \\ \sigma_2^2(x) \\ 1 \\ x^2\sigma_2^2(x) \\ \mu_2(x) \\ x\sigma_2^2(x) \end{pmatrix} = \begin{pmatrix} A & B & C \\ 1 & 0 & 0 \\ D & E & F \\ 0 & 0 & 1 \\ G & H & J \\ 0 & 1 & 0 \end{pmatrix} \begin{pmatrix} \sigma_2^2(x) \\ x\sigma_2^2(x) \\ x^2\sigma_2^2(x) \end{pmatrix}, \tag{3.13}$$

$$
\begin{pmatrix}
\sigma_1^2(y) \\
\mu_1^2(y) - 2v(y)\sigma_1^2(y) \\
-y^2\sigma_1^2(y) \\
1 \\
2y\sigma_1^2(y) \\
-2\mu_1(y)
\end{pmatrix}
=
\begin{pmatrix}
1 & 0 & 0 \\
K & L & M \\
0 & 0 & -1 \\
N & P & Q \\
0 & 2 & 0 \\
R & S & T
\end{pmatrix}
\begin{pmatrix}
\sigma_1^2(y) \\
y\sigma_1^2(y) \\
y^2\sigma_1^2(y)
\end{pmatrix},
\qquad (3.14)
$$

where

$$
\begin{pmatrix}
A & 1 & D & 0 & G & 0 \\
B & 0 & E & 0 & H & 1 \\
C & 0 & F & 1 & J & 0
\end{pmatrix}
\begin{pmatrix}
1 & 0 & 0 \\
K & L & M \\
0 & 0 & -1 \\
N & P & Q \\
0 & 2 & 0 \\
R & S & T
\end{pmatrix}
= 0,
\qquad (3.15)
$$

and A, B, C, D, E, F, G, H, and J are constants. Equation (3.15) is equivalent to

$$
\begin{aligned}
A &= -K, & L &= -2G, & M &= D, & B &= -R, & 2H &= -S, \\
E &= T, & C &= -N, & P &= -2J, & Q &= F.
\end{aligned}
\qquad (3.16)
$$

Substitution of (3.16) into (3.13) and (3.14) leads to the following expressions for the functions involved in (3.8) :

$$
\mu_1(y) = -\frac{B/2 + Hy - Ey^2/2}{C + 2Jy - Fy^2},
\qquad (3.17)
$$

$$
\sigma_1^2(y) = \frac{-1}{C + 2Jy - Fy^2},
\qquad (3.18)
$$

$$
\mu_2(x) = \frac{G + Hx + Jx^2}{D + Ex + Fx^2},
\qquad (3.19)
$$

$$
\sigma_2^2(x) = \frac{1}{D + Ex + Fx^2},
\qquad (3.20)
$$

$$
u(x) = \frac{1}{2}\left[A + Bx + Cx^2 + \frac{(G + Hx + Jx^2)^2}{D + Ex + Fx^2}\right],
\qquad (3.21)
$$

$$
v(y) = \frac{1}{2}\left[A + 2Gy - Dy^2 - \frac{(B/2 + Hy - Ey^2/2)^2}{C + 2Jy + Fy^2}\right],
\qquad (3.22)
$$

$$
f_1(x) = \frac{\exp\left\{\frac{1}{2}\left[A + Bx + Cx^2 + \frac{(G + Hx + Jx^2)^2}{D + Ex + Fx^2}\right]\right\}}{\sqrt{(D + Ex + Fx^2)}},
\qquad (3.23)
$$

$$
f_2(y) = \frac{\exp\left\{\frac{1}{2}\left[A + 2Gy - Dy^2 - \frac{(B/2 + Hy - Ey^2/2)^2}{C + 2Jy - Fy^2}\right]\right\}}{\sqrt{(-C - 2Jy + Fy^2)}},
\qquad (3.24)
$$

and for the joint density $f_{X,Y}(x,y)$:

$$f_{X,Y}(x,y) = \frac{1}{\sqrt{2\pi}}$$

$$\times \exp\{\tfrac{1}{2}[A + Bx + 2Gy + Cx^2 - Dy^2 + 2Hxy + 2Jx^2y - Exy^2 - Fx^2y^2]\}. \tag{3.25}$$

At this point it is convenient to introduce a new parametrization which will extend naturally to higher dimensions and will be consistent with the notation used for more general exponential families in Chapter 4. Thus, instead of expression (3.25), we write

$$f_{X,Y}(x,y) = \exp\left\{(1,x,x^2)\begin{pmatrix} m_{00} & m_{01} & m_{02} \\ m_{10} & m_{11} & m_{12} \\ m_{20} & m_{21} & m_{22} \end{pmatrix}\begin{pmatrix} 1 \\ y \\ y^2 \end{pmatrix}\right\}. \tag{3.26}$$

where

$$\begin{array}{lll} m_{00} = A/2, & m_{01} = G, & m_{02} = -D/2, \\ m_{10} = B/2, & m_{11} = H, & m_{12} = -E/2, \\ m_{20} = C/2, & m_{21} = J, & m_{22} = -F/2, \end{array} \tag{3.27}$$

and (3.17)–(3.24) transform to

$$E(X|Y = y) = \mu_1(y) = -\frac{m_{12}y^2 + m_{11}y + m_{10}}{2(m_{22}y^2 + m_{21}y + m_{20})}, \tag{3.28}$$

$$\text{var}(X|Y = y) = \sigma_1^2(y) = \frac{-1}{2(m_{22}y^2 + m_{21}y + m_{20})}, \tag{3.29}$$

$$E(Y|X = x) = \mu_2(x) = -\frac{m_{21}x^2 + m_{11}x + m_{01}}{2(m_{22}x^2 + m_{12}x + m_{02})}, \tag{3.30}$$

$$\text{var}(Y|X = x) = \sigma_2^2(x) = \frac{-1}{2(m_{22}x^2 + m_{12}x + m_{02})}, \tag{3.31}$$

$$f_1(x) = \frac{\exp\left\{\tfrac{1}{2}\left[2(m_{20}x^2 + m_{10}x + m_{00}) - \dfrac{(m_{21}x^2 + m_{11}x + m_{01})^2}{2(m_{22}x^2 + +m_{12}x + m_{02})}\right]\right\}}{\sqrt{-2(m_{22}x^2 + m_{12}x + m_{02})}}, \tag{3.32}$$

$$f_2(y) = \frac{\exp\left\{\tfrac{1}{2}\left[2(m_{02}y^2 + 2m_{01}y + m_{00}) - \dfrac{(m_{12}y^2 + m_{11}y + m_{10})^2}{2(m_{22}y^2 + m_{21}y + m_{20})}\right]\right\}}{\sqrt{-2(m_{22}y^2 + m_{21}y + m_{20})}}, \tag{3.33}$$

It remains only to determine appropriate conditions on the constants m_{ij}, $i,j = 0,1,2$, in (3.26) to ensure nonnegativity of $f_{X,Y}(x,y)$ and its marginals and the integrability of those marginals. In all cases the constant m_{00} will be a function of the other parameters chosen so that one (and hence all) of the densities (3.23), (3.24), (3.25), and (3.26) integrate to 1.

In order to guarantee that the marginals in (3.32) and (3.33) are non-negative (or, equivalently, to guarantee that for each fixed x, $f_{X,Y}(x,y)$ is integrable with respect to y and for each fixed y it is integrable with respect to x), the coefficients in (3.26) must satisfy one of the two sets of conditions

$$m_{22} = m_{12} = m_{21} = 0, \quad m_{20} < 0, \quad m_{02} < 0, \tag{3.34}$$

$$m_{22} < 0, \quad 4m_{22}m_{02} > m_{12}^2, \quad 4m_{20}m_{22} > m_{21}^2. \tag{3.35}$$

If (3.34) holds then we need to assume in addition that

$$m_{11}^2 < 4m_{02}m_{20}, \tag{3.36}$$

in order to guarantee that (3.23) and (3.24) and hence (3.25) and (3.26) are integrable. Note that (3.34) and (3.36) yield the classical bivariate normal model.

If (3.35) holds then $(m_{22}x^2 + m_{12}x + m_{02})$ is bounded away from zero and the function within square brackets in (3.32) will, for large values of $|x|$, behave like $x^2(4m_{20}m_{22} + m_{21}^2)/2m_{22}$ and consequently (3.32) will be integrable.

It is interesting to note that in his paper in which he sought characterizations of the classical bivariate normal, Bhattacharyya (1943) derived an equation essentially equivalent to (3.26). He assumed differentiability of $f_{X,Y}(x,y)$ in his derivation and he did not completely determine what conditions on the parameters were needed to guarantee integrability. He did express interest in making a detailed study of the density equivalent to (3.26) at some future time.

3.4   Properties of the Normal Conditionals Distribution

The normal conditionals distribution has joint density of the form (3.26) where the constants, the m_{ij}'s, satisfy one of the two sets of conditions:

(i) $m_{22} = m_{12} = m_{21} = 0$, $m_{20} < 0$, $m_{02} < 0$, $m_{11}^2 < 4m_{02}m_{20}$;

(ii) $m_{22} < 0$, $4m_{22}m_{02} > m_{12}^2$, $4m_{20}m_{22} > m_{21}^2$.

Models satisfying conditions (i) are classic bivariate normal with normal marginals, normal conditionals, linear regressions, and constant conditional variances (see Figures 3.1 and 3.2).

More interesting are the models satisfying conditions (ii). These models have normal conditionals distributions but have distinctly nonnormal marginal densities (see (3.32) and (3.33)). The regression functions are given by (3.28) and (3.30). These are either constant or nonlinear. Each

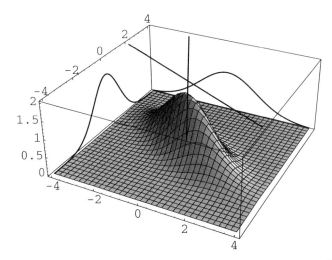

FIGURE 3.1. Density function of a bivariate normal $(1.27324e^{0.241564-x^2/2-xy-y^2})$ showing its regression lines (top projections) and its marginal densities (right and left projections).

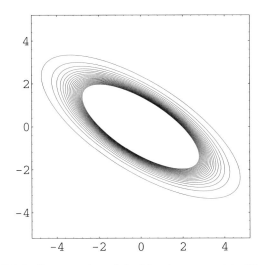

FIGURE 3.2. Contour plot of the bivariate normal in Figure 3.1.

regression function is bounded (a distinct contrast to the classical bivariate normal model). The conditional variance functions are also bounded. They are given by (3.29) and (3.31).

The point or points of intersection of the two regression curves determine the mode(s) of the bivariate distribution in both cases (i) and (ii). The bell-shape of the classical bivariate normal is well known. The form of typical normal conditionals densities satisfying conditions (ii) is not as familiar nor

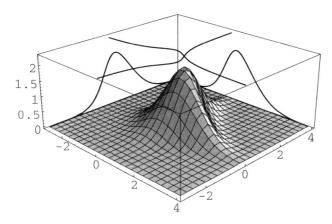

FIGURE 3.3. Example of a normal conditionals density showing its regression lines (top projection) and its marginal densities (right and left projections).

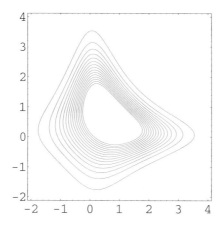

FIGURE 3.4. Contour plot of the normal conditionals density in Figure 3.3.

as easy to visualize. Figures 3.3 and 3.4 illustrate a representative density satisfying conditions (ii). The nonnormal marginals are shown in the backdrop of the figure. Their nonnormal character is evident. The picture is somewhat akin to what we might find if we left a classical Gaussian heap of sand out in the wind overnight. Three other illustrative cases of densities satisfying (ii) are provided by Gelman and Meng (1991), including one which is bimodal (see Figures 3.5 and 3.6 below and the reparametrization discussion immediately preceding them).

What if we require normal conditionals and independent marginals? Referring to (3.26) the requirement of independence translates to the following functional equation:

$$\left(1, x, x^2\right) \begin{pmatrix} m_{00} & m_{01} & m_{02} \\ m_{10} & m_{11} & m_{12} \\ m_{20} & m_{21} & m_{22} \end{pmatrix} \begin{pmatrix} 1 \\ y \\ y^2 \end{pmatrix} = r(x) + s(y), \tag{3.37}$$

which is of the form (2.24). Its solution using Theorem 2.3 eventually leads us to the conclusion that, for independence, we must have

$$m_{11} = m_{21} = m_{12} = m_{22} = 0. \tag{3.38}$$

This shows that independence is only possible within the classical bivariate normal model.

As consequences of the above discussion, Castillo and Galambos (1989) derive the following interesting conditional characterizations of the classical bivariate normal distribution:

Theorem 3.1 $f_{X,Y}(x,y)$ *is a classical bivariate normal density if and only if all conditional distributions, both of X given Y and of Y given X, are normal and any one of the following properties hold:*

(i) $\sigma_2^2(x) = var(Y|X = x)$ or $\sigma_1^2(y) = var(X|Y = y)$ is constant;

(ii) $\lim_{y\to\infty} y^2\sigma_1^2(y) = \infty$ or $\lim_{x\to\infty} x^2\sigma_2^2(x) = \infty$;

(iii) $\underline{\lim}_{y\to\infty} \sigma_1(y) \neq 0$ or $\underline{\lim}_{x\to\infty} \sigma_2(x) \neq 0$; and

(iv) $E(Y|X = x)$ or $E(X|Y = y)$ is linear and nonconstant.

$\square$

Proof.

(i) If $\sigma_2^2(x)$ is constant, from (3.29) we get $m_{21} = m_{22} = 0$ which implies classical bivariate normality.

(ii) If $y^2\sigma_1^2(y) \to \infty$, from (3.29) we get $m_{22} = 0$ which implies classical bivariate normality.

(iii) If $\underline{\lim} \sigma_1(y) \neq 0$, then from (3.29) we get $m_{21} = m_{22} = 0$ which implies classical bivariate normality.

(iv) If $E(Y|X = x)$ is linear, from (3.30) we get $m_{22} = 0$ which implies classical bivariate normality. $\square$

One final reparametrization merits mention. Following Gelman and Meng (1991), if in (3.26) we make the following linear change of variables:

$$u = \frac{x - a}{b},$$

$$v = \frac{y - c}{d}, \tag{3.39}$$

where

$$a = -\frac{m_{12}}{2m_{22}}, \tag{3.40}$$

$$b = 2\sqrt{\frac{-m_{22}}{4m_{20}m_{22} - m_{21}^2}}, \tag{3.41}$$

$$c = -\frac{m_{21}}{2m_{22}}, \tag{3.42}$$

$$d = 2\sqrt{\frac{-m_{22}}{4m_{02}m_{22} - m_{12}^2}}, \tag{3.43}$$

and we rename u and v as x and y, respectively, we obtain the density function

$$f(x, y) \propto \exp\left(\alpha x^2 y^2 - x^2 - y^2 + \beta xy + \gamma x + \delta y\right), \tag{3.44}$$

where α, β, γ, and δ are the new parameters which are functions of the old m_{ij} parameters.

In this parametrization, the conditional distributions are

$$X|Y = y \sim N\left(-\frac{\beta y + \gamma}{2(\alpha y^2 - 1)}, -\frac{1}{2(\alpha y^2 - 1)}\right), \tag{3.45}$$

$$Y|X = x \sim N\left(-\frac{\beta x + \delta}{2(\alpha x^2 - 1)}, -\frac{1}{2(\alpha x^2 - 1)}\right). \tag{3.46}$$

The only constraints for this parametrization are

$$\alpha \leq 0 \quad \text{and} \quad \text{if } \alpha = 0 \quad \text{then} \quad |\beta| < 2. \tag{3.47}$$

An advantage of this Gelman and Meng parametrization is that in some cases it renders it easy to recognize multimodality. Bimodality of a distribution with normal conditionals is perhaps a surprising development. It is, of course, retrospectively obvious that the conditional mode curves (which correspond to the conditional mean curves (3.28) and (3.30)) can intersect at more than one point.

Since modes are at the intersection of regression lines, from (3.45) and (3.46) the coordinates of the modes satisfy the system of equations

$$x = -\frac{\beta y + \gamma}{2(\alpha y^2 - 1)},$$
$$y = -\frac{\beta x + \delta}{2(\alpha x^2 - 1)}. \tag{3.48}$$

Substituting the first into the second we get

$$4\alpha^2 y^5 - 2\alpha^2 \delta y^4 - 8\alpha y^3 + \alpha(4\delta - \beta\gamma)y^2 + (4 - \beta^2 - \alpha\gamma^2)y - 2\delta - \beta\gamma = 0, \tag{3.49}$$

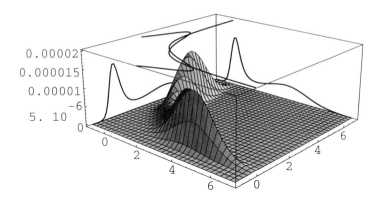

FIGURE 3.5. Example of a normal conditionals density with two modes showing its regression lines and its marginal densities.

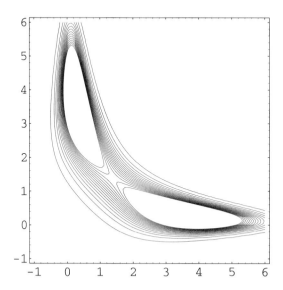

FIGURE 3.6. Contour plot of the normal conditionals density in Figure 3.5 (the mask of Zorro?).

which is a polynomial of degree 5. When this polynomial has a unique real root, the density is unimodal, if it has three distinct real roots the density is bimodal.

But can life get even more complicated? Can there exist three modes? It turns out that (3.49) can have five distinct real roots, though diligent searching of the parameter space is required before encountering such cases. In addition, one mode may be a molehill while the others are mountains. Such is the case in the example described below. But modes they all three are, even though we have to resort to graphing $f^{1/200}(x, y)$ in order to make them more evident in our picture.

Example 3.1 (Two-modes). Figures 3.5 and 3.6 show one example of a density satisfying conditions (ii) with two modes. Its marginals (plotted on the right and left projections), which are not normal and the corresponding nonlinear regression lines (plotted on top of the joint density), clearly indicate that the model is nonclassical. Observe that in this example the regression curves intersect three times, corresponding to two modes and a saddle point. Note also the unexpected bimodality of the corresponding marginal densities. □

Example 3.2 (Three modes). Consider the model in (3.44) with parameters

$$\alpha = -30, \quad \beta = -100, \quad \gamma = -22, \quad \delta = 20.$$

For this case the fifth degree polynomial (3.49) becomes

$$-2240 + 4524y + 63600y^2 + 240y^3 - 36000y^4 + 3600y^5 \qquad (3.50)$$

with roots

$$y = -1.199, \quad y = -0.231, \quad y = 0.1563, \quad y = 1.466, \quad y = 9.808,$$

which correspond to three relativa maxima (modes) and two saddle points. Figure 3.7 shows the 200th root of the corresponding probability density function (i.e., $[f(x, y)]^{1/200}$) and Figure 3.8 shows its associated contour plot together with the two regression lines and the five critical points identified above. □

3.5 The Centered Model

Consider the bivariate random variable (X, Y) with joint density function

$$f_{X,Y}(x, y) = k(c) \frac{1}{2\pi\sigma_1\sigma_2} \exp\left\{-\frac{1}{2}[(x/\sigma_1)^2 + (y/\sigma_2)^2 + c(x/\sigma_1)^2(y/\sigma_2)^2]\right\} \tag{3.51}$$

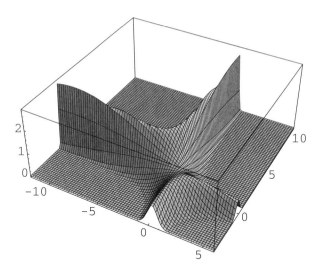

FIGURE 3.7. Transformed probability density of the three modes example.

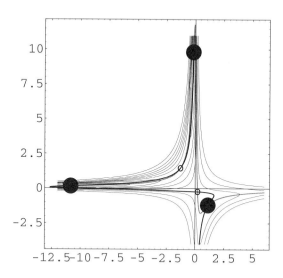

FIGURE 3.8. Contour plot of the transformed probability density of the three modes example in Figure 3.7. The regression curves are also shown in the diagram as are the three modes (marked by solid dots) and the two other critical points (marked by open dots).

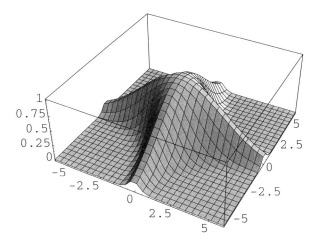

FIGURE 3.9. Bivariate centered model density plot for $\sigma_1^2 = \sigma_2^2 = 10$ and $c = 20$.

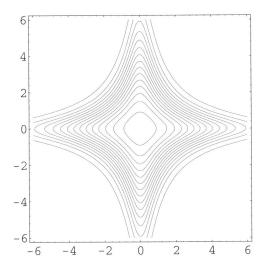

FIGURE 3.10. Contour plot of the centered normal conditionals density for $\sigma_1^2 = \sigma_2^2 = 10$ and $c = 20$.

defined on the whole plane, where σ_1, $\sigma_2 > 0$, $c \geq 0$, and $k(c)$ is a normalizing constant. With different parameters, the density function (3.51)
may be recognized as one of those which Bhattacharyya (1943) identified
as nonstandard models with normal conditional distributions. More particularly, (3.51) corresponds to the most general density function which
has normal conditional distributions with zero means. Consider (X, Y), a

bivariate random variable, such that

$$X|Y = y \sim N(0, \sigma_1^2(y)), \tag{3.52}$$

$$Y|X = x \sim N(0, \sigma_2^2(x)), \tag{3.53}$$

where $\sigma_1^2(y)$, $\sigma_2^2(x)$ are unknown functions such that $\sigma_1^2(y) > 0$, $\sigma_2^2(x) > 0$ for all x, y. Then, it can be proved that (Castillo and Galambos (1989)): $\sigma_1^2(y) = \sigma_1^2/(1 + c(y/\sigma_2)^2)$, $\sigma_2^2(x) = \sigma_2^2/(1 + c(x/\sigma_1)^2)$, and the density function of (X, Y) is given by (3.51). Equivalently, the density can be identified as that obtainable from (3.26) upon setting $m_{01} = m_{10} = m_{21} = m_{11} = 0$. A typical representation of the form of such densities is provided in Figures 3.9 and 3.10. The centered distribution can be used to model bivariate data which are uncorrelated yet nonindependent. An example of this, provided by Arnold and Strauss (1991), involves 30 shots at a target, under slow firing conditions, in a pistol training session.

By integrating (3.51) with respect to y, using the normal density function, we obtain the marginal density of X,

$$f_X(x) = k(c)\frac{1}{\sigma_1\sqrt{2\pi}}\left[1 + c(x/\sigma_1)^2\right]^{-1/2}\exp\left[-\frac{1}{2}(x/\sigma_1)^2\right] \tag{3.54}$$

and, similarly,

$$f_Y(y) = k(c)\frac{1}{\sigma_2\sqrt{2\pi}}\left[1 + c(y/\sigma_2)^2\right]^{-1/2}\exp\left[-\frac{1}{2}(y/\sigma_2)^2\right]. \tag{3.55}$$

Except when $c = 0$, (3.54) and (3.55) are not normal.

The normalizing constant $k(c)$ in (3.51) can be obtained by the representation of the integral of the confluent hypergeometric function given by $(a > 0, z > 0)$,

$$U(a, b, z) = \frac{1}{\Gamma(a)}\int_0^\infty e^{-tz}t^{a-1}(1 + t)^{b-a-1}\,dt \tag{3.56}$$

(Abramowitz and Stegun (1964), eq. 13.2.5). Now, integrating (3.54) and making the indicated change of variables we have

$$\int_{-\infty}^{+\infty}\frac{1}{\sigma_1\sqrt{2\pi}}\left(1 + c(x/\sigma_1)^2\right)^{-1/2}e^{-(x/\sigma_1)^2/2}\,dx$$

$$= 2\int_0^{+\infty}\frac{1}{\sigma_1\sqrt{2\pi}}\left(1 + c(x/\sigma_1)^2\right)^{-1/2}e^{-(x/\sigma_1)^2/2}\,dx$$

$$= \frac{1}{\sqrt{2c\pi}}\int_0^{+\infty}t^{-1/2}(1 + t)^{-1/2}e^{-t/2c}\,dt = \frac{1}{\sqrt{2c}}U\left(\frac{1}{2}, 1, \frac{1}{2c}\right),$$

where we have made $\left[c(x/\sigma_1)^2 = t\right]$. Therefore, the value of k is given by

$$k(c) = \frac{\sqrt{2c}}{U(1/2,\ 1,\ 1/2c)}. \tag{3.57}$$

3.5.1   Distribution Theory

Suppose that (X, Y) is a bivariate random variable with a centered normal conditionals distribution (CNC) with density function (3.51). The parameters σ_1 and σ_2 are scale parameters, and c is the dependence parameter, where $c = 0$ corresponds to independence between X and Y. Notice that the correlation coefficient is always zero. From (3.54) and (3.55) the following unconditional associations between X and Y may be verified:

$$(X/\sigma_1)\left[1 + c(Y/\sigma_2)^2\right]^{1/2} \sim N(0, 1), \tag{3.58}$$

$$(Y/\sigma_2)\left[1 + c(X/\sigma_1)^2\right]^{1/2} \sim N(0, 1). \tag{3.59}$$

Notice that the variable defined in (3.58) is independent of Y and the variable in (3.59) is independent of X. Using (3.58) and (3.59), the following relations among the moments of X and Y can be obtained:

$$E\left\{(X/\sigma_1)^n \left[1 + c(Y/\sigma_2)^2\right]^{n/2}\right\} = E\left\{(Y/\sigma_2)^n \left[1 + c(X/\sigma_1)^2\right]^{n/2}\right\} = E(Z^n), \tag{3.60}$$

where for n even $E(Z^n) = (n-1)(n-3)\cdots 1$, since Z denotes an $N(0, 1)$ random variable. Because the random variable $Z = (X/\sigma_1)(1 + c(Y/\sigma_2)^2)^{1/2}$ is $N(0, 1)$ and independent of Y, if $V = \sigma_1(1 + c(Y/\sigma_1)^2)^{-1/2}$, it follows that $E(Z^n V^n) = E(Z^n)E(V^n)$ or, equivalently,

$$\frac{E\left[(X/\sigma_1)^n\right]}{E\left\{\left[1 + c(Y/\sigma_2)^2\right]^{-n/2}\right\}} = \frac{E\left[(Y/\sigma_2)^n\right]}{E\left\{\left[1 + c(X/\sigma_1)^2\right]^{-n/2}\right\}} = E(Z^n). \tag{3.61}$$

The second relation can be obtained by symmetry. From the marginal density functions (3.54) and (3.55) we can obtain some particular moments related to X and Y,

$$E\left\{(X/\sigma_1)^n \left[1 + c(X/\sigma_1)^2\right]^{1/2}\right\} = E\left\{(Y/\sigma_2)^n \left[1 + c(Y/\sigma_2)^2\right]^{1/2}\right\}$$

$$= k(c)E(Z^n). \tag{3.62}$$

Alternatively, we can obtain the moments by using a moment generating function. Since only the moments of even order are different from zero, we shall calculate the moment generating function of (X^2, Y^2),

$$M_{X^2, Y^2}(s, t) = E(e^{sX^2 + tY^2})$$

$$= \int_{-\infty}^{+\infty} \int_{-\infty}^{+\infty} \frac{k(c)}{2\pi\sigma_1\sigma_2} e^{-1/2\left[(\sigma_1^{-2} - 2s)x^2 + (\sigma_2^{-2} - 2t)y^2 + c(\frac{xy}{\sigma_1\sigma_2})^2\right]} \, dx \, dy \qquad (3.63)$$

$$= \frac{(1 - 2\sigma_1^2 s)^{-1/2}(1 - 2\sigma_2^2 t)^{-1/2}k(c)}{k[c(1 - 2\sigma_1^2 s)^{-1}(1 - 2\sigma_2^2 t)^{-1}]},$$

where $s < 1/(2\sigma_1^2)$, $t < 1/(2\sigma_2^2)$ in order to guarantee the convergence of the integrand. When $c = 0$, (3.63) is recognized as the generating function of two independent rescaled chi-square variables. Most of the moments can be expressed in terms of the function

$$\delta(c) = \frac{d \log k(c)}{dc} = \frac{k'(c)}{k(c)}. \qquad (3.64)$$

From the moment generating function (3.63), in conjunction with (3.60), we have

$$E(X^2) = \sigma_1^2[1 - 2c\delta(c)], \qquad V(X^2) = \sigma_1^4\left[1 + 2\delta(c) - 4c^2\delta^2(c)\right], \qquad (3.65)$$

$$E(Y^2) = \sigma_2^2[1 - 2c\delta(c)], \qquad V(Y^2) = \sigma_2^4\left[1 + 2\delta(c) - 4c^2\delta^2(c))\right], \qquad (3.66)$$

$$E(X^2 Y^2) = 2\sigma_1^2\sigma_2^2\delta(c), \qquad (3.67)$$

$$\rho(X^2, Y^2) = \frac{1 - 2\delta(c) - 4c\delta(c) + 4c^2\delta^2(c)}{-1 - 2\delta(c) + 4c^2\delta^2(c)}. \qquad (3.68)$$

If we calculate $E(X^2 Y^2)$ by means of the generating function, and equate the result to (3.67) which was obtained using (3.60), we find that the function δ must satisfy the following differential equation:

$$1 - (2 + 8c)\delta(c) + 4c^2\delta^2(c) - 4c^2\delta'(c) = 0. \qquad (3.69)$$

By means of (3.69), the moments of higher order can be expressed as functions of δ. As an alternative, the values of k and δ can be expressed as

$$k^{-1}(c) = E\left[(1 + cZ^2)^{-1/2}\right], \qquad (3.70)$$

$$\delta(c) = \frac{1}{2}\frac{E\left[Z^2 \left(1 + cZ^2\right)^{-3/2}\right]}{E\left[(1 + cZ^2)^{-1/2}\right]}, \qquad (3.71)$$

where $Z \sim N(0, 1)$. These expressions allow straightforward numerical approximation using draws from an $N(0, 1)$ distribution. Expression (3.70) also makes transparent the fact that $k(c)$ is a monotone increasing function of c. Representative values of $k(c)$ and $\delta(c)$ are displayed in Figures 3.11 and 3.12.

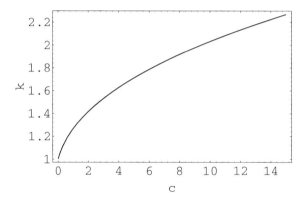

FIGURE 3.11. Plot of $k(c)$ versus c for the normal centered model.

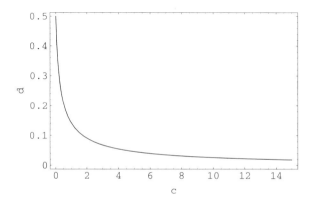

FIGURE 3.12. Plot of $\delta(c)$ versus c for the normal centered model.

3.5.2   Marginal Distribution Theory

Now we shall take a closer look at the marginal density functions of X and Y, given by (3.54) and (3.55). These are symmetric unimodal random variables with less kurtosis than the normal distribution (refer to (3.61)). The tails of these distributions are less heavy than those of the $N(0,1)$ distribution. The distribution function of X is not simple, since (3.54) has no closed form integral. However, a simple approximation for small values of c is possible. If $\Phi_c(x) = P(X \le x)$ and $x \ge 0$, when we expand in a power series (justifiably differentiating under the integral sign) we obtain

$$\Phi_c(x) = \Phi(x/\sigma_1) + \left\{ \frac{1}{2}\left[\Phi(x/\sigma_1) - \frac{1}{2}\right] - \frac{1}{4}I(\frac{3}{2}, \frac{1}{2}(x/\sigma_1)^2) \right\} c + o(c),$$
(3.72)

where $I(a,x) = \int_0^x e^{-t}t^{a-1}\,dt/\Gamma(a)$ represents the incomplete gamma function and Φ represents the standard normal distribution function. As an alternative to the methods used in the previous section, the moments of X

can be expressed in terms of the confluent hypergeometric function. If n is an even number, it may be verified that

$$E(X^n) = k(c)\frac{1}{\sqrt{2\pi}}\sigma_1^n c^{-(n+1)/2}\Gamma\left(\frac{n+1}{2}\right)U\left(\frac{n+1}{2}, \frac{n}{2}+1, \frac{1}{2c}\right).$$
(3.73)

Since the expressions for the moments seem to be complex, some kind of recurrence relation would be useful between them. By using recurrence properties of the confluent hypergeometric function (Abramowitz and Stegun (1964), eqs. 13.4.21 and 13.5.27) we obtain

$$\frac{\sigma_1^2(n-1)}{c}E(X^{n-2}) = \left(\frac{1}{c}-n\right)E(X^n) + \frac{1}{\sigma_1^2}E(X^{n+2}),$$
(3.74)

where n is an even number. Analogously, it may be verified that

$$\frac{\sigma_1^2(n-1)}{c}E(X^{n-2}Y^m) = \left[\frac{1}{c}-(n-m)\right]E(X^nY^m) + \frac{1}{\sigma_1^2}E(X^{n+2}Y^m)$$
(3.75)

from a formula similar to (3.73) for $E(X^nY^m)$.

3.6   Bibliographic Notes

The key references for material in this chapter are Bhattacharyya (1943) and Castillo and Galambos (1987a, 1989). Gelman and Meng (1991) provide interesting illustrations and an alternative parametrization. Further discussion of multiple modes may be found in Arnold, Castillo, Sarabia and González-Vega (1999b). The centered model is discussed in Sarabia (1995).

Exercises

3.1 Solve functional equation (3.37).

3.2 Let (X, Y) be a bivariate random variable with pdf given by

$$f_{X,Y}(x, y) \propto \exp\left[-\left(x^2y^2 + x^2 + y^2 - 2\delta x - 2\delta y\right)/2\right]$$

with δ a real constant.

(a) Prove that $E(X|Y)/\mathrm{var}(X|Y) = E(Y|X)/\mathrm{var}(Y|X) = \delta$.

(b) Prove that $f_{X,Y}(x, y)$ is bimodal if and only if $|\delta| > 2$.

(Gelman and Meng (1991).)

3.3 Obtain the most general bivariate distributions (X, Y) with lognormal conditionals $X|Y = y$, $\forall y$ and $Y|X = x$, $\forall x$.

Note: A random variable X is said to be lognormal, iff $\log X \sim N(\mu, \sigma^2)$.

3.4 Consider the model (4.44) written in terms of the parametrization

$$f(x, y) = \frac{k(a, b, c, d)}{2\pi} \exp\left[-\left(ax^2 y^2 + x^2 + y^2 + bxy + cx + dy\right)/2\right].$$

with $a > 0$ and b, c, d real numbers.

(a) If U, V are $N(0, 1)$ and independent random variables, prove that

$$k^{-1}(a, b, c, d) = E\left\{\exp\left[-(aU^2 V^2 + bUV + cU + dV)/2\right]\right\}.$$

(b) Give an expression for the correlation coefficient in terms of the function $k(a, b, c, d)$ and its partial derivatives.

3.5 Compute the coefficient of linear correlation (3.68) for some values of $c \geq 0$.

3.6 Let (X, Y) be a bivariate random variable such that $X|Y = y$ is normal for all y and $Y|X = x$ is normal for all x and $X \overset{d}{=} Y$. Does X, Y have a classical bivariate normal distribution?

3.7 Let (X, Y) be a bivariate random variable such that X and $Y|X = x$ for all x are normal distributions. Does X, Y have a classical bivariate normal distribution?
Hint: Consider the joint pdf:

$$f(x, y) \propto \sqrt{1 + x^2} \exp\left[-(x^2 y^2 + x^2 + y^2)/2\right].$$

(Hamedani (1992).)

3.8 Is it possible to have (X_1, X_2) with a normal conditionals distribution such that

$$E(X_1|X_2 = x_2) = \text{var}(X_1|X_2 = x_2), \quad \forall x_2,$$

and

$$E(X_2|X_1 = x_1) = \text{var}(X_2|X_1 = x_1), \quad \forall x_1?$$

(Gelman and Meng (1991).)

3.9 If (X_1, X_2) has a normal conditionals distribution then any one of the following conditions is sufficient to ensure that (X_1, X_2) has a classical bivariate normal distribution:

(a) X_1 is normally distributed.

(b) The contours of the joint density are similar concentric ellipses.

4
Conditionals in Exponential Families

4.1 Introduction

Following our careful analysis of the normal conditionals example in Chapter 3 and our brief mention of the exponential conditionals distribution in Chapter 1, it is natural to seek out more general results regarding distributions whose conditionals are posited to be members of quite general exponential families. Indeed the discussion leading up to Theorem 1.3, suggests that things should work well when conditionals are from exponential families. The key reference for the present chapter is Arnold and Strauss (1991). However it should be mentioned that results due to Besag (1974), in a stochastic process setting, anticipate some of the observations in this chapter.

4.2 Distributions with Conditionals in Given Exponential Families

In this section we consider the important case of exponential families.

Definition 4.1 (Exponential family). An ℓ_1-parameter family of densities $\{f_1(x; \underline{\theta}) : \underline{\theta} \in \Theta\}$, with respect to μ_1 on D_1, of the form

$$f_1(x; \underline{\theta}) = r_1(x)\beta_1(\underline{\theta}) \exp\left\{ \sum_{i=1}^{\ell_1} \theta_i q_{1i}(x) \right\}, \tag{4.1}$$

is called an exponential family of distributions.

Here Θ is the natural parameter space and the $q_{1i}(x)$'s are assumed to be linearly independent. Frequently, μ_1 is Lebesgue measure or counting measure and often D_1 is some subset of Euclidean space of finite dimension.   □

Consider the exponential family in (4.1) and let $\{f_2(y; \underline{\tau}) : \underline{\tau} \in T\}$ denote another ℓ_2-parameter exponential family of densities with respect to μ_2 on D_2, of the form

$$f_2(y; \underline{\tau}) = r_2(y)\beta_2(\underline{\tau}) \exp \left\{ \sum_{j=1}^{\ell_2} \tau_j q_{2j}(y) \right\}, \tag{4.2}$$

where T is the natural parameter space and as is customarily done, the $q_{2j}(y)$'s are assumed to be linearly independent.

Our goal is the identification of the class of bivariate densities $f(x, y)$ with respect to $\mu_1 \times \mu_2$ on $D_1 \times D_2$ for which conditional densities $f(x|y)$ and $f(y|x)$ are well defined and satisfy:

(i) for every y for which $f(x|y)$ is defined, this conditional density belongs to the family (4.1) for some $\underline{\theta}$ which may depend on y; and

(ii) for every x for which $f(y|x)$ is defined, this conditional density belongs to the family (4.2) for some $\underline{\tau}$ which may depend on x.

The general class of such bivariate distributions is described in the following result:

Theorem 4.1 *Let $f(x, y)$ be a bivariate density whose conditional densities satisfy*

$$f(x|y) = f_1(x; \underline{\theta}(y)) \tag{4.3}$$

and

$$f(y|x) = f_2(y; \underline{\tau}(x)) \tag{4.4}$$

for some function $\underline{\theta}(y)$ and $\underline{\tau}(x)$ where f_1 and f_2 are defined in (4.1) and (4.2). It follows that $f(x, y)$ is of the form

$$f(x, y) = r_1(x) r_2(y) \exp\{\underline{q}^{(1)}(x)' M \underline{q}^{(2)}(y)\}, \tag{4.5}$$

where

$$\underline{q}^{(1)}(x) = (q_{10}(x), q_{11}(x), q_{12}(x), \ldots, q_{1\ell_1}(x)),$$

$$\underline{q}^{(2)}(y) = (q_{20}(y), q_{21}(y), q_{22}(y), \ldots, q_{2\ell_2}(y)).$$

where $q_{10}(x) = q_{20}(y) \equiv 1$ and M is a matrix of constants parameters of appropriate dimensions (i.e., $(\ell_1 + 1) \times (\ell_2 + 1)$) subject to the requirement that

$$\int_{D_1} \int_{D_2} f(x, y) \, d\mu_1(x) \, d\mu_2(y) = 1.$$

For convenience we can partition the matrix M as follows:

$$M = \begin{pmatrix} m_{00} & | & m_{01} & \cdots & m_{0\ell_2} \\ -- & + & -- & -- & -- \\ m_{10} & | & & & \\ \cdots & | & & \tilde{M} & \\ m_{\ell_1 0} & | & & & \end{pmatrix}. \tag{4.6}$$

Note that the case of independence is included; it corresponds to the choice $\tilde{M} \equiv \mathbf{0}$. □

Proof. Consider a joint density with conditionals in the given exponential families. Denote the marginal densities by $g(x), x \in S(X) = \{x : r_1(x) > 0\}$ and $h(y), y \in S(Y) = \{y : r_2(y) > 0\}$, respectively. Write the joint density as a product of a marginal and a conditional density in two ways to obtain the relation

$$g(x)r_2(y)\beta_2(\underline{\tau}(x))\exp[\underline{\tau}(x)'\underline{\tilde{q}}^{(2)}(y)] = h(y)r_1(x)\beta_1(\underline{\theta}(y))\exp\{\underline{\theta}(y)'\underline{\tilde{q}}^{(1)}(x)\} \tag{4.7}$$

for $(x, y) \in S(X) \times S(Y)$ where

$$\underline{\tilde{q}}^{(1)}(x) = (q_{11}(x), q_{12}(x), \ldots, q_{1\ell_1}(x)),$$

$$\underline{\tilde{q}}^{(2)}(y) = (q_{21}(y), q_{22}(y), \ldots, q_{2\ell_2}(y)).$$

Now define

$$\tau_0(x) = \log[g(x)\beta_2(\underline{\tau}(x))/r_1(x)],$$

$$\theta_0(y) = \log[h(y)\beta_1(\underline{\theta}(y))/r_2(y)],$$

and then (4.7) can be written in the form

$$r_1(x)r_2(y)\exp\left[\sum_{j=0}^{\ell_2}\tau_j(x)q_{2j}(y)\right] = r_1(x)r_2(y)\exp\left[\sum_{i=0}^{\ell_1}\theta_i(y)q_{1i}(x)\right]. \tag{4.8}$$

Note both sides of (4.8) represent $f(x, y)$. If we cancel $r_1(x)r_2(y)$ in (4.8) we are left with a functional equation to which Theorem 2.4 applies directly. It follows that

$$\sum_{j=0}^{\ell_2}\tau_j(x)q_{2j}(y) = \sum_{i=0}^{\ell_1}\theta_i(y)q_{1i}(x) = \underline{q}^{(1)'}(x)M\underline{q}^{(2)}(y). \tag{4.9}$$

We then obtain (4.5) by substituting (4.9) in (4.8). □

An alternative perhaps more elementary proof involves taking logarithms on both sides of (4.8) and then taking differences with respect to x and y (see Arnold and Strauss (1991).)

The factor $e^{m_{00}}$ in (4.5) is a normalizing constant. The other elements of M are constrained to be such that

$$\psi(M) = \int\int_{D_1 \times D_2} e^{-m_{00}} f(x, y)\, d\mu_1(x)\, d\mu_2(y) < \infty. \qquad (4.10)$$

The normalizing constant is then necessarily given by $1/\psi(M)$, so that the joint density integrates to 1. The normalizing constant frequently must be evaluated numerically. As a consequence, the likelihood function associated with samples from conditionals in exponential families distributions are often intractably complicated. Standard maximum likelihood techniques are, at best, difficult to implement. The picture is not completely bleak however. As we shall see in Chapter 9, pseudo-likelihood and method of moments approaches are feasible and often prove to be quite efficient. We remark also that a lack of explicit knowledge of the normalizing factor does not prevent us from simulating samples from distributions with conditionals in exponential families of distributions (see Appendix A). It will be convenient to introduce the acronym CEF to avoid the necessity of repeating the mouthful "conditionals in exponential families." Thus CEF distributions are those whose densities are of the form (4.5).

We have already met two important classes of CEF distributions; the exponential conditionals class (introduced in Chapter 1) and the normal conditionals class discussed in detail in Chapter 3. In the subsequent catalog of CEF distributions in Sections 4.4 and 4.5, these old friends will be only briefly described. Essentially we will just verify how they need to be reparametrized in order to fit in the formulation given by (4.5).

We remark in passing that the requirement for the density to be integrable may be so restrictive as to rule out any possible model except an independent marginals model. See, for example, the discussion of Planck conditionals in Section 4.9.

4.3   Dependence in CEF Distributions

The exponential conditionals distribution described in Chapter 1 has the following joint density (repeating (1.39))

$$f(x, y) = \exp(m_{00} - m_{10}x - m_{01}y + m_{11}xy)\,, \qquad x > 0, \qquad y > 0,$$

where $m_{10} > 0, m_{01} > 0$, and $m_{11} \leq 0$. For such a joint distribution it is readily verified that

$$P(X > x | Y = y) = \exp[-(m_{10} - m_{11}y)x]. \qquad (4.11)$$

It follows that X is stochastically decreasing in Y. Consequently, applying Theorem 5.4.2 of Barlow and Proschan (1981), X and Y are negatively quadrant dependent and consequently have nonpositive correlation, i.e., $\rho(X, Y) \leq 0$. It is reasonable to ask whether this negative dependence is characteristic of the CEF distributions. As we shall see, in certain families negative dependence is assured, in others positive dependence and in some families a unrestrained spectrum of correlations is encountered. It is valuable to know about any correlation restraints in CEF models since those constraints may help justify or countermand use of a particular model in specific practical applications.

It remains a surprising fact that we cannot have $X|Y = y$ exponential with parameter dependent on y, $Y|X = x$ exponential with parameter dependent on x, and (X, Y) positively correlated. But such is the case, and in any given situation we must sacrifice either exponential conditionals or positive correlation in order to have a proper model. Of course with a normal conditionals model, correlations of either sign are possible.

Let us now focus on CEF distributions given by (4.5). Under what conditions can we assert that positive quadrant dependence obtains and hence that, provided adequate moments exist, nonnegative correlation will be encountered? A convenient sufficient condition for such positive dependence is that the density be totally positive of order 2, i.e., that

$$\begin{vmatrix} f(x_1, y_1) & f(x_1, y_2) \\ f(x_2, y_1) & f(x_2, y_2) \end{vmatrix} \geq 0 \qquad (4.12)$$

for every $x_1 < x_2, y_1 < y_2$ in $S(X)$ and $S(Y)$, respectively. (See Barlow and Proschan's (1981), Theorem 5.4.2.) The determinant (4.12) assumes a particularly simple form if the joint density is of the form (4.5). Substitution into (4.12) yields the following sufficient condition for total positivity of order 2 (abbreviated as TP2):

$$[\tilde{\underline{q}}^{(1)}(x_1) - \tilde{\underline{q}}^{(1)}(x_2)]' \tilde{M}[\tilde{\underline{q}}^{(2)}(y_1) - \tilde{\underline{q}}^{(2)}(y_2)] \geq 0 \qquad (4.13)$$

for every $x_1 < x_2$ in $S(X)$ and $y_1 < y_2$ in $S(Y)$. Thus, for example, if the $q_{1i}(x)$'s and the $q_{2j}(y)$'s are all increasing functions, then a sufficient condition for TP2 and hence for nonnegative correlation is that $\tilde{M} \geq 0$ (i.e., if $m_{ij} \geq 0, \forall i = 1, 2, \ldots, \ell_1, j = 1, 2, \ldots, \ell_2$). If $\tilde{M} \leq 0$ then negative correlation is assured (as was encountered in the exponential conditionals example). If the q_{1i}'s and q_{2j}'s are not monotone then it is unlikely that any choice for $\tilde{M}$ will lead to a TP2 density, and in such settings it is quite possible to encounter both positive and negative correlations (as in the normal conditionals example).

4.4   Exponential Conditionals

In this case, $\ell_1 = \ell_2 = 1, r_1(t) = r_2(t) = I(t > 0)$, and $q_{11}(t) = q_{21}(t) = -t$. The resulting densities are of the form

$$f(x, y) = \exp(m_{00} - m_{10}x - m_{01}y + m_{11}xy), \quad x > 0, \ y > 0. \qquad (4.14)$$

For convergence we must have $m_{10} > 0, m_{01} > 0$, and $m_{11} \leq 0$. It follows from the discussion in Section 4.3, that only nonpositive correlation will be encountered. With different parametrization, this density has been discussed extensively in Arnold and Strauss (1988a). It was also treated by Besag (1974), Abrahams and Thomas (1984), and Inaba and Shirahata (1986).

A more convenient parametrization is the following (see Arnold and Strauss (1988a)):

$$f(x, y) = \frac{k(c)}{\sigma_1 \sigma_2} \exp[-x/\sigma_1 - y/\sigma_2 - cxy/(\sigma_1 \sigma_2)]. \qquad (4.15)$$

The conditional densities are exponential, that is,

$$X|Y = y \sim \text{Exp}[(1 + cy/\sigma_2)/\sigma_1], \qquad (4.16)$$

$$Y|X = x \sim \text{Exp}[(1 + cx/\sigma_1)/\sigma_2]. \qquad (4.17)$$

The marginal densities are

$$f_X(x) = \frac{k(c)}{\sigma_1}(1 + cx/\sigma_1)^{-1}e^{-x/\sigma_1}, \quad x > 0, \qquad (4.18)$$

$$f_Y(y) = \frac{k(c)}{\sigma_2}(1 + cy/\sigma_2)^{-1}e^{-y/\sigma_2}, \quad y > 0. \qquad (4.19)$$

Observe that from (4.16) and (4.17) we have unconditionally

$$X(1 + cY/\sigma_2)/\sigma_1 \sim \text{Exp}(1), \qquad (4.20)$$

$$Y(1 + cX/\sigma_1)/\sigma_2 \sim \text{Exp}(1). \qquad (4.21)$$

Note that the variable defined in (4.20) is independent of Y and the variable defined in (4.21) is independent of X.

Example 4.1 (Exponential conditionals).   Figure 4.1 shows an example of an exponential conditionals distribution of the type in (4.15) corresponding to $c = 2, \sigma_1 = \sigma_2 = 1$. The left figure corresponds to the probability density function, and the right figure to the contours.   □

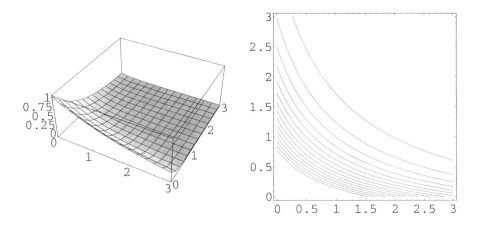

FIGURE 4.1. Example of an exponential conditionals with $c = 2, \sigma_1 = 2, \sigma_2 = 1$ showing (left side) the probability density function, and (right side) the contour plot.

The normalizing constant can be obtained in terms of the classical exponential integral function. More precisely, we have

$$k(c) = \frac{ce^{-1/c}}{[-\mathrm{Ei}(1/c)]}, \qquad (4.22)$$

where

$$-\mathrm{Ei}(u) = \int_u^\infty \frac{e^{-w}}{w}\,dw. \qquad (4.23)$$

The joint moment generating function is given by

$$M_{X,Y}(s,t) = E(e^{sX+tY}) = \frac{(1-\sigma_1 s)^{-1}(1-\sigma_2 t)^{-1}k(c)}{k[c(1-\sigma_1 s)^{-1}(1-\sigma_2 t)^{-1}]}, \qquad (4.24)$$

where $s < 1/\sigma_1$ and $t < 1/\sigma_2$. Thus, we have

$$
\begin{aligned}
E(X) &= \sigma_1[k(c) - 1]/c, & (4.25)\\
E(Y) &= \sigma_2[k(c) - 1]/c, & (4.26)\\
\mathrm{var}(X) &= \sigma_1^2 k(c)[1 + c - k(c)]/c^2, & (4.27)\\
\mathrm{var}(Y) &= \sigma_2^2 k(c)[1 + c - k(c)]/c^2, & (4.28)\\
\mathrm{cov}(X,Y) &= \sigma_1\sigma_2[k(c) - k^2(c) + c]/c^2. & (4.29)
\end{aligned}
$$

Consequently, the coefficient of correlation is

$$\rho(X,Y) = \frac{c + k(c) - k^2(c)}{k(c)[1 + c - k(c)]}. \qquad (4.30)$$

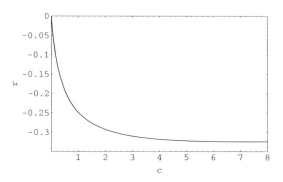

FIGURE 4.2. Coefficient of correlation ρ as a function of c for an exponential conditionals density.

Figure 4.2 shows the coefficient of correlation ρ as a function of c for the exponential conditionals family. Note that it is always negative and bounded from below by the value -0.32.

4.5   Normal Conditionals

Assuming unknown mean and variance, we are dealing with two-parameter exponential families here, i.e., $\ell_1 = \ell_2 = 2$. We have $r_1(t) = r_2(t) = 1$ and we may parametrize in such a fashion that

$$\underline{q}^{(1)}(t) = \underline{q}^{(2)}(t) = \begin{pmatrix} 1 \\ t \\ t^2 \end{pmatrix},$$

yielding a bivariate density of the form

$$f(x,y) = \exp\left\{\underline{q}^{(1)\prime}(x)M\underline{q}^{(2)}(y)\right\} = \exp\left\{(1 \quad x \quad x^2)M\begin{pmatrix} 1 \\ y \\ y^2 \end{pmatrix}\right\}.$$
(4.31)

This is a reparametrized version of the density discussed in detail in Chapter 3. Necessary conditions on M to ensure a valid density are found in that chapter. The choice $m_{22} = m_{12} = m_{21} = 0$ yields the classic bivariate normal provided that

$$m_{20} < 0, \quad m_{02} < 0, \quad m_{11}^2 < 4m_{02}m_{20}.$$

Correlations of both signs are possible. The nonclassical normal conditionals models are governed by the following parametric constraints (equivalent to (3.35)):

$$m_{22} < 0, \quad 4m_{22}m_{02} > m_{12}^2, \quad 4m_{22}m_{20} > m_{21}^2.$$

4.6   Gamma Conditionals

The family of gamma distributions with scale and shape parameters forms a two-parameter exponential family of the form

$$f(x; \theta_1, \theta_2) = x^{-1} e^{\theta_1 \log x - \theta_2 x} \theta_2^{\theta_1} \left[\Gamma(\theta_1)\right]^{-1} I(x > 0). \qquad (4.32)$$

If X has its density of the form (4.32) then we write $X \sim \Gamma(\theta_1, \theta_2)$. If we require all conditionals to be in the family (4.32) then Theorem 4.1 may be invoked. In this example, $\ell_1 = \ell_2 = 2, r_1(t) = r_2(t) = t^{-1} I(t > 0)$, and

$$\underline{q}^{(1)}(t) = \underline{q}^{(2)}(t) = \begin{pmatrix} 1 \\ -t \\ \log t \end{pmatrix}.$$

Consequently, the general gamma conditionals class of densities is given by

$$f(x, y) = (xy)^{-1} \exp \left\{ \begin{pmatrix} 1 & -x & \log x \end{pmatrix} M \begin{pmatrix} 1 \\ -y \\ \log y \end{pmatrix} \right\}, \quad x > 0, \ y > 0.$$
$$(4.33)$$

It remains only to determine appropriate values of the parameters M to ensure integrability of this joint density. Such conditions were provided by Castillo, Galambos, and Sarabia (1990) using a different parametrization. For fixed y, the density $f(x, y)$ is of the form $c(y) x^{\alpha(y)-1} e^{-\beta(y)x}$ for suitable $\alpha(y)$ and $\beta(y)$. For this to be integrable $\alpha(y)$ and $\beta(y)$ must both be positive. The conditional distributions corresponding to (4.33) are of the form

$$X|Y = y \sim \Gamma(m_{20} + m_{22} \log y - m_{21} y, m_{10} - m_{11} y + m_{12} \log y) \qquad (4.34)$$

and

$$Y|X = x \sim \Gamma(m_{02} + m_{22} \log x - m_{12} x, m_{01} - m_{11} x + m_{21} \log x). \qquad (4.35)$$

Thus, our parameters must be such that all the expressions on the right-hand sides of (4.34) and (4.35) are positive. In addition, we must verify that the marginal densities thus obtained are themselves integrable. We have

$$f_X(x) = x^{-1} \frac{\Gamma(m_{02} + m_{22} \log x - m_{12} x) e^{m_{00} - m_{10} x + m_{20} \log x}}{(m_{01} - m_{11} x + m_{21} \log x)^{m_{02} + m_{22} \log x - m_{12} x}}, \quad x > 0,$$
$$(4.36)$$

and an analogous expression for $f_Y(y)$. It turns out that under the parametric conditions sufficient to ensure positivity of the gamma parameters in (4.34) and (4.35), the function $f_X(x)$ is bounded in a neighborhood of the origin and, for large x, is bounded by $x^{1/2} e^{-\delta x}$ for some $\delta > 0$. Consequently it is integrable. The requisite conditions for a proper density $f(x, y)$ in (4.33) may be summarized as follows:

MODEL I (in this case, X and Y are independent):

$$m_{11} = 0, \quad m_{12} = 0, \quad m_{21} = 0, \quad m_{22} = 0,$$
$$m_{10} > 0, \quad m_{20} > 0, \quad m_{01} > 0, \quad m_{02} > 0. \tag{4.37}$$

MODEL II:

$$m_{11} < 0, \quad m_{12} = 0, \quad m_{21} = 0, \quad m_{22} = 0,$$
$$m_{10} > 0, \quad m_{20} > 0, \quad m_{01} > 0, \quad m_{02} > 0. \tag{4.38}$$

MODEL IIIA:

$$m_{11} < 0, \quad m_{12} = 0, \quad m_{21} < 0, \quad m_{22} = 0,$$
$$m_{10} > 0, \quad m_{20} > 0, \quad m_{02} > 0, \quad m_{01} > m_{21}\left(1 - \log \frac{m_{21}}{m_{11}}\right). \tag{4.39}$$

MODEL IIIB:

$$m_{11} < 0, \quad m_{12} < 0, \quad m_{21} = 0, \quad m_{22} = 0,$$
$$m_{20} > 0, \quad m_{01} > 0, \quad m_{02} > 0, \quad m_{10} > m_{12}\left(1 - \log \frac{m_{12}}{m_{11}}\right). \tag{4.40}$$

MODEL IV:

$$m_{01} > m_{21}\left(1 - \log \frac{m_{21}}{m_{11}}\right), \quad m_{11} < 0, \quad m_{12} < 0, \quad m_{21} < 0,$$
$$m_{10} > m_{12}\left(1 - \log \frac{m_{12}}{m_{11}}\right), \quad m_{20} > 0, \quad m_{02} > 0, \quad m_{22} = 0. \tag{4.41}$$

and finally MODEL V:

$$\left. \begin{array}{l} m_{11} < 0, \quad m_{10} > m_{12}\left(1 - \log \dfrac{m_{12}}{m_{11}}\right), \\[2mm] m_{12} < 0, \quad m_{20} > m_{22}\left(1 - \log \dfrac{m_{22}}{m_{21}}\right), \\[2mm] m_{21} < 0, \quad m_{01} > m_{21}\left(1 - \log \dfrac{m_{21}}{m_{11}}\right), \\[2mm] m_{22} < 0, \quad m_{02} > m_{22}\left(1 - \log \dfrac{m_{22}}{m_{12}}\right). \end{array} \right\} \tag{4.42}$$

The regression functions for the gamma conditionals distribution are of course generally nonlinear. We have

$$E(X|Y = y) = \frac{m_{20} + m_{22}\log y - m_{21}y}{m_{10} + m_{12}\log y - m_{11}y} \tag{4.43}$$

and

$$E(Y|X = x) = \frac{m_{02} + m_{22}\log x - m_{12}x}{m_{01} + m_{21}\log x - m_{11}x}, \tag{4.44}$$

obtained using (4.34) and (4.35). Expressions for the conditional variances can also be written by referring to (4.34) and (4.35). As a curiosity, we may note that certain fortuitous parametric choices can lead to $E(X|Y = y) = c_1$ and $E(Y|X = x) = c_2$, i.e., $m_{20}/m_{10} = m_{22}/m_{12} = m_{21}/m_{11}$, etc. These will generally not correspond to independent marginals since the corresponding conditional variances will not be constant.

The modes of this distribution are at the intersection of the conditional mode curves, that is, they are the solution of the system

$$x = \frac{m_{20} + m_{22} \log y - m_{21} y - 1}{m_{10} + m_{12} \log y - m_{11} y},$$

$$y = \frac{m_{02} + m_{22} \log x - m_{12} x - 1}{m_{01} + m_{21} \log x - m_{11} x},$$

where we assume that $m_{20} > 1$ and $m_{02} > 1$.

4.6.1 Model II

In this section we give a more detailed analysis of Model II, which can be reparametrized as

$$f(x,y) = \frac{k_{r,s}(c)}{\sigma_1^r \sigma_2^s \Gamma(r) \Gamma(s)} x^{r-1} y^{s-1} \exp\left(-\frac{x}{\sigma_1} - \frac{y}{\sigma_2} - c\frac{xy}{\sigma_1 \sigma_2}\right) I(x > 0, y > 0) \tag{4.45}$$

with $r, s > 0$, $\sigma_1, \sigma_2 > 0$, and $c \geq 0$. Note that r and s are shape parameters, σ_1 and σ_2 scale parameters, and c is a dependence parameter, such that $c = 0$ corresponds to the case of independence.

If a random variable (X, Y) has probability density function (4.45), then we write $(X, Y) \sim \text{GCD}(r, s; \sigma_1, \sigma_2, c)$. It is obvious that (X, Y) has conditionals

$$X|Y = y \sim \Gamma(r, (1 + cy/\sigma_2)/\sigma_1) \tag{4.46}$$

and

$$Y|X = x \sim \Gamma(s, (1 + cx/\sigma_1)/\sigma_2) \tag{4.47}$$

and marginals

$$f_X(x) = \frac{k_{r,s}(c)}{\sigma_1^r \Gamma(r)} (1 + cx/\sigma_1)^{-s} x^{r-1} e^{-x/\sigma_1}, \quad x > 0, \tag{4.48}$$

and

$$f_X(x) = \frac{k_{r,s}(c)}{\sigma_2^s \Gamma(s)} (1 + cy/\sigma_2)^{-r} x^{s-1} e^{-y/\sigma_2}, \quad x > 0. \tag{4.49}$$

It is worthwhile mentioning that only in the case $c = 0$ are these marginals of the gamma form.

From (4.46) and (4.47) we get the relations

$$\frac{X}{\sigma_1}\left(1 + c\frac{Y}{\sigma_2}\right) \sim \Gamma(r, 1) \tag{4.50}$$

and

$$\frac{Y}{\sigma_2}\left(1 + c\frac{X}{\sigma_1}\right) \sim \Gamma(s, 1), \tag{4.51}$$

where the random variable in (4.50) is independent of Y and the random variable in (4.51) is independent of X. Many of the moments of (4.45) can be written in terms of

$$\delta_{r,s}(c) = \frac{\partial}{\partial c}\log k_{r,s}(c). \tag{4.52}$$

One first observation is that the condition $E(\partial \log f(x, y)/\partial c) = 0$ becomes

$$E(XY) = \sigma_1\sigma_2\delta_{r,s}(c). \tag{4.53}$$

The moment generating function of (X, Y) is

$$G_{X,Y}(u, v) = E(e^{uX+vY}) = \frac{(1 - \sigma_1 u)^{-r}(1 - \sigma_2 v)^{-s}k_{r,s}(c)}{k_{r,s}[c(1 - \sigma_1 u)^{-1}(1 - \sigma_2 v)^{-1}]}, \tag{4.54}$$

where $u < 1/\sigma_1$ and $v < 1/\sigma_2$ to ensure the convergence of the integrand. If we make use of the moment generating function to calculate $E(XY)$ and set the result equal to (4.53) we obtain the differential equation

$$rs - c(r + s + 1)\delta_{r,s}(c) + c^2\delta_{r,s}^2(c) - c^2\delta_{r,s}'(c) = \delta_{r,s}(c), \tag{4.55}$$

where $\delta_{r,s}'(c) = \partial\delta_{r,s}(c)/\partial c$. Using this differential equation we can conclude that all high-order moments of (X, Y) can be expressed in terms of (4.52). Now, making use of (4.50), (4.51), (4.52), and (4.55), we obtain the moments

$$
\begin{aligned}
E(X) &= \sigma_1[r - c\delta_{r,s}(c)], & (4.56)\\
E(Y) &= \sigma_2[s - c\delta_{r,s}(c)], & (4.57)\\
\text{var}(X) &= \sigma_1^2[r(1-s)+(c(r + s - 1)+1)\delta_{r,s}(c)-c^2\delta_{r,s}^2(c)], & (4.58)\\
\text{var}(Y) &= \sigma_1^2[s(1-r)+(c(r + s - 1)+1)\delta_{r,s}(c)-c^2\delta_{r,s}^2(c)], & (4.59)\\
\text{cov}(X, Y) &= \sigma_1\sigma_2[(r + s)c\delta_{r,s}(c)-rs+\delta_{r,s}(c)-c^2\delta_{r,s}^2(c)]. & (4.60)
\end{aligned}
$$

The Normalizing Constant

The normalizing constant and the function $\delta_{r,s}(c)$ play an important role in the calculus of the GCD moments. We are interested in some expressions that are relatively easy to use. For example, from (4.48), the function $k_{r,s}(c)$ can be written as

$$k_{r,s}(c) = \frac{c^r}{U(r, r-s+1, 1/c)},\tag{4.61}$$

where $U(a, b, z)$ is the "Confluent Hypergeometric" function, which is defined in (3.56).

Alternatively, using again (4.48), $k_{r,s}$ and $\delta_{r,s}$ can be given as

$$k_{r,s}^{-1}(c) = E[(1+cZ_s)^{-r}] = E[(1+cZ_r)^{-s}],\tag{4.62}$$

$$\delta_{r,s}(c) = r\frac{E[Z_s(1+cZ_s)^{-r-1}]}{E[(1+cZ_s)^{-r}]} = s\frac{E[Z_r(1+cZ_r)^{-s-1}]}{E[(1+cZ_r)^{-s}]},\tag{4.63}$$

where $Z_\alpha \sim \Gamma(\alpha, 1)$. These expressions allow us to obtain $k_{r,s}$ and $\delta_{r,s}$ by simulating samples from a $\Gamma(\alpha, 1)$ distribution. Expression (4.62) shows that $k_{r,s}(c)$ is an increasing function of c.

Higher-Order Moments and Mode

Higher-order moments of X and Y can also be expressed in terms of the function $k_{r,s}(c)$, that is, in terms of (4.61). Writing $X_{\sigma_1} = X/\sigma_1$, $Y_{\sigma_2} = Y/\sigma_2$ and taking into account the fact that $E(X_{\sigma_1}^{n_1} Y_{\sigma_2}^{n_2}) = E[Y_{\sigma_2}^{n_2} E(X_{\sigma_1}^{n_1}|Y)]$ we obtain

$$E(X_{\sigma_1}^{n_1} Y_{\sigma_2}^{n_2}) = \frac{\Gamma(r+n_1)\Gamma(s+n_2)}{\Gamma(r)\Gamma(s)} \times \frac{k_{r,s}(c)}{k_{r+n_1,s+n_2}(c)}.\tag{4.64}$$

Using the relations (4.61) we can then obtain recurrence formulas for the moments. Using relations 13.4.21 and 13.4.27 in Abramowitz and Stegun (1964) it can be shown that

$$(k+r-1)E(X^{k-1}) = -\left(k+r-s-\frac{1}{c}\right)\frac{c}{\sigma_1}E(X^k) + \frac{c}{\sigma_1^2}E(X^{k+1})\tag{4.65}$$

with $k = 1, 2, \ldots$. The associated recurrence relations for the random variable Y are obtained by symmetry. In addition, we have

$$E(\log X) = \psi(r) + \log\sigma_1 - \frac{\partial}{\partial r}\log k_{r,s}(c),\tag{4.66}$$

where ψ is the digamma function.

If $r, s > 1$, then the mode of the density (4.45) is

$$x_0 = \frac{\sigma_1}{2c}\left[-(1+cs-cr) + \sqrt{(1+cs-cr)^2 + 4c(s-1)}\right],\tag{4.67}$$

$$y_0 = \frac{\sigma_2}{2c}\left[-(1+cr-cs) + \sqrt{(1+cr-cs)^2 + 4c(r-1)}\right].\tag{4.68}$$

If $c = 0$ the mode becomes $(x_0, y_0) = (\sigma_1(r-1), \sigma_2(s-1))$ for $r, s > 1$.

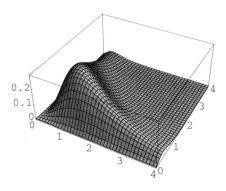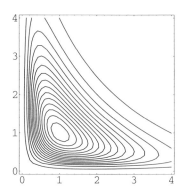

FIGURE 4.3. Example of a gamma conditionals distribution in (4.45) with $c = 1, \sigma_1 = \sigma_2 = 1, r = s = 3$ showing (left side) the probability density function, and (right side) the contour plot.

Example 4.2 (Gamma conditionals Model II). Figure 4.3 shows an example of a gamma conditionals distribution of the type in (4.45) corresponding to $c = 1, \sigma_1 = \sigma_2 = 1, r = s = 3$. The left figure shows the probability density function, and the right figure the associated contours. $\square$

4.7   Weibull Conditionals

In general, the two-parameter Weibull distribution does not form an exponential family and consequently more detailed discussion of this model will be deferred to Chapter 5. However, if the power parameter is held fixed, the Weibull becomes a one-parameter exponential family. In fact, we may view a Weibull random variable W as a positive power of an exponential random variable, i.e.,

$$W = X^c,$$

where X is exponential. We may introduce the notation Weibull(c) to denote the class of Weibull distributions with fixed power c and arbitrary scale. It follows readily, since the functions x^{c_1} and x^{c_2} are invertible on $(0, \infty)$, that the class of all bivariate distributions with $W_1|W_2 = w_2 \sim$ Weibull(c_1) for every w_2, and $W_2|W_1 = w_1 \sim$ Weibull(c_2) for every w_1 is merely given by taking (X, Y) to have a general exponential conditionals distribution given by (4.14) and defining

$$(W_1, W_2) = (X^{c_1}, Y^{c_2}).$$

Of course, analogous distributions could be generated by other invertible functions $\phi_1(X)$ and $\phi_2(Y)$ instead of X^{c_1} and Y^{c_2}.

4.8 Gamma–Normal Conditionals

In the study of the stochastic behavior of ocean waves, Longuet-Higgins (1975) were led to a model in which squared wave amplitude and wave period had a joint distribution with gamma and normal marginals. However, since the theoretical derivation was performed assuming a narrow energy spectrum, Castillo and Galambos (1987a) suggest that it is perhaps more reasonable to assume a model with gamma and normal conditional distributions. Such a model might be applicable even in the wide energy spectrum case. Models of this type may be identified as CEF distributions as follows.

The class of bivariate distributions with $X|Y = y$ having a gamma distribution for all y and $Y|X = x$ having a normal distribution for all x will be given by (4.5) with the following choices for the r's and q's:

$$r_1(x) = x^{-1}I(x > 0), \quad r_2(y) = 1,$$

$$q_{11}(x) = -x, \qquad\qquad q_{21}(y) = y, \qquad\qquad (4.69)$$

$$q_{12}(x) = \log x, \qquad\qquad q_{22}(y) = y^2.$$

Then, the joint density can be written as

$$f(x, y) = (x)^{-1} \exp \left\{ (1 \quad -x \quad \log x) M \begin{pmatrix} 1 \\ y \\ y^2 \end{pmatrix} \right\}, \quad x > 0, \ y \in \mathbb{R}. \quad (4.70)$$

Provided that the parameters are suitably constrained to yield a proper joint density, the specific forms of the conditional distributions will be as follows:

$$X|Y = y \sim \Gamma(m_{20} + m_{21}y + m_{22}y^2, m_{10} + m_{11}y + m_{12}y^2) \qquad (4.71)$$

and

$$Y|X = x \sim N(\mu(x), \sigma^2(x)), \qquad\qquad (4.72)$$

where

$$\mu(x) = \frac{m_{01} - m_{11}x + m_{21}\log x}{2(-m_{02} + m_{12}x - m_{22}\log x)}, \qquad\qquad (4.73)$$

$$\sigma^2(x) = \frac{1}{2}(-m_{02} + m_{12}x - m_{22}\log x)^{-1}. \qquad\qquad (4.74)$$

Thus three necessary conditions for a proper joint distribution are

$$m_{20} + m_{21}y + m_{22}y^2 > 0, \quad \forall y \in \mathbb{R}, \qquad\qquad (4.75)$$

$$m_{10} + m_{11}y + m_{12}y^2 > 0, \quad \forall y \in \mathbb{R}, \qquad\qquad (4.76)$$

and

$$-m_{02} + m_{12}x - m_{22} \log x > 0, \quad \forall x > 0. \tag{4.77}$$

Provided our parameters are constrained to ensure that (4.75)–(4.77) hold, the only further constraint is that one (and hence both) of the resulting marginal densities be integrable. In this manner we are led to the following set of parametric constraints for valid gamma-normal models.

MODEL I (in this case, X and Y are independent):

$$m_{12} = 0, \quad m_{22} = 0, \quad m_{02} < 0, \quad m_{11} = 0,$$
$$m_{21} = 0, \quad m_{10} > 0, \quad m_{20} > 0, \quad m_{01} \in \mathbb{R}. \tag{4.78}$$

MODEL II:

$$m_{22} = 0, \quad m_{12} > 0, \quad m_{02} < 0, \quad m_{11}^2 < 4m_{10}m_{12},$$
$$m_{20} > 0, \quad m_{21} = 0, \quad m_{01} \in \mathbb{R}. \tag{4.79}$$

MODEL III:

$$m_{12} > 0, \quad m_{02} < m_{22} \left(1 - \log \frac{m_{22}}{m_{12}} \right),$$
$$m_{22} > 0, \quad m_{11}^2 < 4m_{10}m_{12}, \tag{4.80}$$
$$m_{01} \in \mathbb{R}, \quad m_{21}^2 < 4m_{20}m_{22}.$$

The regression function of X on Y is biquadratic, while the regression of Y on X is a bilinear function of y and $\log y$.

Example 4.3 (Gamma-Normal conditionals). Figure 4.4 shows an example of a normal-gamma conditionals distribution of type II corresponding to $m_{02} = -1, m_{01} = 1, m_{10} = 1.5, m_{12} = 2, m_{11} = 1, m_{20} = 3, m_{22} = m_{21} = 0$. The left figure corresponds to the probability density function, and the right figure to the contours. □

The conditional mode functions can intersect more than once (similar to the case for normal conditionals distributions). In Figure 4.5, a bimodal gamma-normal conditionals density is shown. It corresponds to the parameter values $m_{01} = 0.7, m_{02} = 1.4, m_{10} = 1.8, m_{11} = 1.3, m_{12} = 2.3, m_{20} = 3.7, m_{21} = 4.8, m_{22} = 3.3$.

4.9 Power-Function and Other Weighted Distributions as Conditionals

The family of power-function densities

$$f(x; \theta) = \theta x^{\theta - 1}, \quad 0 < x < 1, \tag{4.81}$$

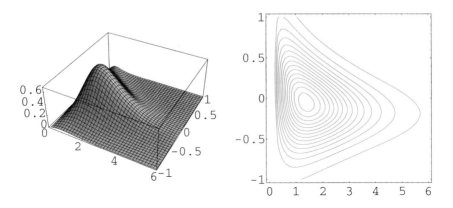

FIGURE 4.4. Example of a normal-gamma conditionals distribution with $m_{02} = -1, m_{01} = 1, m_{10} = 1.5, m_{12} = 2, m_{11} = 1, m_{20} = 3, m_{22} = m_{21} = 0$ showing (left side) the probability density function, and (right side) the contour plot.

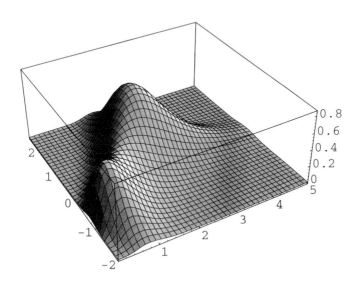

FIGURE 4.5. An example of a gamma-normal conditionals distributions with two modes.

where $\theta \in (0, \infty)$ is clearly an exponential family and modeling situations might suggest use of joint densities with conditionals in this family. Rather than describe this specific class of joint distributions it is convenient to recognize that (4.81) is a special case of a general class of weighted distributions on $\mathbb{R}^+$ (see, for example, Patil and Rao (1977) for a good introduction to weighted distributions and their potential for applications). For any nonnegative function $A(x)$ on $\mathbb{R}^+$ we may define the corresponding class of g weighted distributions (g again is a nonnegative function) by

$$f(x; \theta) = [g(x)]^\theta A(x) B(\theta), \quad x > 0. \tag{4.82}$$

If we let

$$q(x) = \log g(x),$$

then we see immediately that (4.82) is a one-parameter exponential family. Denote the natural parameter space of this family by the interval

$$(\theta_1(g, A), \theta_2(g, A)). \tag{4.83}$$

Of course different choices of g and/or A will lead to different distributions with possibly different natural parameter spaces. We will say that X has a (g, A) weighted distribution if (4.82) holds. We may then ask what is the nature of all bivariate distributions for which $X|Y = y$ is a (g_1, A_1) weighted distribution for all y and $Y|X = x$ is a (g_2, A_2) weighted distribution for every x? Using (4.5) it is apparent that the form of the joint density must be as follows:

$$f(x, y) = A_1(x) A_2(y) e^{m_{00}} [g_1(x)]^{m_{10}} [g_2(y)]^{m_{01}} e^{m_{11} (\log g_1(x))(\log g_2(y))},$$
$$x, y > 0 \tag{4.84}$$

where $m_{10}, m_{01},$ and m_{11} must satisfy

$$\theta_1(g_1, A_1) < m_{10} + m_{11} \log g_2(y) < \theta_2(g_1, A_1), \quad \forall y \quad \text{with } A_2(y) > 0,$$

and

$$\theta_1(g_2, A_2) < m_{01} + m_{11} \log g_1(x) < \theta_2(g_2, A_2), \quad \forall x \quad \text{with } A_1(x) > 0.$$

In some cases we will also need to impose the additional conditions that

$$\theta_1(g_1, A_1) < m_{10} < \theta_2(g_1, A_1)$$

and

$$\theta_1(g_2, A_2) < m_{01} < \theta_2(g_2, A_2)$$

to ensure that the joint density is integrable. This would be necessary, for example, in the case in which $A_1(x) = 1, A_2(y) = 1, g_1(x) = e^{-x}$, and $g_2(y) = e^{-y}$. This of course is yet another representation of the exponential conditionals distribution.

In some cases, severe parametric restrictions are necessary in order to have a valid distribution. An example is provided by the Planck distribution. It is a weighted distribution with $g(x) = x$ and $A(x) = (e^x - 1)^{-1}$, $x > 0$. Thus

$$f(x; \theta) = x^\theta / [(e^x - 1)\Gamma(\theta + 1)\xi(\theta + 1)], \quad x > 0,$$

where $\theta > -1$ (here ξ is the Riemann zeta function). If a bivariate density is to have all conditionals in the Planck family we must have a representation like (4.84) and for legitimate Planck conditionals of X given y we will require

$$m_{11} \log y + m_{01} > -1, \quad \forall y > 0.$$

Evidently this is only possible if $m_{11} = 0$. If follows that the only Planck conditionals model that can be constructed is the trivial one with independent Planck marginals.

4.10   Beta Conditionals

The beta conditionals model is associated with the following choices for the r's and q's in (4.5):

$$
\begin{aligned}
r_1(x) &= [x(1 - x)]^{-1} I(0 < x < 1), \\
r_2(y) &= [y(1 - y)]^{-1} I(0 < y < 1), \\
q_{11}(x) &= \log x, \\
q_{21}(y) &= \log y, \\
q_{12}(x) &= \log(1 - x), \\
q_{22}(y) &= \log(1 - y).
\end{aligned}
\tag{4.85}
$$

This yields a joint density of the form

$$
\begin{aligned}
f(x, y) = [x(1 - x)y(1 - y)]^{-1} \exp\{ &m_{11} \log x \log y + m_{12} \log x \log(1 - y) \\
&+ m_{21} \log(1 - x) \log y + m_{22} \log(1 - x) \log(1 - y) \\
&+ m_{10} \log x + m_{20} \log(1 - x) \\
&+ m_{01} \log y + m_{02} \log(1 - y) \\
&+ m_{00}\} I(0 < x, y < 1).
\end{aligned}
\tag{4.86}
$$

In order that the associated beta conditionals will have parameters in the natural parameter space of the beta exponential family we require that

$$
\begin{aligned}
m_{10} + m_{11} \log y + m_{12} \log(1 - y) &> 0, \quad \forall y \in (0, 1), \\
m_{20} + m_{21} \log y + m_{22} \log(1 - y) &> 0, \quad \forall y \in (0, 1), \\
m_{01} + m_{11} \log x + m_{21} \log(1 - x) &> 0, \quad \forall x \in (0, 1), \\
m_{02} + m_{12} \log x + m_{22} \log(1 - x) &> 0, \quad \forall x \in (0, 1).
\end{aligned}
\tag{4.87}
$$

Evidently (4.87) cannot be true if any m_{ij} with i and $j \geq 1$ is positive. So $m_{ij} \leq 0, i = 1, 2, j = 1, 2$. In order to guarantee integrability of the

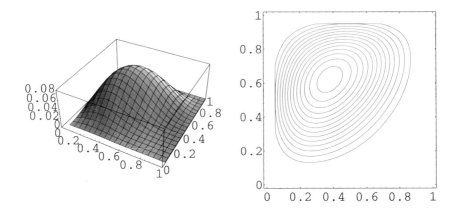

FIGURE 4.6. Example of a beta conditionals distribution with $m_{01} = 1, m_{02} = 1, m_{10} = 1, m_{11} = -1, m_{12} = -1, m_{20} = 1, m_{21} = -3, m_{22} = -1$ showing (left side) the probability density function, and (right side) the contour plot.

marginal distributions we require $m_{10} > 0, m_{20} > 0, m_{01} > 0, m_{02} > 0$. Referring to (4.87) to read off the appropriate parameters for the conditional distributions we may readily write conditional means and variances. For example

$$E(X|Y = y) = \frac{m_{10} + m_{11} \log y + m_{12} \log(1 - y)}{(m_{10} + m_{20}) + (m_{11} + m_{21}) \log y + (m_{12} + m_{22}) \log(1 - y)}.$$
$$(4.88)$$

As usual $\tilde{M} \equiv 0$ corresponds to independent marginals. Both negative and positive correlations are possible in this family.

Example 4.4 (Beta conditionals). Figure 4.6 shows an example of a beta conditionals distribution corresponding to (4.86) with $m_{01} = 1, m_{02} = 1, m_{10} = 1, m_{11} = -1, m_{12} = -1, m_{20} = 1, m_{21} = -3, m_{22} = -1$. The left figure corresponds to the probability density function, and the right figure to the contours. □

4.11   Inverse Gaussian Conditionals

We say that X has an inverse Gaussian distribution if its density is of the form

$$f(x) = \sqrt{\frac{\eta_2}{\pi}} e^{2\sqrt{\eta_1 \eta_2}} e^{-\eta_1 x - \eta_2 x^{-1}} I(x > 0). \qquad (4.89)$$

In such a case we write $X \sim \text{IG}(\eta_1, \eta_2)$. The parameters η_1, η_2 are constrained to be positive in order to have a proper density.

The inverse Gaussian conditionals model corresponds to the following choices for the r's and q's in (4.5):

$$
\begin{aligned}
r_1(x) &= x^{-3/2}I(x > 0), \\
r_2(y) &= y^{-3/2}I(y > 0), \\
q_{11}(x) &= -x, \\
q_{21}(y) &= -y, \\
q_{12}(x) &= -x^{-1}, \\
q_{22}(y) &= -y^{-1}.
\end{aligned}
\tag{4.90}
$$

The joint density is of the form

$$
\begin{aligned}
f(x,y) = (xy)^{-3/2}\exp\{ & m_{11}xy + m_{12}xy^{-1} + m_{21}x^{-1}y \\
& + m_{22}x^{-1}y^{-1} - m_{10}x - m_{20}x^{-1} \\
& - m_{01}y - m_{02}y^{-1} + m_{00}\}I(x > 0, y > 0).
\end{aligned}
\tag{4.91}
$$

In order to have proper inverse Gaussian conditionals distributions we require that

$$
\begin{aligned}
m_{10} - m_{11}y - m_{12}y^{-1} &> 0, \quad \forall y > 0, \\
m_{20} - m_{21}y - m_{22}y^{-1} &> 0, \quad \forall y > 0, \\
m_{01} - m_{11}x - m_{21}x^{-1} &> 0, \quad \forall x > 0, \\
m_{02} - m_{12}x - m_{22}x^{-1} &> 0, \quad \forall x > 0.
\end{aligned}
\tag{4.92}
$$

Clearly then, $m_{ij} \leq 0$, $i = 1, 2$, $j = 1, 2$. In addition we require that

$$
\begin{aligned}
m_{10} > -2\sqrt{m_{11}m_{12}}, \quad m_{20} > -2\sqrt{m_{21}m_{22}}, \\
m_{01} > -2\sqrt{m_{11}m_{21}}, \quad m_{02} > -2\sqrt{m_{12}m_{22}},
\end{aligned}
\tag{4.93}
$$

in order to guarantee that (4.92) holds and that the resulting marginal densities are integrable. Reference may be made to (4.92) to determine the relevant parameter values in the conditional distributions. For example,

$$
X|Y = y \sim \mathrm{IG}(m_{10} - m_{11}y - m_{12}y^{-1}, m_{20} - m_{21}y - m_{22}y^{-1})
$$

and, consequently,

$$
E(X|Y = y) = \sqrt{\frac{m_{20} - m_{21}y - m_{22}y^{-1}}{m_{10} - m_{11}y - m_{12}y^{-1}}}.
\tag{4.94}
$$

4.12   Three Discrete Examples (Binomial, Geometric, and Poisson)

Definition 4.2 (Binomial distribution).   A random variable has a binomial(n, p) distribution if $P(X = x) = \dbinom{n}{x}p^x(1-p)^{n-x}, x = 0, 1, \ldots, n,$ $0 < p < 1$.   □

A two-dimensional random vector (X, Y) will have

$$X|Y = y \sim \text{binomial } (n_1, p_1(y))$$

for each y and $Y|X = x \sim \text{binomial}(n_2, p_2(x))$ for each x if its joint density is of the form

$$f_{X,Y}(x, y) = k_B(p_1, p_2, t) \binom{n_1}{x} p_1^x (1 - p_1)^{n_1 - x} \binom{n_2}{y} p_2^y (1 - p_2)^{n_2 - y} t^{xy},$$

$$x = 0, 1, \ldots, n_1, \quad y = 0, 1, \ldots, n_2.$$
(4.95)

Here $p_1 \in (0, 1), p_2 \in (0, 1)$, and $t > 0$. This is merely a reparametrized version of (4.5) with the appropriate choices of r's and q's. The case $t = 1$, corresponds to independence. The correlation is positive if $t > 1$ and negative if $t < 1$. Conditional means and variances are readily written down since $X|Y = y \sim \text{binomial}(n_1, [p_1 t^y / (1 - p_1 + p_1 t^y)])$ and $Y|X = x \sim \text{binomial}(n_2, [p_2 t^x / (1 - p_2 + p_2 t^x)])$. Thus

$$E(X|Y = y) = n_1 p_1 t^y / (1 - p_1 + p_1 t^y),$$
(4.96)

etc.

Definition 4.3 (Geometric distribution). A random variable has a geometric (q) distribution if $P(X = x) = (1 - q)q^x, x = 0, 1, 2, \ldots$, where $0 < q < 1$.   □

Since this is a one-parameter exponential family of distributions, a two-dimensional random vector with geometric conditionals can be readily written using (4.5). After convenient reparametrization we obtain

$$f_{X,Y}(x, y) = k_G(q_1, q_2, q_3) q_1^x q_2^y q_3^{xy}, \quad x = 0, 1, 2, \ldots, \quad ; y = 0, 1, 2, \ldots.$$
(4.97)

For convergence and legitimate geometric conditionals we require that $q_1 \in (0, 1), q_2 \in (0, 1)$ while $0 < q_3 \leq 1$. The case $q_3 = 1$ corresponds to independent marginals. Otherwise, the correlation between X and Y is negative. Since $X|Y = y \sim \text{geometric}(q_1 q_3^y)$ and $Y|X = x \sim \text{geometric}(q_2 q_3^x)$, we may determine conditional moments such as

$$E(X|Y = y) = q_1 q_3^y / (1 - q_1 q_3^y).$$
(4.98)

Our final discrete example is the Poisson conditionals distribution.

Definition 4.4 (Poisson distribution). A random variable X has a Poisson(λ) distribution if $P(X = x) = e^{-\lambda} \lambda^x / x!, x = 0, 1, 2, \ldots, \lambda > 0$.   □

Since this is a one-parameter exponential family we may use (4.5) to describe the class of Poisson conditionals distributions. When reparametrized, we obtain

$$f_{X,Y}(x, y) = k_P(\lambda_1, \lambda_2, \lambda_3)\lambda_1^x \lambda_2^y \lambda_3^{xy}/(x!\, y!),$$
$$x = 0, 1, \ldots, \quad y = 0, 1, \ldots \, . \tag{4.99}$$

In order to have summability in (4.99) and legitimate Poisson conditional distributions the parameters are constrained to satisfy $\lambda_1 > 0, \lambda_2 > 0$, and $0 < \lambda_3 \le 1$. The conditional distribution of X given $Y = y$ is Poisson($\lambda_1 \lambda_3^y$) while $Y|X = x \sim$ Poisson($\lambda_2 \lambda_3^x$). Consequently,

$$E(X|Y = y) = \lambda_1 \lambda_3^y, \tag{4.100}$$

etc. If $\lambda_3 = 1, X$ and Y are independent. If $0 < \lambda_3 < 1, X$ and Y are negatively correlated.

We also have

$$E(X) = \lambda_1 \frac{k_P(\lambda_1, \lambda_2, \lambda_3)}{k_P(\lambda_1, \lambda_2 \lambda_3, \lambda_3)},$$

$$E(Y) = \lambda_2 \frac{k_P(\lambda_1, \lambda_2, \lambda_3)}{k_P(\lambda_1 \lambda_3, \lambda_2, \lambda_3)}, \tag{4.101}$$

$$E(XY) = \lambda_1 \lambda_2 \lambda_3 \frac{k_P(\lambda_1, \lambda_2, \lambda_3)}{k_P(\lambda_1 \lambda_3, \lambda_2 \lambda_3, \lambda_3)}.$$

Just as we cannot have exponential conditionals with positive correlation, we cannot have Poisson (or geometric) conditionals with positive correlation.

The Poisson conditionals distribution, (4.99), is also known as Obrechkoff's distribution (Obrechkoff (1963)).

4.13   Poisson–Gamma Conditionals

Consider a random variable (X, Y) such that

$$X|Y = y \sim \text{ Poisson}(y), \tag{4.102}$$

where we assume that the random variable Y is a gamma variable with density function

$$Y \sim \Gamma(\alpha, \lambda). \tag{4.103}$$

The resulting distribution of X is known as a compound Poisson distribution. It is well known that the unconditional distribution of X is negative binomial with probability mass density

$$f_X(x) = \frac{\Gamma(x + \alpha)}{x!\, \Gamma(\alpha)} \left(\frac{\lambda}{\lambda + 1} \right)^\alpha \left(\frac{1}{\lambda + 1} \right)^x, \quad x = 0, 1, \ldots \, . \tag{4.104}$$

From (4.102) and (4.103) it easy to prove that the conditional density of Y given $X = x$ is gamma, that is,

$$Y|X = x \sim \Gamma(x + \alpha, \lambda + 1). \tag{4.105}$$

As a generalization of this scenario we can ask for the most general distribution of a random variable (X, Y) such that its conditional distributions are of the Poisson and gamma type. This will be a CEF distribution as in (4.5) with $\ell_1 = 1$ and $\ell_2 = 2$ and,

$$\underline{q}^{(1)}(x) = \begin{pmatrix} 1 \\ x \end{pmatrix},$$

$$\underline{q}^{(2)}(y) = \begin{pmatrix} 1 \\ -y \\ \log y \end{pmatrix},$$

and with $r_1(x) = 1/x!$ and $r_2(y) = y^{-1}I(y > 0)$. In this way, we obtain the joint density

$$f(x, y) = \frac{1}{x! \, y} \exp \left\{ \begin{pmatrix} 1 & x \end{pmatrix} M \begin{pmatrix} 1 \\ -y \\ \log y \end{pmatrix} \right\}, \quad x = 0, 1, 2, \ldots, \ y > 0. \tag{4.106}$$

which is equivalent to

$$f(x, y) = \frac{1}{x! \, y} \exp(m_{00} + m_{10}x - m_{01}y - m_{11}xy + m_{02} \log y + m_{12}x \log y),$$
$$x = 0, 1, 2, \ldots, y > 0. \tag{4.107}$$

where

$$m_{01} > 0, \quad m_{02} > 0, \quad m_{11} \geq 0, \quad m_{12} \geq 0. \tag{4.108}$$

Some extra conditions are required to ensure the integrability of (4.107). The case $m_{11} = 0$ and $m_{12} = 1$ corresponds to the compound Poisson distribution, and the case $m_{11} = m_{12} = 0$ to the case in which X and Y are independent. The conditional densities of the new model are

$$X|Y = y \sim \text{Poisson}(e^{m_{10} - m_{11}y + m_{12} \log y}), \tag{4.109}$$
$$Y|X = x \sim \Gamma(m_{02} + m_{12}x, m_{01} + m_{11}x). \tag{4.110}$$

The marginal density of X in (4.107) is a generalization of the negative binomial distribution. Its density function is

$$f_X(x) = \frac{\Gamma(m_{02} + m_{12}x)}{(m_{01} + m_{11}x)^{m_{02} + m_{12}x}} \times \frac{e^{m_{00} + m_{10}x}}{x!}, \quad x = 0, 1, \ldots . \tag{4.111}$$

4.14   Bibliographic Notes

Section 4.2 is based on Arnold and Strauss (1991). Section 4.3 is drawn from Arnold (1988a). Convenient references for the examples in later sections are as follows:

(i) Exponential: Arnold and Strauss (1988a);

(ii) Normal: Castillo and Galambos (1987a);

(iii) Gamma: Castillo, Galambos, and Sarabia (1990);

(iv) Weibull: Castillo and Galambos (1990);

(v) Gamma-Normal: Castillo and Galambos (1987a);

(vi) Power function and other weighted distributions: this material is new;

(vii) Beta: Arnold and Strauss (1991) and Castillo and Sarabia (1990a);

(viii) Inverse Gaussian: this material is new;

(ix) Binomial, geometric, Poisson: Arnold and Strauss (1991);

(x) Poisson-gamma: this material is new.

A good source of information on the various univariate exponential families discussed in this chapter is Johnson, Kotz, and Kemp (1992) and Johnson, Kotz, and Balakrishnan (1994, 1995).

Exercises

4.1 Consider the distribution with exponential conditionals given by (4.15).

    (a) If $k(c)$ is the normalizing constant and if we define

$$\psi(c) = \frac{1}{ck(c^{-1})},$$

    prove that

$$\psi'(c) = \psi(c) - \frac{1}{c}.$$

    (b) Using the joint moment generating function (4.24) and (a), obtain the moment expressions (4.26)–(4.29).

    (Arnold and Strauss (1988a).)

4.2 Consider the model with Poisson conditionals given by (4.99).

(a) Obtain the marginal distributions of X and Y.

(b) If $0 < \phi \le 1$, prove that

$$\sum_{n=0}^{\infty} \frac{5^n}{n!} \exp(7\phi^n) = \sum_{n=0}^{\infty} \frac{7^n}{n!} \exp(5\phi^n).$$

4.3 Obtain the modal value of the distribution with Poisson conditionals given by (4.99).

4.4 Identify the class of bivariate distributions for which all conditional distributions of X given Y and of Y given X are in the family of gamma distributions with scale and shape parameters equal to each other.

4.5 A random variable X is said to be a Laplace random variable and is denoted by $X \sim \text{Lap}(\mu, \sigma)$, if its pdf is given by

$$f(x; \mu, \sigma) = \frac{1}{2\sigma} \exp\left(-\left|\frac{x-\mu}{\sigma}\right|\right), \quad -\infty < x < \infty,$$

with $\mu \in \mathbb{R}$ and $\sigma \in \mathbb{R}^+$.

(a) Find the most general random variable (X, Y) such that $X|Y = y \sim \text{Lap}(\mu_1, \sigma_1(y))$ and $Y|X = x \sim \text{Lap}(\mu_2, \sigma_2(x))$. Find their marginal probability density functions.

(b) For the centered case ($\mu_1 = \mu_2 = 0$), obtain an expression for the normalizing constant.

(c) Solve the problem in the general case, i.e., find the most general random variable such that $X|Y = y \sim \text{Lap}(\mu_1(y), \sigma_1(y))$ and $Y|X = x \sim \text{Lap}(\mu_2(x), \sigma_2(x))$.

4.6 Discuss the sign of the correlation coefficient for the gamma-normal conditionals model with joint pdf given by (4.70).

4.7 Consider the bivariate weighted distributions conditionals with joint pdf given by (4.84).

(a) Consider some particular cases.

(b) Discuss the sign of the correlation coefficient.

(c) Find the regression lines $x = E(X|Y = y)$ and $y = E(Y|X = x)$.

(d) Characterize some bivariate distributions with weighted conditionals and a particular regression function.

(Sarabia and Castillo (1991).)

4.8 Characterize the bivariate distributions (X, Y) with beta conditionals satisfying:

(a) The conditional expectations $E(X|Y = y)$ and $E(Y|X = x)$ are constant.

(b) The two conditional mode functions are constant.

(c) The conditional variances are constant (homoscedastic).

4.9 A random variable X is said to be a negative binomial random variable and is denoted by $X \sim \mathrm{NB}(\alpha, p)$, iff

$$P(X = x) = \frac{\Gamma(x + \alpha)}{\Gamma(x + 1)\Gamma(\alpha)} p^\alpha q^x, \quad x = 0, 1, 2, \ldots,$$

with $\alpha > 0$, $0 < p < 1$, and $q = 1 - p$.

(a) Obtain the most general random variable (X, Y) such that $X|Y = y \sim \mathrm{NB}(\alpha_1, p_1(y))$ and $Y|X = x \sim \mathrm{NB}(\alpha_2, p_2(x))$. Use the most convenient parametrization.

(b) Solve the general case with nonconstant α_i, $i = 1, 2$.

4.10 Obtain the most general bivariate distribution (X, Y) such that the conditional distributions of $X|X + Y$ and $X + Y|X$ are exponential distributions.

4.11 Consider the gamma conditionals distribution (4.33). Determine sufficient conditions on the parameters of this model to guarantee that X has a gamma distribution and Y has a Pareto distribution.

4.12 For $n = 1, 2, \ldots$ consider the extreme order statistics $X_{1:n}, Y_{1:n}, X_{n:n}, Y_{n:n}$ based on a sample of independent random vectors (X_i, Y_i), $i = 1, 2, \ldots, n$, with common exponential conditionals distributions (4.14).

(a) Verify that $X_{1:n}$ and $Y_{1:n}$ are asymptotically independent and determine the limiting distribution of $(X_{1:n}, Y_{1:n})$.

(b) Now consider $X_{n:n}$ and $Y_{n:n}$.

(Angus (1989).)

5
Other Conditionally Specified Families

5.1 Introduction

Of course, not every conditionally specified model involves exponential families. The present chapter surveys a variety of conditionally specified models not fitting into the exponential family paradigm. No general theorem analogous to Theorem 4.1 is available and results are obtained on a case by case basis. The key tools are of course Theorems 1.3 and 1.4 which permit us to solve the functional equations characterizing many conditionally specified models.

We shall treat first the case of distributions with Pareto conditionals, and two extensions involving Pearson type VI and generalized Pareto distributions. These three models, together with the Dagum type I distribution, which can be derived from them, can be used for modeling data involving income for related individuals or income from different sources. Next we describe distributions with Cauchy and Student-t conditionals, heavy tailed alternatives to the normal conditionals model. Then, the most general cases of distributions with uniform and exponential conditionals, in the sense of having nonrectangular support, are analyzed. In this study we shall use the compatibility conditions described in Chapter 1, which are specially suited for distributions with a support depending on the parameters. Finally, we shall study the distributions with scaled (first type) beta conditionals with nonrectangular support. In particular, those subject to the condition $X + Y \leq 1$ have applications in simulating random proportions.

The plot thickens a little when we turn to study Weibull and logistic conditionals models. As we shall see, there is no trouble developing the corresponding functional equations. The general solutions however are not easy to obtain and even when, as is the case with Weibull conditionals, we can solve the equations, or prove they have a solution, it is not easy to write down the form of the resulting joint density. In fact, at the present time, we are not able to write an explicit form for a nontrivial Weibull conditionals density (i.e., one which does not have independent marginals and is distinct from those obtained by power transformations of exponential conditionals models as described in Section 4.7). Trivial logistic conditionals densities are related to certain Pareto conditionals distributions and have constant conditional scale parameters. Nontrivial logistic conditionals models have evaded discovery. In this case even existence is open to question.

Our final section of this chapter deals with mixtures. A time honored device for constructing bivariate densities with specified marginals involves mixtures of densities with suitable independent marginals. Can mixtures of suitable conditionally specified distributions yield useful models with conditionals in specified families? Not quite, as it turns out, but the mixture device may be useful in suggesting possible forms of well-behaved conditionally specified models.

5.2   Bivariate Distributions with Pareto Conditionals

Definition 5.1 (Pareto type II distribution). We say that a random variable has a Pareto type II distribution, according to the hierarchy of Pareto distributions described in Arnold (1983), if its probability density function is

$$f(x, \alpha) = \frac{\alpha}{\sigma} \left(1 + \frac{x}{\sigma}\right)^{-(\alpha+1)} I(x > 0), \tag{5.1}$$

where $\alpha, \sigma > 0$. In the following this distribution will be denoted by $P(\sigma, \alpha)$. □

This law is closely related to the one introduced by Pareto in 1895, in his famous polemic against the French and Italian socialists who were pressing for institutional reforms to reduce inequality in the distribution of income. Pareto analyzed the characteristics of regularity and permanence in observed income distributions, which indicated that the income elasticity of the survival distribution function was constant, that is,

$$d \log P(X > x)/d \log x = \alpha, \tag{5.2}$$

where X is the income variable with range (x_0, ∞).

The value of α (which is called the Pareto index), is normally close to 1.5, even though it changes with time and from population to population. It is often interpreted as an inequality measure. It may be noted that the Gini index corresponding to (5.2) is given by $(2\alpha - 1)^{-1}$. The distributions described by (5.1) and (5.2) differ only by translation. Recently, the Pareto laws have been applied to model stochastic phenomena from other areas such as health care, telephone circuitry, oil explorations, queue service disciplines, and reliability investigations. For more details about generalizations, properties, and applications of Pareto distributions, see Arnold (1983).

In this section, we look for all bivariate densities $f_{X,Y}(x, y)$, such that all conditional distributions are of the type (5.1), with constant Pareto index α. Then the conditional distributions are

$$X|Y = y \sim P(\sigma_1(y), \alpha),$$

$$Y|X = x \sim P(\sigma_2(x), \alpha),$$

where $\sigma_1(y)$, $\sigma_2(x)$ are positive functions with positive arguments. Thus, writing the joint density as a product of marginal and conditional densities in both ways, we get the functional equation

$$\frac{g_1(y)}{\sigma_1(y) + x} = \frac{g_2(x)}{\sigma_2(x) + y}, \tag{5.3}$$

where

$$g_1(y) = (\alpha \, \sigma_1(y)^\alpha f_Y(y))^{1/(\alpha+1)}, \quad g_2(x) = (\alpha \, \sigma_2(x)^\alpha f_X(x))^{1/(\alpha+1)}. \tag{5.4}$$

Equation (5.3) can be readily rearranged to be of the form (1.31) and consequently we may find the following general solution:

$$\sigma_1(y) = \frac{\lambda_{00} + \lambda_{01}y}{\lambda_{10} + \lambda_{11}y}, \quad \sigma_2(x) = \frac{\lambda_{00} + \lambda_{10}x}{\lambda_{01} + \lambda_{11}x}, \tag{5.5}$$

$$g_1(y) = \frac{1}{\lambda_{10} + \lambda_{11}y}, \quad g_2(x) = \frac{1}{\lambda_{01} + \lambda_{11}x}, \tag{5.6}$$

where the λ_{ij}'s are arbitrary constants. From (5.3)–(5.6) we get the following joint and marginal densities:

$$f_{X,Y}(x, y) \propto \frac{1}{(\lambda_{00} + \lambda_{10}x + \lambda_{01}y + \lambda_{11}xy)^{\alpha+1}}, \tag{5.7}$$

$$f_X(x) \propto \frac{1}{(\lambda_{01} + \lambda_{11}x)(\lambda_{00} + \lambda_{10}x)^\alpha}, \quad f_Y(y) \propto \frac{1}{(\lambda_{10} + \lambda_{11}y)(\lambda_{00} + \lambda_{01}y)^\alpha}. \tag{5.8}$$

It remains now to identify constraints on the λ_i's to ensure that (5.7) and (5.8) are nonnegative and integrable. Clearly for nonnegativity we must assume all λ_{ij}'s are ≥ 0. It turns out that to ensure integrability, we must distinguish three cases.

Case (i): $0 < \alpha < 1$. In this case we must have $\lambda_{10} > 0, \lambda_{01} > 0, \lambda_{11} > 0$ and $\lambda_{00} \geq 0$,

Case (ii): $\alpha = 1$. Then we need $\lambda_{10} > 0, \lambda_{01} > 0, \lambda_{11} > 0$, and $\lambda_{00} > 0$.

Case (iii): $\alpha > 1$. Here we must have $\lambda_{00} > 0, \lambda_{10} > 0$, and $\lambda_{01} > 0$ while $\lambda_{11} \geq 0$.

It will be noted that in case (iii), the choice $\lambda_{11} = 0$ leads to a distribution with Pareto marginals and Pareto conditionals (this special case is the bivariate Pareto introduced by Mardia (1962)). In order to calculate the normalizing constant in (5.7) we need to evaluate the integral

$$I = \int_0^\infty \int_0^\infty \frac{dx \, dy}{(\lambda_{00} + \lambda_{10}x + \lambda_{01}y + \lambda_{11}xy)^{\alpha+1}} \tag{5.9}$$

with the above conditions. For this we shall distinguish three cases:

Case (i): $\lambda_{11} = 0$. In this case elementary calculus yields

$$I = \alpha(\alpha - 1)/(\lambda_{00}^{\alpha-1}\lambda_{10}\lambda_{01}). \tag{5.10}$$

Case (ii): $\lambda_{00} = 0$. In this case, define $\phi = \lambda_{11}/(\lambda_{10}\lambda_{01})$ and make a simple change of variables to obtain

$$I = \frac{\pi\phi^{\alpha-1}}{\alpha\lambda_{10}\lambda_{01}\sin(\alpha\pi)}. \tag{5.11}$$

Case (iii): $\lambda_{00} \neq 0$. Setting $\lambda_{10}^* = (\lambda_{10}/\lambda_{00}), \lambda_{01}^* = (\lambda_{01}/\lambda_{00}), \lambda_{11}^* = (\lambda_{11}/\lambda_{00}), \phi^* = \lambda_{11}^*/(\lambda_{10}^*\lambda_{01}^*)$, and using a technique similar to that given by Arnold (1987), we get (5.9) in terms of hypergeometric functions:

$$I = \begin{cases} \dfrac{F(\alpha, 1; \alpha + 1; 1 - 1/\phi^*)}{\alpha^2 \lambda_{00}^{\alpha+1} \lambda_{10}^* \lambda_{01}^* \phi^*} = \dfrac{\sum_{k=0}^\infty \dfrac{\alpha}{\alpha + k}\left(1 - \dfrac{1}{\phi^*}\right)^k}{\alpha^2 \lambda_{00}^{\alpha+1} \lambda_{10}^* \lambda_{01}^* \phi^*} \\ \qquad\qquad\qquad\qquad\qquad\qquad \text{if } \phi^* > \dfrac{1}{2}, \\[2em] \dfrac{\phi^{*\alpha+1} F(\alpha, \alpha; \alpha + 1; 1 - \phi^*)}{\alpha^2 \lambda_{00}^{\alpha+1} \lambda_{10}^* \lambda_{01}^*} = \dfrac{\phi^{*\alpha+1} \sum_{k=0}^\infty \dfrac{\Gamma(\alpha + k)}{(\alpha + k)k!}(1 - \phi^*)^k}{\alpha \lambda_{00}^{\alpha+1} \lambda_{10}^* \lambda_{01}^* \Gamma(\alpha)} \\ \qquad\qquad\qquad\qquad\qquad\qquad \text{if } 0 < \phi^* \leq \dfrac{1}{2}. \end{cases}$$

$$\tag{5.12}$$

Expression (5.12) simplifies considerably in the case when $\alpha = 1$ or 2. We find

$$[\alpha = 1], \quad I = \frac{1}{\lambda_{10}\lambda_{01}} \frac{-\log[(\lambda_{00}\lambda_{11})/(\lambda_{10}\lambda_{01})]}{\left(1 - \dfrac{\lambda_{00}\lambda_{11}}{\lambda_{10}\lambda_{01}}\right)} \tag{5.13}$$

and

$$[\alpha = 2], \quad I = \frac{1}{2\lambda_{00}\lambda_{10}\lambda_{01}} \frac{\phi^*}{(1-\phi^*)^2} \left[\frac{1}{\phi^*} + \log \phi^* - 1\right]. \tag{5.14}$$

The quantity I defined in (5.12) clearly deserves the title of an awkward normalizing constant and parameters estimation techniques will need to be carefully chosen to sidestep the problem of repeatedly evaluating I. Such techniques will be discussed in Chapter 9. In order to study the dependence between the random variables X and Y in the model (5.7), we compute

$$\frac{d}{dy}P(X > x|Y = y) = -\alpha\left(1 + \frac{x}{\sigma_1(y)}\right)^{-(\alpha+1)} \frac{[\lambda_{00}\lambda_{11} - \lambda_{10}\lambda_{01}]}{(\lambda_{00} + \lambda_{01}y)^2}x$$

and observe that its sign depends on the sign of $\lambda_{00}\lambda_{11} - \lambda_{10}\lambda_{01}$. Thus, according to Barlow and Proschan (1981), we conclude that X is stochastically increasing or decreasing with Y. Consequently, for values of α and the λ_i's such that the correlation coefficient exists, we have

$$\text{sign}\rho(X,Y) = \text{sign}(\lambda_{10}\lambda_{01} - \lambda_{00}\lambda_{11}). \tag{5.15}$$

In particular, the bivariate densities with $\lambda_{00} = 0$, or $\lambda_{11} = 0$ always have a positive coefficient of correlation.

By using some changes of variable in (5.7) other interesting bivariate distributions with given conditionals can be obtained. For example, if $\alpha = 1$ in (5.7) and if we let $U = \log X, V = \log Y$, then (U, V) has all its conditional distributions of the logistic form with unit scale parameters. These are the trivial logistic conditionals distributions referred to in the Introduction. Nontrivial logistic conditionals densities will be discussed in Section 5.10.

5.3   Pearson Type VI Conditionals

In this section we consider a generalization of the above Pareto families. Specifically we treat Pearson type VI laws, which are also called beta distributions of the second kind. The densities associated with Pearson VI distributions are

$$f(x; p, q) = \frac{\sigma^q}{B(p, q)} x^{p-1}(\sigma + x)^{-(p+q)}, \quad x > 0, \tag{5.16}$$

TABLE 5.1. Reciprocals of the normalizing constants for the Pearson type VI models.

$\lambda_{00} = 0$	$J = \dfrac{B(p,q)B(p-q,q)}{\lambda_{10}^q \lambda_{01}^q \lambda_{11}^{p-q}}$
$\lambda_{11} = 0$	$J = \dfrac{B(p,q)B(p,q-p)}{\lambda_{00}^{q-p} \lambda_{10}^q \lambda_{01}^p}$
$\lambda_{00}, \lambda_{11} > 0$	$J = \dfrac{B(p,q)^2}{\lambda_{00}^{q-p} \lambda_{10}^p \lambda_{01}^p} F\left(p,p; p+q, 1 - \dfrac{1}{\theta}\right)$

where $p, q, \sigma > 0$. In the following discussion, this family will be denoted by $B2(p, q, \sigma)$.

We wish to determine the most general class of bivariate random variables (X, Y) such that $X|Y = y \sim$ beta $2(p, q, \sigma_1(y))$ and $Y|X = x \sim$ beta $2(p, q, \sigma_2(x))$. Using a similar technique to that used in the Pareto case, we get

$$f_{X,Y}(x,y) \propto \frac{x^{p-1}y^{p-1}}{(\lambda_{00} + \lambda_{10}x + \lambda_{01}y + \lambda_{11}xy)^{p+q}}, \qquad (5.17)$$

where $\lambda_{00}, \lambda_{11} \geq 0$, and $\lambda_{10}, \lambda_{01} > 0$. If $\lambda_{00} = 0$, then $q < p$ and $\lambda_{11} \neq 0$, and if $\lambda_{11} = 0$, then $1 < p < q$ and $\lambda_{00} \neq 0$. Note that (5.17) reduces to (5.7) when $p = 1$. The marginal densities corresponding to (5.17) are

$$f_X(x) \propto \frac{x^{p-1}}{(\lambda_{01} + \lambda_{11}x)^p(\lambda_{00} + \lambda_{10}x)^q}, \quad f_Y(y) \propto \frac{y^{p-1}}{(\lambda_{10} + \lambda_{11}y)^p(\lambda_{00} + \lambda_{01}y)^q}.$$
$$(5.18)$$

To determine the appropriate normalizing constant we must evaluate the integral

$$J = \int_0^\infty \int_0^\infty \frac{x^{p-1}y^{p-1}}{(\lambda_{00} + \lambda_{10}x + \lambda_{01}y + \lambda_{11}xy)^{p+q}} \, dx \, dy. \qquad (5.19)$$

This can be accomplished by methods similar to those used in the Pareto case. We thus obtain the entries in Table 5.1.

If $\lambda_{10}\lambda_{01} = \lambda_{00}\lambda_{11}$, we have the trivial case of independence where $X \sim$ beta $2(p, q; \lambda_{00}/\lambda_{10})$ and $Y \sim$ beta $2(p, q; \lambda_{00}/\lambda_{01})$. The density (5.17) with $\lambda_{11} = 0$ is an extension of the bivariate Pareto distribution introduced by Mardia (1962).

In direct analogy to the univariate beta(2) distribution, the class of distributions (5.17) is closed under reciprocation, i.e. if we denote (5.17) by $B2C(\lambda_{00}, \lambda_{10}, \lambda_{01}, \lambda_{11}; p, q)$, then if $(X, Y) \sim B2C(\lambda_{00}, \lambda_{10}, \lambda_{01}, \lambda_{11}; p, q)$, we have $(1/X, 1/Y) \sim B2C(\lambda_{11}, \lambda_{01}, \lambda_{10}, \lambda_{00}; q, p)$. In addition, it is easy to prove that

$$E(X^k | Y = y) = \frac{B(p+k, q-k)}{B(p, q)} \left(\frac{\lambda_{00} + \lambda_{01} y}{\lambda_{10} + \lambda_{11} y} \right)^k, \qquad (5.20)$$

$$E(Y^k | X = x) = \frac{B(p+k, q-k)}{B(p, q)} \left(\frac{\lambda_{00} + \lambda_{01} x}{\lambda_{10} + \lambda_{11} x} \right)^k, \qquad (5.21)$$

provided that $q > k$. Thus the conditional moments are rational functions of the conditioned variable. In a similar way, it is easy to prove that if the coefficient of correlation exists, then we have

$$\text{sign} \rho(X, Y) = \text{sign}(\lambda_{10} \lambda_{01} - \lambda_{00} \lambda_{11}),$$

just as in the Pareto case.

5.4 Bivariate Distributions with Generalized Pareto Conditionals

A hierarchy of Pareto distributions was introduced in Arnold (1983). The generalized Pareto distribution that we are considering in the present book was called a Pareto (IV) distribution in that monograph. Since we will not consider other distributions in the hierarchy we will merely refer to the distribution as generalized Pareto (it is also known as a Burr XII distribution).

Definition 5.2 (Generalized Pareto distribution). We say that X has a generalized Pareto distribution and write $X \sim \mathcal{GP}(\sigma, \delta, \alpha)$ if its survival function is of the form

$$P(X > x) = \left[1 + \left(\frac{x}{\sigma} \right)^\delta \right]^{-\alpha} I(x > 0), \qquad (5.22)$$

where σ, δ and α are positive. □

Our goal is to identify all bivariate distributions with the property that all of their conditional distributions are members of the family (5.22). We will outline analogous multivariate extensions in Chapter 8.

We wish to identify the class of all bivariate random variables (X, Y) such that for every $y > 0$ we have

$$X | Y = y \sim \mathcal{GP}(\sigma(y), \delta(y), \alpha(y)), \qquad (5.23)$$

and for every $x > 0$ we have

$$Y|X = x \sim \mathcal{GP}(\tau(x), \gamma(x), \beta(x)). \tag{5.24}$$

If we denote the corresponding marginal densities of X and Y by $f(x)$ and $g(y)$, respectively, and write the joint density of (X, Y) as the product of marginal and conditional densities in both possible ways ($f_X(x)f_{Y|X}(y|x)$ and $f_Y(y)f_{X|Y}(x|y)$) we conclude that the following functional equation must hold:

$$\frac{f(x)\beta(x)\gamma(x)y^{\gamma(x)-1}}{[\tau(x)]^{\gamma(x)}[1 + [\frac{y}{\tau(x)}]^{\gamma(x)}]^{\beta(x)+1}} = \frac{g(y)\alpha(y)\delta(y)x^{\delta(y)-1}}{[\sigma(y)]^{\delta(y)}[1 + [\frac{x}{\sigma(y)}]^{\delta(y)}]^{\alpha(y)+1}}, x, y > 0. \tag{5.25}$$

If we introduce the following new notation:

$$\begin{aligned}
a_1(x) &= xf(x)\beta(x)\gamma(x)/[\tau(x)]^{\gamma(x)}, \\
b_1(x) &= \tau(x)^{-\gamma(x)}, \\
c_1(x) &= -[\beta(x) + 1], \\
a_2(y) &= yg(y)\alpha(y)\delta(y)/[\sigma(y)]^{\delta(y)}, \\
b_2(y) &= \sigma(y)^{-\delta(y)}, \\
c_2(y) &= -[\alpha(y) + 1],
\end{aligned}$$

we may rewrite (5.25) in a simplified form.

$$a_1(x)y^{\gamma(x)}[1 + b_1(x)y^{\gamma(x)}]^{c_1(x)} = a_2(y)x^{\delta(y)}[1 + b_2(y)x^{\delta(y)}]^{c_2(y)}, \quad x, y > 0. \tag{5.26}$$

Since the functions $f, \gamma, \tau, \beta, g, \delta, \sigma$, and α in (5.25) are only constrained to be always positive, it follows that the functions $a_1, b_1, \gamma, a_2, b_2$, and δ in (5.26) must be positive, while c_1 and c_2 are constrained to assume values in the interval $(-\infty, -1)$. Subject only to these constraints we need to solve (5.26). In principle this can be done as follows. Take the logarithm of both sides of (5.26). Differentiate successively three times with respect to y (or take differences if you wish to avoid assuming differentiability). Set $y = 1$ in all four equations. This gives four horrendous equations to solve for a_1, b_1, c_1, and γ. Finally, constraints may need to be imposed on the free parameters in the solution to ensure integrability of the corresponding joint density. The situation is completely analogous to that encountered in the search for the general class of distributions with Weibull conditionals (as described in Arnold, Castillo, and Sarabia (1992), pp. 73–75).

Two special cases of (5.26) are tractable and will be discussed below.

Suppose that (5.23) and (5.24) hold with $\gamma(x) = \gamma$ and $\delta(y) = \delta$. The functional equation to be solved (i.e., (5.26)) now assumes the form

$$a_1(x)y^{\gamma}[1 + b_1(x)y^{\gamma}]^{c_1(x)} = a_2(y)x^{\delta}[1 + b_2(y)x^{\delta}]^{c_2(y)}. \tag{5.27}$$

Now make the charge of variables $u = x^\delta$, $v = y^\gamma$ and introduce the new functions

$$\tilde{a}_1(u) = u^{-1} a_1(u^{1/\delta}), \quad \tilde{a}_2(v) = v^{-1} a_2(v^{1/\gamma}),$$
$$\tilde{b}_1(u) = b_1(u^{1/\delta}), \qquad \tilde{b}_2(v) = b_2(v^{1/\gamma}),$$
$$\tilde{c}_1(u) = c_1(u^{1/\delta}), \qquad \tilde{c}_2(v) = c_2(v^{1/\gamma}).$$

These functions must satisfy the equation

$$\tilde{a}_1(u)[1 + \tilde{b}_1(u)v]^{\tilde{c}_1(u)} = \tilde{a}_2(v)[1 + \tilde{b}_2(v)u]^{\tilde{c}_2(v)}. \tag{5.28}$$

However, this functional equation is equivalent to one solved earlier in Castillo and Galambos (1987a) (their equation (5.23)). The regularity conditions are slightly different but this does not affect the form of the general solution. Two families of solutions to (5.28) exist. In case 1, there exist constants $\lambda_1, \lambda_2, \lambda_3, \lambda_4$, and λ_5 such that

$$\tilde{a}_1(u)[1 + \tilde{b}_1(u)v]^{\tilde{c}_1(u)} = [\lambda_1 + \lambda_2 u + \lambda_3 v + \lambda_4 uv]^{\lambda_5}, \tag{5.29}$$

while in case 2, there exist constants $\theta_1, \theta_2, \theta_3, \theta_4, \theta_5$, and θ_6 such that

$$\tilde{a}_1(u)[1 + \tilde{b}_1(u)v]^{\tilde{c}_1(u)}$$
$$= \exp\left[\theta_1 + \theta_2 \log(\theta_5 + u) + \theta_3 \log(\theta_6 + v) + \theta_4 \log(\theta_5 + u) \log(\theta_6 + v)\right]. \tag{5.30}$$

Tracing these expressions back through the changes of variables and redefinitions of functions used in the derivation, we are led to the following two classes of joint densities with generalized Pareto marginals with constant $\gamma(x)$ and $\delta(y)$.

Model I:

$$f_{X,Y}(x,y) = x^{\delta-1} y^{\gamma-1} [\lambda_1 + \lambda_2 x^\delta + \lambda_3 y^\gamma + \lambda_4 x^\delta y^\gamma]^{\lambda_5}, \quad x, y > 0, \tag{5.31}$$

and

Model II:

$$f_{X,Y}(x,y) = x^{\delta-1} y^{\gamma-1} \exp\{\theta_1 + \theta_2 \log(\theta_5 + x^\delta) + \theta_3 \log(\theta_6 + y^\gamma) \\ + \theta_4 \log(\theta_5 + x^\delta) \log(\theta_6 + y^\gamma)\}, \quad x, y > 0. \tag{5.32}$$

In (5.31) we require that $\lambda_5 < -1$ while $\lambda_1 \geq 0, \lambda_2 > 0, \lambda_3 > 0, \lambda_4 \geq 0$. This family coincides with the family described in detail in Arnold, Castillo, and Sarabia (1992) pp. 56–60). The common constant value for $\alpha(y)$ and $\beta(x)$ is $[-\lambda_5 - 1]$. Both negative and positive correlations are possible in this model. When second moments exist (i.e., when $\min(2/\delta, 2/\gamma) + \lambda_5 + 1 < 0$) the sign of the correlation is determined by sign $(\lambda_2 \lambda_3 - \lambda_1 \lambda_4)$ (cf. Arnold, Castillo, and Sarabia (1992), p. 59 with slightly different notation).

Densities of the form (5.25) have marginals given by

$$f_X(x) = \frac{1}{\delta(-1-\lambda_5)} x^{\delta-1} (\lambda_3 + \lambda_4 x^\delta)^{-1} (\lambda_1 + \lambda_2 x^\delta)^{\lambda_5+1}, \quad x > 0,$$

and

$$f_Y(y) = \frac{1}{\gamma(-1-\lambda_5)} y^{\gamma-1} (\lambda_2 + \lambda_4 y^\gamma)^{-1} (\lambda_1 + \lambda_3 y^\gamma)^{\lambda_5+1}, \quad y > 0.$$

These marginals will be generalized Pareto distributions only if $\lambda_4 = 0$.

A joint density described by (5.31) will be unbounded as x or y approaches 0 if $\delta < 1$ or, respectively, $\gamma < 1$. If $\delta \geq 1$ and $\gamma \geq 1$, the density is bounded and is unimodal. One representative three-dimensional example, including the associated contour plot, of the density is displayed in Figure 5.1.

The second family, (5.32), is distinct from the family described in Arnold, Castillo and Sarabia. The parameters θ_5 and θ_6 must be positive, the parameters θ_2 and θ_3 must be less than -1, and the parameter θ_4 must be ≤ 0. In this family the functions $\sigma(y)$ and $\tau(x)$ are constants; specifically $\sigma(y) = \theta_5^{1/\delta}$ and $\tau(x) = \theta_6^{1/\gamma}$. In this family only nonpositive correlations are possible. The fact that $\theta_4 \leq 0$ ensures that the density of (X, Y) is totally negative of order 2 which, when the correlation exists, determines that its sign be nonpositive (cf. Section 4.3). The marginal density of X, corresponding to a joint density of the form (5.32), is given by

$$f_X(x) = \frac{-x^{\delta-1}}{1 + \theta_3 + \theta_4 \log(\theta_5 + x^\delta)}$$

$$\times \exp[\theta_1 + (1 + \theta_3) \log \theta_6 + (\theta_2 + \theta_4 \log \theta_6) \log(\theta_5 + x^\delta)], \quad x > 0.$$

An analogous expression can be written for $f_Y(y)$. One representative three-dimensional example, including the associated contour plot, of the density (5.32) is displayed in Figure 5.2.

In summary, we have identified two classes of distributions with generalized Pareto conditionals with $\gamma(x)$ and $\delta(y)$ constant: (i) those with a common constant value for $\alpha(y)$ and $\beta(x)$, (5.31); and (ii) those with constant values for $\sigma(y)$ and $\tau(x)$, (5.32). It is not possible to have a model in which $\gamma(x)$ and $\delta(y)$ are constant and both $\alpha(y)$ and $\sigma(y)$ are nonconstant. Nor is it possible to have $\gamma(x)$ and $\delta(y)$ both constant and $\beta(x)$ and $\tau(x)$ both nonconstant. Next we consider the models that can arise when we require that $\alpha(y)$ and $\beta(x)$ be possibly different constants.

Suppose that (5.23) and (5.24) hold with $\alpha(y) = \alpha$ and $\beta(x) = \beta$. The functional equation to be solved (i.e., (5.26)) now assumes the form

$$a_1(x) y^{\gamma(x)} [1 + b_1(x) y^{\gamma(x)}]^{c_1} = a_2(y) x^{\delta(y)} [1 + b_2(y) x^{\delta(y)}]^{c_2}. \tag{5.33}$$

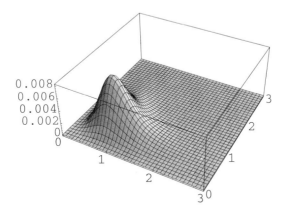

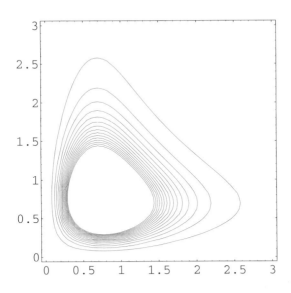

FIGURE 5.1. Probability density function of a generalized Pareto conditionals distribution Model I with $\delta = 3.5, \gamma = 3.5, \lambda_1 = \lambda_2 = \lambda_3 = 2, \lambda_4 = 3, \lambda_5 = -2.5$.

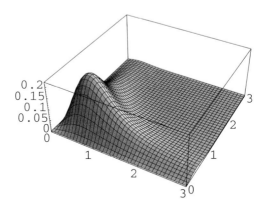

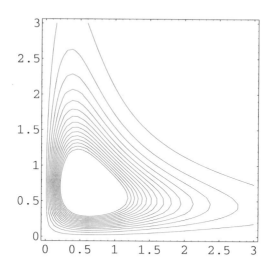

FIGURE 5.2. Probability density function of a generalized Pareto conditionals distribution Model I with $\delta = 2.2, \gamma = 2.5, \theta_1 = \theta_5 = \theta_6 = 1, \theta_2 = \theta_3 = -2, \theta_4 = -1.5$.

The general solution is obtainable by techniques analogous to those used previously. The constants c_1 and c_2 must be equal to or less than -1. Reparametrizing, the corresponding family of joint densities is of the form

$$f(x, y) = \frac{(xy)^{-1} \exp[\theta_0 + \theta_2 \log x + \theta_3 \log y + \theta_4 \log x \log y]}{[1 + \exp\{\theta_1 + \theta_2 \log x + \theta_3 \log y + \theta_4 \log x \log y\}]^{\alpha+1}}$$

$$x > 0, \quad y > 0$$
$$(5.34)$$

where $\alpha > 0$. To ensure that the functions $\gamma(x)$ and $\delta(y)$ are uniformly positive we must set $\theta_4 = 0$ in (5.34). With this modification the resulting distribution has $\gamma(x)$ and $\delta(y)$ constant and is thus a special case of Model I in (5.31) with $\lambda_2 = \lambda_3 = 0$.

5.4.1   Conjectures About More General Cases

In light of the results obtained above, we conjecture that there will not exist solutions to (5.25) with more than one of the functions $\gamma(x), \tau(x)$ and $\beta(x)$ nonconstant. If this proves to be true then the totality of generalized Pareto conditionals models will be subsumed by Models I and II ((5.31) and (5.32)).

5.4.2   Related Distributions

We could consider the addition of a location parameter in our generalized Pareto model. We will write $X \sim \mathcal{GP}^*(\mu, \sigma, \delta, \alpha)$ if its survival function is of the form

$$P(X > x) = \left[1 + \left(\frac{x - \mu}{\sigma}\right)^\delta\right]^{-\alpha}, \quad x > \mu. \tag{5.35}$$

We would then seek to identify all bivariate random variables (X, Y) with

$$X|Y = y \sim \mathcal{GP}^*(\mu(y), \sigma(y), \delta(y), \alpha(y))$$

for every possible value y of Y and

$$Y|X = x \sim \mathcal{GP}^*(\nu(x), \tau(x), \gamma(x), \beta(x))$$

for every possible value x of X. The trivial case in which $\mu(y) \equiv \mu$ and $\nu(x) \equiv \nu$ leads to translated versions of Models I and II. We conjecture that no nontrivial solutions will be found (i.e., solutions with nonconstant $\mu(x)$ and $\nu(y)$). Compatibility conditions on the conditional distribution will require that $\nu(y) = \mu^{-1}(y)$ if ν and μ are nonconstant functions (cf. Section 5.8) where translated exponential conditionals are discussed.

The family of densities (5.31) is a generalization of the bivariate Burr type XII distribution given by Takahasi (1965). Some dependent distributions with Burr marginals exist in the family (5.31). An important advantage of this family with respect to the Takahasi family is that the correlation coefficient may be of either sign. Other bivariate extensions of the Burr type XII have been given by Durling, Owen, and Drane (1970). See also Arnold (1990).

By means of the change of variable $U = 1/X, V = 1/Y$ in (5.31) we can obtain the most general distribution with Dagum type I conditionals, i.e. with cumulative distribution functions of the form

$$F(x) = \begin{cases} 0 & \text{if } x < 0, \\ 1/[1 + 1/(\lambda x^q)]^p & \text{if } x \geq 0. \end{cases} \tag{5.36}$$

A close relative of the family (5.22) is the family of survival functions

$$\bar{F}(x; \alpha, k) = \left(1 - \frac{kx}{\alpha}\right)^{1/k}, \qquad 0 < x < \frac{\alpha}{\max(0, k)}, \tag{5.37}$$

where $k \in \mathbb{R}, \alpha > 0$. We will call distributions of the form (5.37), Pickands-deHaan distributions. If X has survival function (5.37) we write $X \sim PdH(k, \alpha)$. The case $k = 0$ (evaluated by taking the limit as $k \to 0$ in (5.37)), corresponds to the exponential distribution. The case $k < 0$ corresponds to (5.1).

Bivariate distributions with conditionals in the Pickands-deHaan family are discussed in Arnold, Castillo, and Sarabia (1995a).

5.5   Bivariate Distributions with Cauchy Conditionals

Definition 5.3 (Location-scale Cauchy distribution). A random variable X has a Cauchy (μ, σ) distribution if its density is of the form

$$f_X(x) = \left[\pi\sigma\left(1 + (\frac{x - \mu}{\sigma})^2\right)\right]^{-1}, \qquad -\infty < x < \infty; \ \sigma > 0, \ \mu \in \mathbb{R}. \tag{5.38}$$

□

The Cauchy model is usually selected when we need a density function with tails heavier than those of the normal distribution. In the following it will be denoted by $C(\mu, \sigma)$.

In our search for the most general Cauchy conditionals distribution we assume $Y|X = x \sim C(\mu_2(x), \sigma_2(x)), X|Y = y \sim C(\mu_1(y), \sigma_1(y))$, where

$\sigma_2(x), \sigma_1(x) > 0$ for all x. This leads to the functional equation

$$\frac{f_X(x)}{\sigma_2(x)\left(1 + \left(\dfrac{y - \mu_2(x)}{\sigma_2(x)}\right)^2\right)} = \frac{f_Y(y)}{\sigma_1(y)\left(1 + \left(\dfrac{x - \mu_1(y)}{\sigma_1(y)}\right)^2\right)}. \qquad (5.39)$$

Setting

$$g_1(x) = f_X(x)/\sigma_2(x), \quad g_2(y) = f_Y(y)/\sigma_1(y), \qquad (5.40)$$

equation (5.39) becomes equivalent to

$$g_1(x) + g_1(x)\left(\frac{x - \mu_1(y)}{\sigma_1(y)}\right)^2 = g_2(y) + g_2(y)\left(\frac{y - \mu_2(x)}{\sigma_2(x)}\right)^2, \qquad (5.41)$$

which can be easily converted to one of the form (1.31). Thus, the general solution may be expressed as

$$\begin{bmatrix} g_1(x) \\ g_1(x)x \\ g_1(x)x^2 \\ 1 + \dfrac{a^2(x)}{b^2(x)} \\ \dfrac{\mu_2(x)}{b^2(x)} \\ \dfrac{1}{b^2(x)} \end{bmatrix} = A \begin{bmatrix} g_1(x) \\ g_1(x)x \\ g_1(x)x^2 \end{bmatrix}; \qquad \begin{bmatrix} 1 + \dfrac{c^2(y)}{d^2(y)} \\ -2\dfrac{\mu_1(y)}{d^2(y)} \\ \dfrac{1}{d^2(y)} \\ -g_2(y) \\ 2g_2(y)y \\ -g_2(y)y^2 \end{bmatrix} = B \begin{bmatrix} g_2(y) \\ g_2(y)y \\ g_2(y)y^2 \end{bmatrix},$$

where:

$$A' = \begin{pmatrix} 1 & 0 & 0 & a_{11} & a_{21} & a_{31} \\ 0 & 1 & 0 & a_{12} & a_{22} & a_{32} \\ 0 & 0 & 1 & a_{13} & a_{23} & a_{33} \end{pmatrix}, \quad B' = \begin{pmatrix} b_{11} & b_{21} & b_{31} & -1 & 0 & 0 \\ b_{12} & b_{22} & b_{32} & 0 & 1 & 0 \\ b_{13} & b_{23} & b_{33} & 0 & 0 & -1 \end{pmatrix},$$

with

$$A'B = 0,$$

from which $a_{ij} = (-1)^{j+1}b_{ij}$ $(i, j = 1, 2, 3)$.

If for $k = 1, 2, 3$ we define

$$P_k(x) = a_{k1} + a_{k2}x + a_{k3}x^2, \quad Q_k(y) = a_{1k} - 2a_{2k}y + a_{3k}y^2, \qquad (5.42)$$

the solution of (5.41) becomes

$$\mu_2(x) = \frac{P_2(x)}{P_3(x)}, \qquad\qquad \mu_1(y) = -\frac{Q_2(y)}{2Q_3(y)}, \qquad (5.43)$$

$$\sigma_1^2(x) = \frac{P_1(x)P_3(x) - P_2^2(x)}{P_3^2(x)}, \qquad \sigma_2^2(y) = \frac{4Q_1(y)Q_3(y) - Q_2^2(y)}{4Q_3^2(y)}, (5.44)$$

$$g_1(x) = \frac{P_3(x)}{P_1(x)P_3(x) - P_2^2(x)}, \qquad g_2(y) = \frac{4Q_3(y)}{4Q_1(y)Q_3(y) - Q_2^2(y)}. (5.45)$$

If we introduce the notation

$$
\begin{array}{lll}
m_{00} = a_{11}, & m_{10} = a_{12}, & m_{01} = -2a_{21}, \\
m_{20} = a_{13}, & m_{02} = a_{31}, & m_{11} = -2a_{22}, \\
m_{12} = a_{32}, & m_{21} = -2a_{23}, & m_{22} = a_{33}.
\end{array}
\tag{5.46}
$$

the joint density of (X, Y) finally becomes

$$
f_{X,Y}(x, y) \propto \left[(\begin{array}{ccc} 1 & x & x^2 \end{array}) M \begin{pmatrix} 1 \\ y \\ y^2 \end{pmatrix} \right]^{-1},
\tag{5.47}
$$

where $M = (m_{ij}), i, j = 0, 1, 2$, is a matrix of arbitrary constants. It remains to determine constraints on these parameters to guarantee that (5.47) is nonnegative and integrable over the plane.

We write:

$$
\begin{aligned}
f_{X,Y}(x, y) &= \left[a_1(y) + b_1(y)x + c_1(y)x^2 \right]^{-1}, & (5.48) \\
&= \left[\tilde{a}_1(x) + \tilde{b}_1(x)y + \tilde{c}_1(x)y^2 \right]^{-1}, & (5.49)
\end{aligned}
$$

where

$$
\begin{aligned}
a_1(y) &= m_{00} + m_{01}y + m_{02}y^2, & (5.50) \\
b_1(y) &= m_{10} + m_{11}y + m_{12}y^2, & (5.51) \\
c_1(y) &= m_{20} + m_{21}y + m_{22}y^2, & (5.52) \\
\tilde{a}_1(x) &= m_{00} + m_{10}x + m_{20}x^2, & (5.53) \\
\tilde{b}_1(x) &= m_{01} + m_{11}x + m_{21}x^2, & (5.54) \\
\tilde{c}_1(x) &= m_{02} + m_{12}x + m_{22}x^2. & (5.55)
\end{aligned}
$$

If the marginal density functions exist, they are of the form

$$
\begin{aligned}
f_X(x) &\propto \left[4\tilde{a}_1(x)\tilde{c}_1(x) - \tilde{b}_1^2(x) \right]^{-1/2}, & (5.56) \\
f_Y(y) &\propto \left[4a_1(y)c_1(y) - b_1^2(y) \right]^{-1/2}. & (5.57)
\end{aligned}
$$

First, we analyze the positivity of (5.48). We discuss different cases. If $c_1(y) = 0$, then $b_1(y) = 0$ and $a_1(y) > 0$. This case is not possible since it leads to nonintegrability. Consequently, $c_1(y) \neq 0$. Then, the following three conditions must hold for all y:

$$
\begin{aligned}
a_1(y) &> 0, & (5.58) \\
c_1(y) &> 0, & (5.59) \\
4a_1(y)c_1(y) - b_1^2(y) &> 0. & (5.60)
\end{aligned}
$$

From (5.58) we have two possible cases

$$m_{02} = m_{01} = 0, \quad m_{00} > 0, \tag{5.61}$$

or

$$m_{00} > 0, \quad m_{02} > 0, \quad m_{01}^2 - 4m_{00}m_{02} < 0. \tag{5.62}$$

Similarly, from (5.59) we have another two cases

$$m_{22} = m_{21} = 0, \quad m_{20} > 0, \tag{5.63}$$

or

$$m_{22} > 0, \quad m_{20} > 0, \quad m_{21}^2 - 4m_{20}m_{22} < 0. \tag{5.64}$$

Consideration of (5.49) leads to similar relations as those in (5.58) to (5.64).

Now we can divide the class of possible Cauchy conditionals distributions into two classes:

(i) The class with $m_{22} = 0$. Using (5.63) and its counterpart derived from (5.49) we have $m_{21} = m_{12} = 0$. If there were a joint density corresponding to such a parametric configuration, we would have

$$f_{X,Y}(x,y) \propto (m_{00} + m_{10}x + m_{01}y + m_{20}x^2 + m_{02}y^2 + m_{11}xy)^{-1}. \tag{5.65}$$

A natural and convenient reparametrization permits rewriting this in the form

$$f_{X,Y}(x,y) \propto [1 + (x - \mu_1, y - \mu_2)\Lambda(x - \mu_1, y - \mu_2)']^{-1}, \tag{5.66}$$

where $\mu_1, \mu_2 \in \mathbb{R}$ and Λ is positive definite. The condition that Λ be positive definite will guarantee that (5.66) is non-negative. Unfortunately, (5.66) is not integrable over the plane. Thus (5.66) is a species of improper model of the type to be discussed further in Chapter 6.

(ii) The class with $m_{22} > 0$. First, conditions (5.61), (5.64), with (5.60), together with their counterparts derived from (5.49), are not possible because the resulting pdf is not integrable. Thus, only conditions (5.62), (5.64), (5.60) and their counterparts are possible. Using standard results, we can easily prove the integrability. Thus, in summary, the conditions on the parameters to ensure that for $m_{22} > 0$ the joint density of the form (5.48) be well defined with marginals (5.56) and (5.57) are

$$m_{00}, m_{20}, m_{02}, m_{22} > 0 \tag{5.67}$$

and

$$4m_{20}m_{22} - m_{21}^2 > 0, \tag{5.68}$$

$$4m_{02}m_{22} - m_{12}^2 > 0, \tag{5.69}$$
$$4m_{02}m_{00} - m_{01}^2 > 0, \tag{5.70}$$
$$4m_{20}m_{00} - m_{10}^2 > 0, \tag{5.71}$$
$$4a_1(y)c_1(y) - b_1^2(y) > 0, \quad \forall y \tag{5.72}$$
$$4\tilde{a}_1(x)\tilde{c}_1(x) - \tilde{b}_1^2(x) > 0, \quad \forall x. \tag{5.73}$$

Note that the left-hand sides of (5.72) and (5.73) are fourth degree polynomials in y and x, respectively. Conditions (5.72) and (5.73) are equivalent to conditions for the coefficients of a fourth degree polynomial to be strictly positive. These conditions, though not simple, can be explicitly given. The implementation using a personal computer is easy. The proof of the conditions is based in the Sturm-Habicht sequence and can be found in González-Vega (1998). Consider a monic fourth degree polynomial

$$P(\underline{a}, x) = x^4 + a_3 x^3 + a_2 x^2 + a_1 x + a_0. \tag{5.74}$$

Then, $\forall x$, $P(\underline{a}, x) > 0$ if and only if $a_0 > 0$ and $\underline{a} \in H_4$ where

$$
\begin{aligned}
H_4 &= \{\underline{a} \in \mathbb{R}^4 : S_2 < 0, S_1 \neq 0, S_0 > 0\} \\
&\cup \{\underline{a} \in \mathbb{R}^4 : S_2 = 0, S_1 \leq 0, S_0 > 0\} \\
&\cup \{\underline{a} \in \mathbb{R}^4 : S_2 > 0, S_1 < 0, S_0 > 0\} \\
&\cup \{\underline{a} \in \mathbb{R}^4 : S_2 > 0, S_1 = 0\}
\end{aligned}
$$

and

$$
\begin{aligned}
S_2 &= 3a_3^2 - 8a_2, \\
S_1 &= 2a_2^2 a_3^2 - 8a_2^3 + 32a_2 a_0 + 28a_1 a_2 a_3 - 12a_3^2 a_0 - 6a_1 a_3^3 - 36a_1^2, \\
S_0 &= -27a_1^4 - 4a_3^3 a_1^3 + 18a_2 a_3 a_1^3 - 6a_3^2 a_0 a_1^2 + 144a_2 a_0 a_1^2 + a_2^2 a_3^2 a_1^2 \\
&\quad -4a_2^3 a_1^2 - 192a_3 a_0^2 a_1 + 18a_0 a_2 a_3^3 a_1 - 80a_0 a_2^2 a_3 a_1 + 256a_0^3 \\
&\quad -27a_3^4 a_0^2 + 144a_2 a_3^2 a_0^2 - 128a_2^2 a_0^2 - 4a_2^3 a_3^2 a_0 + 16a_2^4 a_0.
\end{aligned}
$$

As a final comment, we point out that the model with $m_{22} > 0$ does include, as a special case, densities with Cauchy marginals. The sufficient condition for having Cauchy marginals is that, for some $\gamma > 0$ and $\delta > 0$ with $\gamma \neq 1$ and $\delta \neq 1$, the following hold:

$$
\begin{aligned}
4\tilde{a}_1(x)\tilde{c}_1(x) &= \gamma \tilde{b}_1^2(x), \\
4a_1(y)c_1(y) &= \delta b_1^2(y).
\end{aligned}
$$

One special submodel may be of interest. If we insist on null location parameters in the conditional densities (i.e., that $m_{10} = m_{01} = m_{11} = m_{12} = m_{21} = 0$), then the model reduces to

$$f_{X,Y}(x, y) \propto \frac{1}{m_{00} + m_{20}x^2 + m_{02}y^2 + m_{22}x^2 y^2}, \tag{5.75}$$

$$f_X(x) \propto \frac{1}{\sqrt{(m_{00} + m_{20}x^2)(m_{02} + m_{22}x^2)}}, \tag{5.76}$$

$$f_Y(y) \propto \frac{1}{\sqrt{(m_{00} + m_{02}y^2)(m_{20} + m_{22}y^2)}}, \tag{5.77}$$

where $m_{00}, m_{20}, m_{02}, m_{22} > 0$. The conditional scale parameters in (5.75) are given by

$$\sigma_2(x) = \sqrt{\frac{m_{00} + m_{20}x^2}{m_{02} + m_{22}x^2}}, \quad \sigma_1(y) = \sqrt{\frac{m_{00} + m_{02}y^2}{m_{20} + m_{22}y^2}}. \tag{5.78}$$

The normalizing constant in (5.75) can be obtained by evaluating the integral

$$I = \int_{-\infty}^{\infty} \int_{-\infty}^{\infty} \frac{dx \; dy}{m_{00} + m_{20}x^2 + m_{02}y^2 + m_{22}x^2y^2}. \tag{5.79}$$

Without loss of generality we can assume $m_{00}m_{22} < m_{02}m_{20}$ and then

$$I = \frac{2\pi}{\sqrt{m_{20}m_{02}}} \int_0^{\pi/2} \frac{d\theta}{(1 - \sin^2 \alpha \sin^2 \theta)^{1/2}} = \frac{2\pi}{\sqrt{m_{20}m_{02}}} F\left(\frac{\pi}{2} \Big/ \alpha\right), \tag{5.80}$$

where

$$\sin^2 \alpha = \frac{m_{20}^2 m_{02}^2 - m_{00}^2 m_{22}^2}{m_{20}^2 m_{02}^2} \tag{5.81}$$

and $F(\pi/(2\alpha))$ is the Complete Elliptic Integral of the First Kind, which has been tabulated in Abramowitz and Stegun (1964), pp. 608–611. An approximation is given by formula 17.3.33 in Abramowitz and Stegun:

$$
\begin{aligned}
F(\pi/2/\alpha) =\; & (a_0 + a_1(1 - \alpha) + a_2(1 - \alpha)^2) \\
& - (b_0 + b_1(1 - \alpha) + b_2(1 - \alpha)^2) \log(1 - \alpha) + \epsilon(\alpha),
\end{aligned} \tag{5.82}
$$

where $|\epsilon(\alpha)| \le 3 \times 10^{-5}$, and

$$
\begin{aligned}
a_0 &= 1.3862944, & a_1 &= 0.1119723, & a_2 &= 0.0725296, \\
b_0 &= 0.5, & b_1 &= 0.1213478, & b_2 &= 0.0288729.
\end{aligned}
$$

Similarly to the Cauchy density function, the densities (5.76) and (5.77) do not possess finite moments or cumulants.

Observe that imposition of a condition requiring constant conditional scale parameters is equivalent to requiring Cauchy marginals. As remarked earlier this can be achieved with both independent and dependent marginals.

The transformation $u = \log x, v = \log y$ applied to the density (5.75) leads us to the hyperbolic secant conditionals distribution with density of the form

$$f_{U,V}(u,v) \propto (\alpha e^{-x-y} + \beta e^{-x+y} + \gamma e^{x-y} + \delta e^{x+y})^{-1}. \tag{5.83}$$

In (5.83), the parameters α, β, γ, and δ are all required to be positive.

5.6   Bivariate Distributions with Student-t Conditionals

As a generalization of the models in the previous section we consider now the most general distribution with Student-t conditionals.

Definition 5.4 (Student-t distribution). A random varaible U_α is said to follow a Student-t distribution if its pdf is given by

$$f_{U_\alpha}(u) = \frac{\Gamma\left[(\alpha+1)/2\right]}{(\alpha\pi)^{1/2}\Gamma(\alpha/2)}\left(1 + \frac{u^2}{\alpha}\right)^{-(\alpha+1)/2} \qquad \text{if } -\infty < u < \infty. \quad (5.84)$$

with $\alpha > 0$.   □

Setting $\alpha = 1$ in (5.84), we obtain the standard Cauchy distribution. If $\alpha > 1$ then $E(U_\alpha) = 0$, and if $\alpha > 2$, then $\operatorname{var}(U_\alpha) = \alpha/(\alpha - 2)$.

Now we are interested in the most general distribution of (X, Y) with conditionals

$$X|Y = y \sim \mu_1(y) + \sigma_1(y)U_\alpha \quad (5.85)$$

and

$$Y|X = x \sim \mu_2(x) + \sigma_2(x)U_\alpha, \quad (5.86)$$

with $\sigma_i(x) > 0, i = 1, 2, \ \forall x$.

Using similar arguments to those used for the Cauchy case, we find that the joint pdf of the random variable (X, Y) is

$$f_{(X,Y)}(x, y) = \left[(1 \quad x \quad x^2)M(1 \quad y \quad y^2)'\right]^{-(\alpha+1)/2}. \quad (5.87)$$

Now we will use the definitions of $a_1(y)$, $b_1(y)$, $c_1(y)$, $\tilde{a}_1(x)$, $\tilde{b}_1(x)$, and $\tilde{c}_1(x)$ given in (5.50)–(5.55). The location and scale parameters for the conditional densities are given by

$$\mu_1(y) = -\frac{1}{2} \times \frac{b_1(y)}{c_1(y)}, \quad (5.88)$$

$$\mu_2(x) = -\frac{1}{2} \times \frac{\tilde{b}_1(x)}{\tilde{c}_1(x)}, \quad (5.89)$$

and

$$\sigma_1^2(y) = \frac{4a_1(y)c_1(y) - b_1^2(y)}{4\alpha c_1^2(y)}, \quad (5.90)$$

$$\sigma_2^2(x) = \frac{4\tilde{a}_1(x)\tilde{c}_1(x) - \tilde{b}_1^2(x)}{4\alpha\tilde{c}_1^2(x)}. \quad (5.91)$$

The marginals, when they exist, are of the form

$$f_X(x) \propto \frac{[\tilde{c}_1(x)]^{(\alpha-1)/2}}{[4\tilde{a}_1(x)\tilde{c}_1(x) - \tilde{b}_1^2(x)]^{\alpha/2}}, \quad (5.92)$$

and

$$f_Y(y) \propto \frac{[c_1(y)]^{(\alpha-1)/2}}{[4a_1(y)c_1(y) - b_1^2(y)]^{\alpha/2}}. \tag{5.93}$$

Setting $\alpha = 1$ in (5.92) and (5.93) we get (5.56) and (5.57) which corresponds to the Cauchy case.

Similarly to the Cauchy case, we can break the class of possible Student t conditionals distributions into two classes:

(i) The class corresponding to $m_{22} = m_{21} = m_{12} = 0$. This class can be parametrized as

$$f_{(X,Y)}(x,y) \propto \left[1 + \frac{1}{\alpha-1}(x - \mu_1 \, y - \mu_2)\Sigma^{-1}(x - \mu_1 \, y - \mu_2)' \right]^{-(\alpha+1)/2}, \tag{5.94}$$

where $\alpha > 1$ to get integrability, and Σ must be positive definite. This model corresponds to the classical bivariate Student-t distribution.

(ii) The class corresponding to $m_{22} > 0$. The development in this case is essentially the same as that provided in the Cauchy case and will not be repeated here. The conditions for integrability are (5.67)–(5.73) together with $\alpha \geq 1$. This model is studied in Sarabia (1994).

5.7   Bivariate Distributions with Uniform Conditionals

Let (X, Y) be a bivariate random variable such that the conditional $X|Y = y \sim U(\phi_1(y), \phi_2(y))$ with $c < y < d, \phi_1(y) \leq \phi_2(y)$, and $Y|X = x \sim U(\psi_1(x), \psi_2(x))$ with $a < x < b, \psi_1(x) \leq \psi_2(x)$. We assume that ϕ_1 and ϕ_2 are either both nondecreasing or both nonincreasing and that the two domains

$$N_\phi = \{(x,y) : \phi_1(y) < x < \phi_2(y), \ c < y < d\}$$

and

$$N_\psi = \{(x,y) : \psi_1(x) < y < \psi_2(x), \ a < x < b\}$$

are coincident, so that the compatibility conditions are satisfied. Note that in principle a or c could be $-\infty$ and b or d could be $+\infty$. If such unbounded domains are considered we will have to make sure that the ϕ's and ψ's are such that the area of the support domain is finite.

Writing the joint density as products of marginals and conditionals gives us the following functional equation:

$$\frac{f_Y(y)}{\phi_2(y) - \phi_1(y)} = \frac{f_X(x)}{\psi_2(x) - \psi_1(x)}, \qquad (x,y) \in N_\psi, \tag{5.95}$$

which implies that both terms must be constant, i.e.,

$$f_X(x) = k[\psi_2(x) - \psi_1(x)], \quad a < x < b,$$

$$f_Y(y) = k[\phi_2(y) - \phi_1(y)], \quad c < y < d, \tag{5.96}$$

where

$$k^{-1} = \text{Area of } N_\psi = \int_a^b [\psi_2(x) - \psi_1(x)]\, dx = \int_c^d [\phi_2(y) - \phi_1(y)]\, dy < \infty.$$

Thus, we get

$$f_{X,Y}(x,y) = \begin{cases} k & \text{if } (x,y) \in N_\psi, \\ 0 & \text{otherwise,} \end{cases} \tag{5.97}$$

and the regression lines

$$E(X|Y = y) = \frac{\phi_2(y) + \phi_1(y)}{2}, \quad c < y < d,$$

$$E(Y|X = x) = \frac{\psi_2(x) + \psi_1(x)}{2}, \quad a < x < b, \tag{5.98}$$

which are monotone. If $\psi_1(x) \equiv 0$ and $\phi_1(y) \equiv 0$ then X and Y will be negatively correlated.

An interesting particular case arises when ϕ_1 and ϕ_2 are invertible; then $\phi_i(x) = \psi_i^{-1}(x), i = 1, 2$.

5.8   Possibly Translated Exponential Conditionals

Consider a random vector (X, Y) with the property that all conditionals are possibly translated exponential distributions. A random variable X has a possibly translated exponential distribution if

$$P(X > x) = e^{-\lambda(x-\alpha)} I(x > \alpha), \tag{5.99}$$

where $\lambda > 0$ and $\alpha \in (-\infty, \infty)$. If (5.99) holds we write $X \sim \exp(\alpha, \lambda)$. We wish to identify the class of all bivariate distributions (X, Y) such that

$$X|Y = y \sim \exp(\alpha(y), \lambda(y)), \quad y \in S(Y),$$

and

$$Y|X = x \sim \exp(\beta(x), \gamma(x)), \quad x \in S(X). \tag{5.100}$$

Clearly, for compatibility, we must assume that

$$D = \{(x,y) : \alpha(y) < x\} = \{(x,y) : \beta(x) < y\}. \tag{5.101}$$

Consequently, $\alpha(\cdot)$ must be nonincreasing and $\beta = \alpha^{-1}$. Then for $(x, y) \in D$, the following functional equation must hold:

$$\lambda(y)\exp\{-\lambda(y)[x - \alpha(y)]\}f_Y(y) = \gamma(x)\exp\{-\gamma(x)[y - \beta(x)]\}f_X(x),$$

$$(x, y) \in D, \tag{5.102}$$

where $f_X(x)$ and $f_Y(y)$ are the marginal pdf's.

Taking logarithms and defining

$$u(x) = \log[\gamma(x)f_X(x)] + \gamma(x)\beta(x), \quad v(y) = \log[\lambda(y)f_Y(y)] + \lambda(y)\alpha(y),$$

we get

$$\gamma(x)y - \lambda(y)x + v(y) - u(x) = 0, \tag{5.103}$$

which is a functional equation of the form (1.31), with general solution

$$\gamma(x) = ax + b, \quad u(x) = cx + d, \quad \lambda(y) = ay - c, \quad v(y) = -by + d. \tag{5.104}$$

Subject to integrability constraints we obtain

$$f_{X,Y}(x, y) \propto \exp(d + cx - by - axy), \quad (x, y) \in D, \tag{5.105}$$

and

$$f_X(x) = \frac{\exp[cx + d - (ax + b)\beta(x)]}{ax + b}, \quad x \in S(X), \tag{5.106}$$

$$f_Y(y) = \frac{\exp[-by + d - (ay - c)\alpha(y)]}{ay - c}, \quad y \in S(Y). \tag{5.107}$$

To ensure that (5.106) and (5.107) are integrable we need to assume that $\inf\{x : (x, y) \in D\} > -\infty$ and $\inf\{y : (x, y) \in D\} > -\infty$. To see the need for such a condition consider the possible choice $D = \{(x, y) : x + y > 0\}$. Everything works well except that in this case (5.105), (5.106), and (5.107) are not integrable.

In the model (5.105), the regression curves are given by

$$E(X|Y = y) = \alpha(y) + (ay - c)^{-1}, \quad y \in S(Y), \tag{5.108}$$

and

$$E(Y|X = x) = \beta(x) + (ax + b)^{-1}, \quad x \in S(X). \tag{5.109}$$

5.9   Bivariate Distributions with Scaled Beta Conditionals

Another interesting model where there is a condition on the range has been given by James (1975). He looked at multivariate distributions with beta conditionals which arise in connection with the generation of distributions for random proportions which do not possess neutrality properties.

Definition 5.5 (Neutrality). Let X and Y be positive continuous random variables such that $X+Y \leq 1$. Then X is said to be neutral if the pairs of random variables X, $Y/(1-X)$ and Y, $X/(1-Y)$ are independent. □

It is well known that the bivariate Dirichlet distribution with density function

$$f_{X,Y}(x,y) = \frac{\Gamma(\alpha + \beta + \gamma)}{\Gamma(\alpha)\Gamma(\beta)\Gamma(\gamma)}x^{\alpha-1}y^{\beta-1}(1-x-y)^{\gamma-1}, \quad x,y > 0, \quad x+y < 1,$$
(5.110)

$(\alpha, \beta, \gamma > 0)$ is characterized by the neutrality character of its marginals. With the purpose of extending the Dirichlet distribution, James assumed first type beta distributions for the conditionals of $X/(1-Y)$ given $Y = y$, and of $Y/(1-X)$ given $X = x$.

More precisely assume that (X, Y) are positive continuous random variables with $P(X + Y < 1) = 1$ such that

$$Y|X = x \sim (1-x)B(\alpha_1(x), \beta_1(x)), \quad 0 < x < 1, \quad (5.111)$$

and

$$X|Y = y \sim (1-y)B(\alpha_2(y), \beta_2(y)), \quad 0 < y < 1. \quad (5.112)$$

Our by now familiar techniques can be used to readily determine the nature of the corresponding joint density. Thus

$$f_{X,Y}(x,y) = \mu x^{\alpha-1}y^{\beta-1}(1-x-y)^{\gamma-1}\exp(\eta \log\ x \log\ y), \\ x,y > 0,\ x+y < 1, \quad (5.113)$$

$$f_X(x) = \mu B(\gamma, \eta\ \log\ x + \beta)x^{\alpha-1}(1-x)^{\eta \log\ x+\beta+\gamma-1}, \quad 0 < x < 1,\ (5.114)$$

where $\alpha, \beta, \gamma > 0$, $\eta \leq 0$, and μ is a suitable normalizing factor. From expressions (5.113) and (5.114), we conclude that the Dirichlet distribution ($\eta = 0$) is characterized by having first kind beta conditionals $X|Y = y$, and $Y|X = x$ (with $X + Y \leq 1$) and at least one of its marginals being of the beta kind (James (1975), pp. 683).

A slight generalization of (5.113) is possible. We can consider scaled beta's with support $\{0 < x < h(y)\}$ instead of $\{0 < x < 1 - y\}$. See the discussion in Section 5.7.

5.10   Weibull and Logistic Conditionals

Definition 5.6 (Weibull distribution). We say that X has a Weibull distribution if

$$P(X > x) = \exp[-(x/\sigma)^\gamma], \quad x > 0, \quad (5.115)$$

where $\sigma > 0$ and $\gamma > 0$.   □

If (5.115) holds we write $X \sim \text{Weibull}(\sigma, \gamma)$. The case $\gamma = 1$, corresponds to the exponential distribution. Our goal is to characterize all bivariate distributions with Weibull conditionals. That is, random variables (X, Y) such that

$$X|Y = y \sim \text{Weibull}(\sigma_1(y), \gamma_1(y)), \quad y > 0, \tag{5.116}$$

and

$$Y|X = x \sim \text{Weibull}(\sigma_2(x), \gamma_2(x)), \quad x > 0. \tag{5.117}$$

Interest centers on nontrivial examples in which $\gamma_1(y)$ and $\gamma_2(x)$ are not constant. If they are constant, the discussion in Section 4.7 provides a solution. Writing the joint density corresponding to (5.116) and (5.117) as products of marginals and conditionals yields the following functional equation, valid for $x, y > 0$:

$$f_Y(y) \frac{\gamma_1(y)}{\sigma_1(y)} \left[\frac{x}{\sigma_1(y)} \right]^{\gamma_1(y)-1} \exp\left[-\left(\frac{x}{\sigma_1(y)} \right)^{\gamma_1(y)} \right]$$

$$= f_X(x) \frac{\gamma_2(x)}{\sigma_2(x)} \left[\frac{y}{\sigma_2(x)} \right]^{\gamma_2(x)-1} \exp\left[-\left(\frac{y}{\sigma_2(x)} \right)^{\gamma_2(x)} \right]. \tag{5.118}$$

For suitably defined functions $\phi_1(y), \phi_2(y), \psi_1(x)$ and $\psi_2(x)$, (5.118) can be written in the form

$$\phi_1(y)x^{\gamma_1(y)} \exp\left[-\phi_2(y)x^{\gamma_1(y)} \right] = \psi_1(x)y^{\gamma_2(x)} \exp\left[-\psi_2(x)y^{\gamma_2(x)} \right]. \tag{5.119}$$

This equation is not as easy to solve as those reducible to the form (2.24). We may indicate the nature of solutions to (5.119) by assuming differentiability (a differencing argument would lead to the same conclusions albeit a bit more painfully). Take logarithms on both sides of (5.119) then differentiate once with respect to y to get a new functional equation, and then again with respect to y to get another functional equation. Now set $y = 1$ in these two equations and in (5.119). This yields the following three equations for $\psi_1(x), \psi_2(x)$ and $\gamma_2(x)$:

$$c_1 x^{\gamma_1} e^{-c_2 x^{\gamma_1}} = \psi_1(x) e^{-\psi_2(x)}, \tag{5.120}$$

$$c_3 + c_4 \log x = -c_5 x^{\gamma_1} - c_6 (\log x) x^{\gamma_1} + \gamma_2(x) \left[1 - \psi_2(x) \right], \tag{5.121}$$

$$c_7 + c_8 \log x = -c_9 x^{\gamma_1} - c_{10} (\log x) x^{\gamma_1} - c_{11} (\log x)^2 x^{\gamma_1}$$
$$- \gamma_2(x) - \psi_2(x) \gamma_2(x) \left[\gamma_2(x) - 1 \right]. \tag{5.122}$$

These three equations may be solved to yield $\gamma_2(x), \psi_1(x)$, and $\psi_2(x)$. See Castillo and Galambos (1990) for details. A similar approach will yield expressions for $\phi_1(y), \phi_2(y)$, and $\gamma_1(y)$. Unfortunately, it appears to be difficult to determine appropriate values for all of the constants appearing in the solutions to guarantee compatibility in a nontrivial situation.

An analogous situation is encountered when we study distributions with logistic conditionals.

Definition 5.7 (Logistic distribution).   We say that X is a logistic (μ, σ) random variable if its density is of the form

$$f_X(x) = \frac{1}{\sigma} e^{(x-\mu)/\sigma} / [1 + e^{(x-\mu)/\sigma}]^2, \qquad (5.123)$$

where $\mu \in \mathbb{R}$ and $\sigma > 0$.   □

We seek distributions for which the conditional distribution of X given $Y = y$ is logistic$(\mu_1(y), \sigma_1(y))$ while the distribution of Y given $X = x$ is logistic$(\mu_2(x), \sigma_2(x))$. Thus we must solve the following functional equation:

$$\frac{f_Y(y)}{\sigma_1(y)} \frac{e^{(x-\mu_1(y))/\sigma_1(y)}}{[1 + e^{(x-\mu_1(y)/\sigma_1(y)}]^2} = \frac{f_X(x)}{\sigma_2(x)} \frac{e^{(y-\mu_2(x))/\sigma_2(x)}}{[1 + e^{(y-\mu_2(x))/\sigma_2(x)}]^2}. \qquad (5.124)$$

If we define

$$\phi(y) = 2\sqrt{\sigma_1(y)/f_Y(y)}$$

and

$$\psi(x) = 2\sqrt{\sigma_2(x)/f_X(x)}$$

we may rewrite (5.124) in the form

$$\phi(y) \cosh\left(\frac{x - \mu_1(y)}{2\sigma_1(y)}\right) = \psi(x) \cosh\left(\frac{y - \mu_2(x)}{2\sigma_2(x)}\right). \qquad (5.125)$$

Despite its attractive simplicity, (5.125) appears to be difficult to solve except in the trivial case when $\sigma_1(y) = \sigma_1$ and $\sigma_2(x) = \sigma_2$ (as described in Section 5.2).

5.11   Mixtures

A suitable scale mixture of exponential densities yields a Pareto (α) density. Thus

$$\int_0^\infty c e^{-cx} \frac{c^{\alpha-1} e^{-c}}{\Gamma(\alpha)} dc = \alpha(1 + x)^{-(\alpha+1)}. \qquad (5.126)$$

If we try a similar scale mixture of exponential conditionals densities, we do not get a simple expression and emphatically do not obtain a Pareto conditionals density. However, a scale mixture of kernels of exponential conditionals densities does yield the kernel of the Pareto conditionals density since

$$\int_0^\infty c e^{-c(m_{10}x+m_{01}y+m_{11}xy)} \frac{c^{\alpha-1} e^{-c}}{\Gamma(\alpha)} dc \qquad (5.127)$$
$$= \alpha(1 + m_{10}x + m_{01}y + m_{11}xy)^{-(\alpha+1)}.$$

An analogous argument involving scale mixtures of normal densities yielding a Cauchy density would have allowed us to guess the form of the Cauchy conditionals density (5.47). Other scale mixtures of exponential families can be treated in a similar fashion.

5.12   Bibliographic Notes

Section 5.2 is based on Arnold (1987) and Castillo and Sarabia (1990a). Section 5.3 covers material from Castillo and Sarabia (1990b). Section 5.4 is based on Arnold, Castillo, and Sarabia (1992). The general Cauchy conditionals material in Section 5.5 has not appeared elsewhere. The expressions for the marginal densities given in Arnold, Castillo, and Sarabia (1992) have been corrected. Section 5.7 is based on Sarabia (1994). Anderson (1990) and Anderson and Arnold (1991) discuss the centered Cauchy conditionals distribution. Uniform conditionals (Section 5.4) were introduced in Arnold (1988a). Section 5.8 on possibly translated exponential conditionals is new. The scaled beta material in Section 5.9 is based on James (1975). In Section 5.10, the Weibull material is based on Castillo and Galambos (1990), while the logistic material is new.

Information on the families of univariate distributions covered in this chapter may be found in Johnson, Kotz, and Balakrishnan (1994, 1995).

Exercises

5.1 Apart from the independent case, is there a joint bivariate density with conditionals of the form:

$$f(x|y) = \lambda(y)e^{-\lambda(y)x}, \quad x > 0,$$

such that $\lambda(y) > 0$, and

$$f(y|x) = \frac{\alpha}{\sigma(x)}\left(1 + \frac{y}{\sigma(x)}\right)^{-(\alpha+1)}, \quad y > 0,$$

with $\sigma(x) > 0$ and $\alpha > 0$?

5.2 The joint pdf,

$$f(x,y) = k(\alpha,\delta)(1 - x - y + \delta xy)^{1/\alpha-1},$$

for $0 < x, y < 1$, and $1 - x - y + \delta xy > 0$ with $\alpha > 0$, is a particular case of the model with conditionals in the Pickands–deHaan family.

(a) Obtain the marginal distributions.

(b) Compute the moments and the normalizing constant.

(Arnold, Castillo, and Sarabia (1995).)

5.3 A random variable X is said to be a Bradford random variable, and is denoted as $X \sim \mathrm{BF}(\theta)$, if its pdf is

$$f(x; \theta) = \frac{\theta}{\log(1 + \theta)(1 + \theta x)}, \quad \text{if } 0 < x < 1,$$

where $\theta > 0$.

(a) Obtain the most general random variable (X, Y) such that $X|Y = y \sim \mathrm{BF}(\theta_1(y))$ and $Y|X = x \sim \mathrm{BF}(\theta_2(x))$.

(b) Write the integral

$$\int_0^1 \frac{\log(1 + ax)}{1 + bx} \, dx$$

in terms of the dilog function

$$\mathrm{dilog}(x) = \int_1^x \frac{\log t}{1 - t} \, dt.$$

Using this fact, obtain the normalizing constant of the bivariate distribution with Bradford conditionals, in some particular cases.

5.4 A random variable X is said to be generalized Poisson with parameters λ and θ, and is denoted as $X \sim \mathrm{GP}(\lambda, \theta)$, if its pdf is

$$P(X = x) = \frac{1}{x!} \lambda(\lambda + \theta x)^{x-1} \exp[-(\lambda + \theta x)], \; x = 0, 1, 2, \ldots,$$

and $P(X = x) = 0$ for $x > m$ when $\theta < 0$, and where $\lambda > 0$, $\max(-1, -\lambda/m) < \theta \le 1$, $m > 4$.

Find the bivariate distributions with conditionals of the generalized Poisson type.

5.5 A random variable is said to be a finite mixture if its pdf is of the form

$$f(x) = \pi_1 f_1(x) + \ldots + \pi_k f_k(x)$$

where $\pi_i \ge 0$ and $\pi_1 + \ldots + \pi_k = 1$ and $f_1(x), \ldots, f_k(x)$ are known linearly independent pdfs.

(a) Obtain the most general pdf $f(x, y)$ with conditionals

$$f(x|y) = \pi_1(y) f_1(x) + \ldots + \pi_k(y) f_k(x)$$

and
$$f(y|x) = \tilde{\pi}_1(x)g_1(y) + \ldots + \tilde{\pi}_r(x)g_r(y),$$

where $\{f_i(x), i = 1, \ldots, k\}$ and $\{g_j(y); i = 1, \ldots, n\}$ are sets of linearly independent pdfs.

(b) Obtain the marginal distributions and the moments.

(c) Characterize the bivariate finite mixture conditionals distributions which are uncorrelated but not independent.

5.6 Suppose that (X_1, X_2) has a Cauchy conditionals distribution (5.47). Show that $(1/X_1, 1/X_2)$ also has a Cauchy conditionals distribution.

5.7 We say that X has a hyperbolic secant distribution if its density is of the form
$$f_X(x) = (\alpha e^{-x/\sigma} + \beta e^{x/\sigma})^{-1}.$$

Investigate the family of bivariate densities with hyperbolic secant conditionals.

6

Improper and Nonstandard Models

6.1 Introduction

In our discussion of compatible conditional densities (Chapter 1) it will be recalled that the key requirements for compatibility were:

(i) $\{(x, y) : f(x|y) > 0\} = \{(x, y) : f(y|x) > 0\}$;

(ii) $f(x|y)/f(y|x) = a(x)b(y)$; and

(iii) $a(x)$ in (ii) must be integrable.

In several potentially interesting situations compatibility fails only because condition (iii) is not satisfied. Such "improper" models may have utility for predictive purposes and in fact are perfectly legitimate models if we relax the finiteness condition in our definition of probability. Many subjective probabilists are willing to make such an adjustment (they can thus pick an integer at random). Another well-known instance in which the finiteness condition is relaxed with little qualm is associated with the use of improper priors in Bayesian analysis. In that setting, both sets of conditional densities (the likelihood and the posterior) are integrable nonnegative densities but one marginal (the prior) and, consequently, both marginals are nonnegative but nonintegrable. For many researchers, then, these "improper" models are perfectly possible. All that is required is that $f(x|y)$ and $f(y|x)$ be nonnegative and satisfy (i) and (ii). Integrability is not a consideration. A simple example (mentioned in Chapter 1) will help visualize the situation.

Suppose we ask for a joint density $f_{X,Y}(x,y)$ such that for each $x > 0$,

$$Y|X = x \sim U\left(0, \frac{1}{x}\right) \tag{6.1}$$

and for each $y > 0$,

$$X|Y = y \sim U\left(0, \frac{1}{y}\right). \tag{6.2}$$

If a corresponding joint density is to exist it clearly must be of the form

$$f_{X,Y}(x,y) = cI(x > 0, y > 0, xy < 1). \tag{6.3}$$

Thus (X, Y) is to have a uniform distribution over a region in the plane of infinite area.

Models such as (6.3) (and the nonintegrable Cauchy conditionals density (5.66)) may correctly summarize our feelings about the relative likelihoods of possible values of (X, Y) and, consequently, may have great predictive utility despite their flawed nature. We do not propose to resolve whether or not such models should be used. We merely provide a modest catalog of "improper" conditionally specified models which have arisen in the literature.

More serious modeling inconsistencies occasionally occur. What if the given functions $f(x|y)$ are $f(y|x)$ are occasionally negative? Or what if the given families of conditional densities are incompatible? Again we will only remark on such modeling anomalies. They do occur. The perils of conditional specification are not always appreciated.

6.2   Logistic Regression

A logistic probability model is often used to relate a binary response variable Y to several predictor variables $X_1, X_2, \ldots, X_k$. The model is of the form

$$P(Y = 1|X_1, X_2, \ldots, X_k) = \frac{1}{\{1 + \exp\{-[\beta_0 + \sum_{j=1}^{p} \beta_j \Phi_j(X_1, \ldots, X_k)]\}\}}, \tag{6.4}$$

where the β_k are unknown constants. In addition, it is often assumed that vector $(X_1, X_2, \ldots, X_k)$ given $Y = 1$ and $Y = 0$ has a specific distribution. For example, these conditional distributions may be posited to be multivariate normal. If we make such a claim then we have a model with conditionals in exponential families (binomial and multivariate normal), and the material in Chapter 4 can be used to completely specify the spectrum of acceptable models. Severe restrictions must be made on the Φ_j's appearing in (6.4). Many of the logistic regression models discussed in the applied literature are questionable in the light of these observations.

6.3   Uniform Conditionals

Let A be a subset of the plane for which each x cross section has finite measure and each y cross section has finite measure. Now define a function on $\mathbb{R}^2$ by

$$f(x, y) = I((x, y) \in A). \qquad (6.5)$$

If A has finite planar measure then this defines a proper joint density with all conditionals uniform (on the appropriate cross-section sets). If A has infinite measure we get "improper" uniform conditionals models analogous to the one defined in (6.3).

6.4   Exponential and Weibull Conditionals

Improper exponential conditionals models may be of either of two kinds (or a combination of both). The first kind are models involving a scale parameter which is a function of the conditioning variable. Recall the general form of the exponential conditionals density ((4.14))

$$f_{X,Y}(x, y) \propto \exp[-m_{10}x - m_{01}y + m_{11}xy]\, I(x > 0)I(y > 0). \qquad (6.6)$$

If $m_{10} > 0$ and $m_{01} > 0$ and $m_{11} \leq 0$, then (6.6) is a perfectly legitimate joint density. If, however, $m_{10}m_{01} = 0$ [i.e., if one or both of m_{10}, m_{01} are zero], then the model still has exponential conditionals but is improper in the sense that (6.6), although nonnegative, no longer is integrable. For example, the specification that $P(X > x|Y = y) = e^{-xy}$ and $P(Y > y|X = x) = e^{-xy}$ is improper in the sense that no integrable joint density can lead to such conditional distributions (it corresponds to the choice $m_{10} = m_{01} = 0, m_{11} = -1$).

The second type of improper exponential conditionals model involves a location parameter which is a function of the conditioning variable. A simple example involves a nonincreasing function $\psi(x)$ defined on the real line with inverse $\psi^{-1}(y)$. In order to have conditional distributions of the form

$$P(X > x|Y = y) = e^{-[x - \psi^{-1}(y)]}I(x > \psi^{-1}(y)) \qquad (6.7)$$

and

$$P(Y > y|X = x) = e^{-[y - \psi(x)]}I(y > \psi(x)), \qquad (6.8)$$

we must consider an improper (nonintegrable) joint density for (X, Y).

Consideration of (X^{c_1}, Y^{c_2}) for $c_1, c_2 > 0$ will lead to improper Weibull conditionals models analogous to (6.6) and (6.7)–(6.8). An example is provided by the function

$$F(x, y) = [1 - \exp(-x^{\gamma}y^{\eta})]I(x > 0)I(y > 0). \qquad (6.9)$$

For each fixed x, (6.9) is a genuine Weibull distribution function for y. Similarly for each fixed y, it is a Weibull distribution function for x. However, there is no integrable joint density over the positive quadrant with conditional distributions given by (6.9).

6.5   Measurement Error Models

In the context of measurement error, we may encounter random variables (X, Y) such that

$$X - Y \sim N(0, \sigma^2). \tag{6.10}$$

Condition (6.10) is compatible with a broad spectrum of joint densities for X and Y, some of which are bivariate normal. In an effort to treat X and Y in an exchangeable fashion the question arises as to whether a joint density for (X, Y) can be found to satisfy (6.10) and to have conditional distributions described as follows:

$$X|Y = y \sim N(y, \sigma_1^2), \tag{6.11}$$

$$Y|X = x \sim N(x, \sigma_2^2). \tag{6.12}$$

Referring to our compatibility theorem (Theorem 2.2) we conclude that we must require $\sigma_1^2 = \sigma_2^2$, but then the ratio $f_{X|Y}(x|y)/f_{Y|X}(y|x)$ will be constant and thus not integrable. Thus no integrable joint density can satisfy (6.11) and (6.12).

Related issues arise when we ask if it is proper to have a valid model in which both forward and backward regression models are valid with normal errors.

6.6   Stochastic Processes and Wöhler Fields

Consider an indexed family of invertible decreasing functions

$$\{y = \phi_z(x) : z \in A\}. \tag{6.13}$$

Let Z be a random variable with range A. For a fixed value of x, we can define a random variable $Y(x)$ as follows:

$$Y(x) = \phi_Z(x). \tag{6.14}$$

Analogously, for a fixed value of y, we may define $X(y)$ by

$$X(y) = \phi_Z^{-1}(y). \tag{6.15}$$

Note that the only randomness in (6.14) and (6.15) is associated with the random variable Z. It is consequently evident that for any x, y,

$$P(Y(x) \leq y) = P(X(y) \leq x). \tag{6.16}$$

The family of curves

$$P(Y(x) \leq y) = c, \quad 0 \leq c \leq 1, \tag{6.17}$$

or, equivalently,

$$P(X(y) \leq x) = c, \quad 0 \leq c \leq 1, \tag{6.18}$$

is called the Wöhler field of the stochastic process. It is important to realize that (6.16) does not deal with conditional distributions. There is no random variable X nor any random variable Y. There are two intimately related stochastic processes (6.14) and (6.15).

Sometimes physical considerations suggest the appropriate form for the densities of $Y(x)$ and $X(y)$. For example, we might posit that they are all of the Gumbel minimum type or perhaps that they are all Weibull.

6.6.1   Compatibility Conditions in Fatigue Models

When modeling fatigue and other similar lifetime data, models are selected mainly because of mathematical tractability, simplicity, and/or concordance with the data. However, models should be derived based on physical and statistical considerations. These considerations require that fatigue models should satisfy the following conditions (see Castillo, Fernández-Canteli, Esslinger, and Thürlimann (1985) and Castillo and Galambos (1987a)):

1. Models should take into account the fact that lifetime is governed by the weakest link principle, that is, the fatigue lifetime of a piece is the fatigue lifetime of its weakest subpiece or link. The weakness of a subpiece is determined by the size of its largest crack and the stress it is subjected to. Thus a piece can fail because it has a large crack and/or a large stress concentration.

2. Models must be stable. For example, because of the weakest link principle, models must be stable with respect to minimum operations. Assume the cdf of the fatigue lifetime of a longitudinal piece belongs to the parametric family $\mathcal{F} = \{F(x; \theta); \theta \in S\}$, where the parameter θ can be assumed, without loss of generality, to be the length of the piece. Then, according to the weakest link principle, the cdf of the first-order statistic of a sample of size n from $F(x; \theta)$ must be the same as the cdf of the lifetime of a piece of length $n\theta$. Thus, the family $\mathcal{F}$ must satisfy the functional equation $F(x; n\theta) = 1 - [1 - F(x; \theta)]^n$ which is the formal statement of stability with respect to minimum operations.

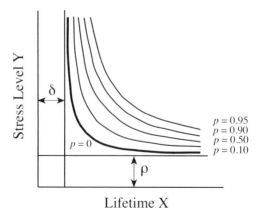

FIGURE 6.1. Wöhler curves for the fatigue model.

3. Models must take into account the positive character of lifetime and stress level.

4. The distributions of lifetime given stress level should be compatible with the distribution of the stress level given lifetime, that is, if $F_X(x; y)$ is the cumulative distribution function (cdf) of X given y, and $F_Y(y; x)$ is the cdf of Y given x, then

$$F_X(x; y) = F_Y(y; x). \tag{6.19}$$

This compatibility equation can be explained as follows. Suppose the curves in Figure 6.1 are in the parametric family $y = \theta_\alpha(x)$ or $x = \theta_\alpha^{-1}(y)$, with the curves heading in the northeast direction as α increases. Think of α as being random with some distribution. For given (x, y), let $\eta(x, y)$ be the value of α such that $\theta_{\eta(x,y)}(x) = y$ or $\theta_{\eta(x,y)}^{-1} = x$. Then

$$\begin{aligned} F_X(x; y) &= P(X \le x; y) = P(\theta_\alpha(y) \le x) = P(\alpha \le \eta(x, y)) \\ &= P(\theta_\alpha^{-1}(x) \le y) = P(Y \le y; x) = F_Y(y; x). \end{aligned} \tag{6.20}$$

Thus when selecting any point on any of the Wöhler percentile curves, the number of curves to the left of the point (i.e., in the X-direction) is equal to the number of curves below the point (i.e., in the Y-direction), that is,

$$F_X(\theta(y); y) = F_Y(\theta^{-1}(x); x), \tag{6.21}$$

which is equivalent to (6.16).

In the following sections we introduce several models. All of them satisfy conditions 1, 2, and 4. In addition, the reversed generalized Pareto model satisfies condition 3.

6.6.2   The Gumbel–Gumbel Model

As in Castillo et al. (1985) and Castillo (1988), suppose that the random variables $X(y), -\infty < y < \infty$, and $Y(x), -\infty < x < \infty$, all have Gumbel distributions. It follows that

$$P(X(y) \leq x) = P(Y(x) \leq y) = 1 - \exp\left\{-\exp\left[\frac{x - a(y)}{b(y)}\right]\right\}$$

$$= 1 - \exp\left\{-\exp\left[\frac{y - c(x)}{d(x)}\right]\right\}, \tag{6.22}$$

where $a(y), b(y), c(x)$, and $d(x)$ are unknown functions such that $b(y) > 0$ and $d(x) > 0$. Equation (6.22) can be written as

$$\frac{x - a(y)}{b(y)} = \frac{y - c(x)}{d(x)} \quad \Leftrightarrow \quad xd(x) - a(y)d(x) - yb(y) + c(x)b(y) = 0, \tag{6.23}$$

which is a functional equation of the form (1.31). Thus, using Theorem 2.3, we get the general solutions

$$a(y) = \frac{Cy - D}{Ay - B}, \quad b(y) = \frac{1}{Ay - B}, \quad c(x) = \frac{Bx - D}{Ax - C}, \quad d(x) = \frac{1}{Ax - C}. \tag{6.24}$$

Substitution into (6.22) leads to

$$P(X(y) \leq x) = P(Y(x) \leq y) = 1 - \exp\{-\exp[Axy - Bx - Cy + D]\}, \quad \forall x, y. \tag{6.25}$$

However, if $A \neq 0$, expression (6.25) is not a monotonic function of x for every y. Thus $A = 0$.

Consequently, the most general model of the Gumbel–Gumbel type is

$$P(X(y) \leq x) = P(Y(x) \leq y)$$

$$= 1 - \exp\{-\exp[-Bx - Cy + D]\}, \quad B, C < 0; \quad \forall x, y. \tag{6.26}$$

Note that the Wöhler field for this model consists of a set of parallel straight lines.

In the context of the discussion in earlier sections of this chapter, we note that (6.26) does provide indexed families of Gumbel distributions but there is no integrable joint density $f_{X,Y}(x, y)$ which has its conditional distributions given by (6.26). Lack of integrability is the only lacuna, otherwise (6.26) would provide a routine example of a conditionally specified distribution with Gumbel marginals.

6.6.3   The Weibull–Weibull Model

Suppose that for each fixed $y > 0, X(y)$ has a Weibull distribution and for each fixed $x > 0, Y(x)$ has a Weibull distribution. From (6.16) we then have

$$P(X(y) \leq x) = P(Y(x) \leq y) \quad = \quad 1 - \exp\left\{-[a(x)y + b(x)]^{c(x)}\right\}$$

$$= \quad 1 - \exp\left\{-[d(y)x + e(y)]^{f(y)}\right\},$$

$$y \geq -\frac{b(x)}{a(x)}, \quad x \geq -\frac{e(y)}{d(y)}, \quad (6.27)$$

where $a(x), b(x), c(x), d(y), e(y)$, and $f(y)$ are unknown positive functions.

Castillo and Galambos (1987b) obtained the following three families of solutions as the general solutions of the functional equation (6.27):

$$P(X(y) \leq x) = P(Y(x) \leq y)$$

$$= 1 - \exp\left\{-\left[E(x-A)^C(y-B)^D \exp[M \log(x-A) \log(y-B)]\right]\right\},$$

$$x > A, \quad y > B,$$
$$(6.28)$$

$$P(X(y) \leq x) = P(Y(x) \leq y)$$

$$= 1 - \exp\{-[C(x - A)(y - B) + D]^E\}, \quad x > A, \ y > B,$$
$$(6.29)$$

and

$$P(X(y) \leq x) = P(Y(x) \leq y)$$

$$(6.30)$$

$$= 1 - \exp\{-E(x - A)^C(y - B)^D\}, \quad x > A, \ y > B.$$

We can change the notation of model (6.30) to get

$$P(X(y) \leq x) = P(Y(x) \leq y)$$

$$(6.31)$$

$$= 1 - \exp\{-\sigma(x - \delta)^\epsilon(y - \rho)^{\epsilon/k}\} \quad x > \delta; \ y > \rho.$$

The physical interpretation of these parameters is given in Section 6.7.

As in the Gumbel–Gumbel case, there do not exist integrable joint densities, $f_{X,Y}(x,y)$, whose conditionals are given by (6.28), (6.29), or (6.30).

6.6.4   The Reversed Generalized Pareto Model

Castillo and Hadi (1995) develop the *reversed generalized Pareto model* (RGPD) which satisfies the above four conditions of compatibility.

Since design values are associated with small probabilities of failure, that is, only with lower tail values it follows that only the left-hand part of the Wöhler field is relevant for design considerations (see Figure 6.2). With this in mind, it is not reasonable to continue fatigue testing after some

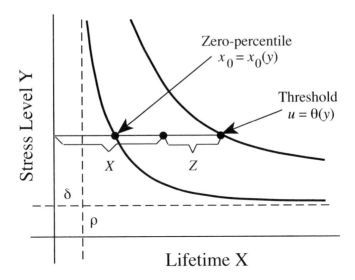

FIGURE 6.2. Wöhler curves for the reeversed generalized Pareto fatigue model.

given time (some standard procedures last until 2×10^6 or 10^7 cycles) unless a very low stress level is used. In some cases these specimens give information about the right tail but give no information about the left tail (see Galambos (1987)). In fact, only fatigue data below a given threshold value u, say, have information about relevant design values. The central idea of the method to be described is to model only the relevant part of the Wöhler field, that is, the small-percentile curves.

To this end, we transform the random variable X by first truncating at a threshold value u and then translating the origin to u, thus getting a new random variable $Z = T_u(X)$, which is defined as $X - u$ given $X \leq u$, (see Figure 6.2). The random variable Z has the cdf

$$H(z; u) = P(Z \leq z) = P\left[X \leq u + z | X \leq u\right] = \frac{F(u + z)}{F(u)}, \quad x_0 - u \leq z \leq 0,$$
$$(6.32)$$

where $F(x)$ is the cdf of the fatigue lifetime X and x_0 is the lower endpoint of F.

If we assume that the distribution of the random variable Z belongs to a parametric family $\mathcal{H} = \{H(x; u); u \in S\}$, then this family cannot be chosen arbitrarily. Since truncating any random variable X at v is the same as truncating $T_u(X)$ at v, for any $v < u$, we have

$$T_v(X) = T_v(T_u(X)).$$
$$(6.33)$$

From (6.32) and (6.33), it follows that

$$H(z; v) = \frac{H(v + z; u)}{H(v; u)}.$$
$$(6.34)$$

This functional equation is the formal statement of compatibility of the family $H(x; u)$ with respect to the above transformation.

It follows from (6.32) that

$$F(x) = F(u)H(x - u; u), \quad x < u. \tag{6.35}$$

Thus, to estimate $F(x)$ for $x < u$, we need to estimate both $F(u)$ and $H(x - u; u)$. In practice, different values of u should be tried out and $F(u)$ can then be estimated by the ratio of the sample data below u to the sample size. To estimate $H(x - u; u)$, we use the following result:

Theorem 6.1 *The RGPD*

$$G(z; \sigma, \alpha) = \begin{cases} \left(1 + \dfrac{\alpha z}{\sigma}\right)^{\frac{1}{\alpha}}, & z > 0, & if \alpha < 0, \\[2mm] \dfrac{-\sigma}{\alpha} < z < 0, & if \alpha > 0, \\[2mm] \exp\left(\dfrac{z}{\sigma}\right), & z < 0, & if \alpha = 0, \end{cases} \tag{6.36}$$

where σ and α are the scale and shape parameters, respectively, is a good approximation of $H(z; u)$, in the sense that

$$\lim_{u \to x_0} \sup_{x_0 - u < z < 0} |H(z; u) - G(z; \sigma(u), \alpha)| = 0, \tag{6.37}$$

for some fixed α and functions $\sigma(u)$, if and only if F is in the minimal domain of attraction of one of the extreme value distributions.

See Pickands (1975) for the proof of the equivalent result for the maximal domain of attraction. Note that the RGPD family satisfies (6.34), that is, it is stable with respect to the transformation $T_u(X)$. Using Theorem 6.1, from (6.35) and (6.36), the proposed model for a given level is

$$F(x) = F(u)G(x - u; \sigma, \alpha), \quad x < u. \tag{6.38}$$

We note, however, that for a good approximation, $F(u)$ need not be small if the specimen length is large and can be close to 1 if it is very large. Thus, the cdf of X for a given stress level y can then be approximated by

$$F_X(x; y) \cong F_X(\theta(y); y)G(x - \theta(y); \sigma(y), \alpha(y)), \quad x < \theta(y), \tag{6.39}$$

where now the parameters α and σ and $u = \theta(y)$ depend on the stress level y. Similarly, the cdf of Y given a lifetime x can be approximated by

$$F_Y(y; x) \cong F_Y(\theta^{-1}(x); x)G\left(y - \theta^{-1}(x); \tau(x), \eta(x)\right), \quad y < \theta^{-1}(x). \tag{6.40}$$

Therefore, from (6.19), (6.39), (6.40), and (6.21) and assuming that (6.39) and (6.40) are identities, we have

$$G(x - \theta(y); \sigma(y), \alpha(y)) = G\left(y - \theta^{-1}(x); \tau(x), \eta(x)\right). \tag{6.41}$$

Then from (6.36), (6.41) becomes

$$H(x;y) = \left[1 - \frac{\alpha(y)}{\sigma(y)}\theta(y) + \frac{\alpha(y)}{\sigma(y)}x\right]^{\frac{1}{\alpha(y)}} = \left[1 - \frac{\eta(x)}{\tau(x)}\theta^{-1}(x) + \frac{\eta(x)}{\tau(x)}y\right]^{\frac{1}{\eta(x)}},$$
(6.42)

for $\alpha \neq 0$, or

$$H(x;y) = \frac{x - \theta(y)}{\sigma(y)} = \frac{y - \theta^{-1}(x)}{\tau(x)},$$
(6.43)

for $\alpha = 0$, where $\sigma(y)$ and $\tau(x)$ are the scale parameters and $\alpha(y)$ and $\eta(x)$ are the shape parameters, for the lifetime given the stress level and for the stress level given the lifetime, respectively. Thus, the compatibility condition has led to the functional equations in (6.42) and (6.43) which impose conditions on the functions $\alpha(y), \sigma(y), \eta(x)$, and $\tau(x)$. The solutions of these functional equations (cf. (5.28)) are given by the following theorem:

Theorem 6.2   *Let $A, B, C, D, \alpha, \kappa$ and ϵ be parameters, then:*

1. *The functional equation (6.42) has only two solutions which are given by*

$$H(x;y) = [Axy + Bx + Cy + D]^{1/\alpha}, \quad 0 \leq Axy + Bx + Cy + D \leq 1,$$
(6.44)

   and

$$H(x;y) = (Ax + B)^{\epsilon}(Cy + D)^{\epsilon/\kappa}, \quad 0 \leq (Ax + B)^{\epsilon}(Cy + D)^{\epsilon/\kappa} \leq 1.$$
(6.45)

2. *The functional equation (6.43) has only one solution which is given by*

$$H(x, y) = \exp\left[Axy + Bx + Cy + D\right], \quad -\infty \leq Axy + Bx + Cy + D \leq 0.$$
(6.46)

Note that in principle the range of α is $(-\infty, +\infty)$. However, we want $H(x, y)$ to be a distribution function for fixed y and for fixed x (Galambos (1987), pp. 53–54), so we must restrict α to be nonnegative. Note also that models (6.44)–(6.46) are derived based on only two assumptions, namely, the compatibility condition (6.19), and the adequacy of the RGPD as an approximation to $H(x; y)$ as given by Theorem 6.1.

We should mention here that $H(x; y)$ does not represent a joint distribution of the random variables X and Y, but it is interpreted as either the distribution of lifetime X for a given stress level y or the distribution of stress level Y associated with a lifetime x. However, since the lifetime cannot be fixed, the second is a nonobservable random variable (an engineer has direct control over the stress level but has no control over the lifetime).

6.7   Physical Interpretations of the Model Parameters

To interpret the parameters of $H(x; y)$, we first note that if $A \neq 0$, model (6.44) can be written as

$$H(x; y) = [\beta + \sigma(x - \delta)(y - \rho)]^{1/\alpha}, \tag{6.47}$$

where
$$\sigma = A, \quad \delta = -C/A, \quad \rho = -B/A, \quad \beta = D - BC/A. \tag{6.48}$$

Also, the parameters A and C in model (6.45) should be different from zero, otherwise $H(x; y)$ is not a distribution function of X given y or of Y given x. Additionally, the Wöhler curves in Figure 6.1 indicate that A, C, κ, and ϵ are positive, in which case model (6.45) can be written as

$$H(x; y) = \sigma(x - \delta)^{\epsilon}(y - \rho)^{\epsilon/\kappa}, \tag{6.49}$$

where $\sigma = A^{\kappa} C^{\epsilon/\kappa}$, $\delta = -B/A$, $\rho = -D/C$. This parametrization has a clear physical interpretation (see Figure 6.1):

- ρ is the endurance limit, that is, the stress level below which failure due to fatigue does not occur;

- δ is a minimum lifetime that can be guaranteed for all specimens;

- σ is a combined scale factor for lifetime and stress level;

- β is associated with the zero-percentile (if β is zero, the zero-percentile curve degenerates to two straight lines parallel to the x- and y-axes); and

- $1/\alpha, \epsilon$, and ϵ/κ are shape parameters.

An interesting special case of model (6.47) is obtained when $\rho = \delta = 0$. In this case there is neither endurance limit nor minimum positive lifetime. Another special case of (6.44) is obtained for $A = 0$, which leads to a set of parallel straight percentile lines.

The zero-percentile curve for model (6.47) is $\beta + \sigma(x - \delta)(y - \rho) = 0$, which implies that at the zero-percentile, the lifetime depends on the stress level. On the other hand, the zero-percentile curve for model (6.49) is $\sigma(x - \delta)^{\epsilon}(y - \rho)^{\epsilon/\kappa} = 0$, which implies that at the zero-percentile $x = \delta$ for all stress levels $y > \rho$. Although specimens having zero lifetime at any stress level are possible, fracture mechanic theory states that fatigue failure is due to the progressive enlargement of existing cracks along the specimen and is governed by the largest crack in the piece. The zero-percentile curve is associated with the behavior of the worst possible specimen. The lifetime of this specimen depends on the stress level because the larger the stress level the faster the progression of its largest crack. For this reason, we would argue that only model (6.47) admits a reasonable physical interpretation.

Several methods for estimation of the parameters of this model have been proposed by Castillo and Hadi (1995).

6.8   Bibliographic Notes

The examples in Sections 6.2, 6.3, 6.4, and 6.5 are taken from or modified from Arnold and Press (1989a). Most of the Wöhler field material in Section 6.6 is based on Castillo et al. (1985) and Castillo (1988).

Exercises

6.1 Discuss the existence of a joint pdf $f(x, y)$ whose conditional densities satisfy
$$f(x|y) \propto y e^{-yx}, \quad 0 < x < A,$$
and
$$f(y|x) \propto x e^{-xy}, \quad 0 < y < A,$$
making a distinction between A finite and A infinite.

6.2 Discuss the existence of bivariate distributions (X, Y) such that
$$X - Y|Y = y \sim N(\mu_1(y), \sigma_1^2(y))$$
and
$$Y - X|X = x \sim N(\mu_2(x), \sigma_2^2(x)).$$

6.3 If $X \sim N(\mu, 1)$ then, from a fiducial viewpoint, the distribution of μ given $X = x$ is $N(x, 1)$. Discuss the implied joint distribution of X and μ in such a setting.

6.4 Consider the uniform conditionals example discussed in Section 6.1 ((6.1) and (6.2)). A Markov Chain with state space $\mathbb{R}$ can be set up as in (1.23) using the conditional densities (6.1) and (6.2). Discuss the long-run behavior of this chain.

6.5 Discuss the Wöhler field for a gamma–gamma model (cf. Section 6.6).

7
Characterizations Involving Conditional Moments

7.1   Introduction

By now we have developed a repertoire of experience in characterizing distributions whose conditionals are required to belong to specified parametric families. Two kinds of unexpected results have been encountered. On the one hand, the class of conditionally specified joint densities might be surprisingly constrained. For example, exponential conditionals models turn out to be always negatively correlated. In some sense, then, specifying the form of the conditional distributions is more restrictive than we might have envisioned. On the other hand, the conditionally specified families often include unexpected models with anomalous properties; for example, the nonclassical normal conditionals models with their unusual rational regression functions (ratios of quadratic functions). Many of the conditionally specified models have high-dimensional parameter spaces. Interesting subfamilies are frequently obtainable by invoking additional conditions on the form of the regression functions or other conditional moments.

One example of the phenomenon in question was provided in Section 3.4. There certain restrictions on the conditional variances and/or regression functions of a normal conditionals distribution were shown to characterize the classical bivariate normal distribution (specifically, see Theorem 3.1). Several other analogous results are described in the following sections.

Instead of being given both families of conditional densities we might be given one family of conditional densities and the other regression function. Interesting characterization problems arise in this context. Attempts to characterize distributions only using conditional moments will also be briefly described.

7.2   Mardia's Bivariate Pareto Distribution

As a bivariate extension of the usual Pareto distribution, Mardia (1962) proposed the family of densities which may be reparametrized in the following form:

$$f_{X,Y}(x,y) = \frac{\alpha(\alpha+1)}{\sigma_1\sigma_2}\left(1 + \frac{x}{\sigma_1} + \frac{y}{\sigma_2}\right)^{-(\alpha+2)} , \quad x > 0, \ y > 0. \qquad (7.1)$$

Such a joint density has Pareto marginals and Pareto conditionals. Using the notation introduced following equation (5.1) we may write

$$\begin{aligned} X &\sim P(\sigma_1, \alpha), \\ Y &\sim P(\sigma_2, \alpha), \end{aligned} \qquad (7.2)$$

and

$$X|Y = y \sim P\left(\sigma_1\left(1 + \frac{y}{\sigma_2}\right), \alpha + 1\right),$$

$$\qquad (7.3)$$

$$Y|X = x \sim P\left(\sigma_2\left(1 + \frac{x}{\sigma_1}\right), \alpha + 1\right).$$

From (7.3) we may compute the corresponding regression functions

$$E(X|Y = y) = \frac{\sigma_1}{\alpha}\left(1 + \frac{y}{\sigma_2}\right) \qquad (7.4)$$

and

$$E(Y|X = x) = \frac{\sigma_2}{\alpha}\left(1 + \frac{x}{\sigma_1}\right). \qquad (7.5)$$

Thus the regressions are linear.

Since $\alpha > 0$ in (7.1), the conditional densities in (7.3) are Pareto with index $\alpha+1$, greater than 1. We may ask whether Pareto conditionals and linear regression functions are enough to characterize the Mardia distribution. Specifically, we ask if

$$X|Y = y \sim P(\sigma_1(y), \alpha + 1) \qquad (7.6)$$

and

$$Y|X = x \sim P(\sigma_2(x), \alpha + 1) \qquad (7.7)$$

and linear regression functions imply that the joint density must be given by (7.1). An affirmative answer is called for. From the discussion in Section

5.2, Pareto conditionals such as (7.6) and (7.7) are only associated with a joint density of the form

$$f_{X,Y}(x, y) \propto (\lambda_0 + \lambda_1 x + \lambda_2 y + \lambda_3 xy)^{-(\alpha+2)}, \tag{7.8}$$

where $\lambda_0 > 0, \lambda_1 > 0, \lambda_2 > 0$, and $\lambda_3 \geq 0$. From (7.8) we conclude that

$$E(X|Y = y) = \frac{1}{\alpha} \left(\frac{\lambda_0 + \lambda_2 y}{\lambda_1 + \lambda_3 y} \right). \tag{7.9}$$

This could be constant (when $\lambda_0/\lambda_1 = \lambda_2/\lambda_3$), which implies X and Y are independent Pareto variables. It will be a non-constant linear function iff $\lambda_3 = 0$ and in this case (7.8) can be readily reparametrized to the form (7.1). Technically we only need to assume Pareto conditionals and <u>one</u> nonconstant linear regression function to conclude that the model is of the Mardia type.

A second characterization of the Mardia distribution is possible. Suppose that (7.3) holds, i.e., Pareto conditionals, and (7.2) holds, i.e., Pareto marginals. Then, again, we may claim that the distribution is of the Mardia type. The argument is as follows. Pareto conditionals implies a joint density of the form (7.8) with marginals given by (5.8). These marginal densities will only be Pareto densities if either $\lambda_3 = 0$ or $\lambda_0/\lambda_1 = \lambda_2/\lambda_3$. The latter yields independent Pareto marginals while the former ($\lambda_3 = 0$) guarantees that the joint density is of the Mardia form. Again, we actually need only assume that one marginal density is Pareto and that we have Pareto conditionals, to draw the conclusion.

A slight generalization of the above observations involves the Pearson type VI or Beta 2 distribution discussed in Section 5.3. By referring to the regression formulas (5.20) and (5.21) we may characterize all distributions with Beta2(p, q) conditionals and regression functions which are polynomials of degree q. They correspond to the choice $\lambda_3 = 0$ in (5.17). We remark in passing that the bivariate Beta 2 density, (5.17), has Beta 2 marginals and conditionals.

7.3   Linear Regressions with Conditionals in Exponential Families

If we have normal conditionals and impose a condition that one regression function is linear and nonconstant, then we characterize the classical bivariate normal density. What happens to other conditionals in exponential families distributions when we add a nonconstant linear regression condition?

If we consider the exponential conditionals family (Section 4.4) we see that the regression function can never be a nonconstant linear function.

Turning to the gamma conditionals family, whose regression functions are given in (4.43) and (4.44), we see that we cannot have either regression function be a nonconstant linear function. As noted in Section 4.6, constant regression functions are indeed possible (in both dependent and independent cases). A similar analysis can be performed for the gamma–normal conditionals model (Section 4.9). Here too it is not possible to have either regression function be a nonconstant linear function. The same observation is correct for the beta conditionals model. We had, in fact, no hope for nonconstant linear regressions in the beta conditionals density. We can only hope to encounter nonconstant linear regressions if, in (4.1), $q_{1i}(x) = x$ for some i or, in (4.2), $q_{2j}(y) = y$ for some j.

If instead of requiring that the conditionals belong to exponential families, we require the marginals to be in exponential families, it is not difficult to construct examples with nonconstant linear regressions.

7.4   Linear Regressions with Conditionals in Location Families

It is sometimes difficult to characterize distributions whose conditionals belong to given location families. Particular cases (the normal and Cauchy, for example) are tractable, others are not. What happens if we put conditions on the regression function? Apparently Narumi (1923) was the first to study this question. He considered all possible bivariate distributions with given regression functions $E(X|Y = y) = a(y)$ and $E(Y|X = x) = b(x)$ with conditionals belonging to unspecified location families. Thus the conditional densities were required to be of the form

$$f_{X|Y}(x|y) = g_1(x - a(y)) \tag{7.10}$$

and

$$f_{X|Y}(y|x) = g_2(y - b(x)) \tag{7.11}$$

(note that (7.10) and (7.11) implicitly include the assumption that conditional variances are constant, a stringent condition). For certain choices of the functions $a(y)$ and $b(x)$, it is possible to determine the nature of the joint distribution associated with (7.10) and (7.11). For example, it is natural to inquire about the case in which $a(y)$ and $b(x)$ are linear. In that case, we will have

$$f_Y(y)g_1(x - a_1y - a_2) = f_X(x)g_2(y - b_1x - b_2). \tag{7.12}$$

Narumi solves (7.12) by taking logarithms of both sides and differentiating, assuming the existence of derivatives up to the third order. He was thus able to conclude that either X and Y were independent, a trivial solution ($a_1 = b_1 = 0$), or $\log g_1$ must be quadratic and eventually (X, Y) must have a classical bivariate normal distribution. Instead of differentiating, one could difference. A third approach involves rewriting (7.12) in terms of functions

$$\tilde{g}_1(u) = g_1(u - a_2)$$

and

$$\tilde{g}_2(u) = g_2(u - b_2).$$

Equation (7.12) then becomes

$$f_Y(y)\tilde{g}_1(x - a_1 y) = f_X(x)\tilde{g}_2(y - b_1 x). \tag{7.13}$$

Lajko (1980) has shown that (7.13) can only hold (for $a_1 \neq 0$ and $b_1 \neq 0$) if the logarithms of all four functions involved are quadratic. Again we are led either to independent marginals or a classical bivariate normal.

More generally, if $a(y)$ and $b(x)$ in (7.10) and (7.11) are known functions we can write the joint density $f_{X,Y}(x, y)$ as a product of marginal and conditional densities in two ways (as we are accustomed to do). Then by considering $\dfrac{\partial^2 \log f_{X,Y}(x, y)}{\partial x \partial y}$ we arrive at Narumi's general functional equation (his equation (iii), with modified notation)

$$a'(y)\psi_1''(x - a(y)) = b'(x)\psi_2''(y - b(x)), \tag{7.14}$$

in which we have defined

$$\psi_i(u) = \log g_i(u), \quad i = 1, 2. \tag{7.15}$$

We were able to solve (7.14) when $a(y)$ and $b(x)$ were linear. What about more general choices of $a(y)$ and $b(x)$? It turns out (Narumi (1923)) that (7.14) actually is sufficiently restrictive to determine the nature of all four functions a, b, ψ_1, and ψ_2. For example, we must have

$$a'(y) = \frac{cm\left(\coth^2 \dfrac{\sqrt{c}\,(y - y_0)}{2} - 1\right)}{\left(\sqrt{c}\,\coth \dfrac{\sqrt{c}\,(y - y_0)}{2} + n\right)^2 - m^2 a} \tag{7.16}$$

for certain choices of the parameters a, c, y_0, m, and n. An analogous expression must hold for $b'(x)$. Thus only very specialized regression functions $a(y)$ and $b(x)$ can appear in (7.10) and (7.11). The following general form of all densities satisfying (7.10) and (7.11) is provided by Narumi, ((1923), p. 214):

$$
\begin{aligned}
f(x, y) \propto e^{\gamma x + \delta y} \Bigg\{ &\left(\cosh\frac{\sqrt{a}(x - x_0)}{2} - \lambda_1 \sinh\frac{\sqrt{a}(x - x_0)}{2}\right) \\
\times &\left(\cosh\frac{\sqrt{c}(y - y_0)}{2} - \lambda_2 \sinh\frac{\sqrt{c}(y - y_0)}{2}\right) \\
&+ \lambda_3 \sinh\frac{\sqrt{a}(x - x_0)}{2} \sinh\frac{\sqrt{c}(y - y_0)}{2} \Bigg\}^{-1}.
\end{aligned}
\tag{7.17}
$$

It includes the classical bivariate normal as a (very) special case and includes a bivariate Makeham distribution.

7.5   Specified Regressions with Conditionals in Scale Families

Suppose instead of (7.10) and (7.11) we require that the conditional densities satisfy

$$f_{X|Y}(x|y) = g_1(x/c(y))/c(y), \tag{7.18}$$

$$f_{Y|X}(y|x) = g_2(y/d(x))/d(x), \tag{7.19}$$

for given positive functions $c(y)$ and $d(x)$. What can be said about the joint density $f_{X,Y}(x,y)$?

Based on the analysis of Narumi for location families, we can expect that (7.18) and (7.19) will only be compatible for very special choices of $c(y)$ and $d(x)$. To begin let us consider particular choices for $c(y)$ and $d(x)$ for which we know there is at least one solution. Recall that the exponential conditionals distribution (1.39) had all its conditionals in scale families and had regressions which were of the form

$$E(Y|X = x) = (c_{21} + c_{22}x)^{-1},$$

$$E(X|Y = y) = (c_{12} + c_{22}y)^{-1}.$$

It is then natural to inquire about the nature of all joint densities whose conditional densities satisfy

$$f_{X|Y}(x|y) = g_1((\alpha + y)x)(\alpha + y), \tag{7.20}$$

$$f_{Y|X}(y|x) = g_2((\beta + x)y)(\beta + x). \tag{7.21}$$

It is reasonable to restrict our search to random variables (X, Y) with support in the positive quadrant. Thus we ask for what functions g_1 and g_2 can we have (7.20) and (7.21) holding for $x > 0$ and $y > 0$? Let $\psi_i(u) = \log g_i(u)$, then we can write $\log f_{X,Y}(x,y)$ in two ways and obtain the relationship

$$\log[(\alpha + y)f_Y(y)] + \psi_1((\alpha + y)x) = \log[(\beta + x)f_X(x)] + \psi_2((\beta + x)y).$$

Now differentiate with respect to x and y to obtain

$$[(\alpha+y)x]\psi_1''((\alpha+y)x)+\psi_1'((\alpha+y)x) = [(\beta+x)y]\psi_2''((\beta+x)y)+\psi_2'((\beta+x)y). \tag{7.22}$$

But this can hold for every x, y if and only if

$$u\psi_1''(u) + \psi_1'(u) = \gamma = v\psi_2''(v) + \psi_2'(v) \tag{7.23}$$

for some constant γ. The differential equation for $\psi_1(u)$ implicit in (7.23) has as its general solution

$$\psi_1(u) = c_0 + c_1 u + c_2 \log u$$

so that

$$g_1(u) = u^{c_2} e^{c_0 + c_1 u}. \tag{7.24}$$

Analogously $g_2(v)$ is a gamma density. Thus (X, Y) has gamma conditionals. Referring back to Section 4.6, we conclude that the class of all solutions to (7.20) and (7.21) coincides with the MODEL II gamma conditionals class (described by (4.33) with parametric constraints (4.38)).

It is possible to derive a general solution to (7.18)–(7.19) defined on the positive quadrant analogous to Narumi's general solution to the system (7.10)–(7.11). This is true because the functional equation associated with the system (7.18)–(7.19), namely

$$g_1(x/c(y)) f_Y(y)/c(y) = g_2(y/d(x)) f_X(x)/d(x), \quad x > 0, \quad y > 0, \tag{7.25}$$

is equivalent to the functional equation

$$\tilde{g}_1(u - a(v)) \tilde{f}_Y(v) = \tilde{g}_2(v - b(u)) \tilde{f}_X(u), \quad -\infty < u < \infty, \quad -\infty < v < \infty, \tag{7.26}$$

where

$$\tilde{g}_1(u) = g_1(e^u), \qquad\qquad \tilde{g}_2(v) = g_2(e^v),$$
$$\tilde{f}_X(u) = f_X(e^u) e^{-b(u)}, \quad \tilde{f}_Y(v) = f_Y(e^v) e^{-a(v)},$$

and $u = \log x, v = \log y$.

Since (7.26) is identical in form to the functional equation associated with the system (7.10)–(7.11), a solution to (7.18)–(7.19) over the positive quadrant can be obtained by suitable substitution in (7.17).

7.6   Conditionals in Location-Scale Families with Specified Moments

We now seek joint densities whose conditional densities satisfy

$$f_{X|Y}(x|y) = g_1 \left(\frac{x - a(y)}{c(y)} \right) \frac{1}{c(y)}, \tag{7.27}$$

$$f_{Y|X}(y|x) = g_2 \left(\frac{y - b(x)}{d(x)} \right) \frac{1}{d(x)}, \tag{7.28}$$

for given functions $a(y)$ and $b(x)$ and given positive functions $c(y)$ and $d(x)$. In principle, we could search for the general solution to (7.27)–(7.28) since from our discussion in Sections 7.3 and 7.4 (based on Narumi's work), we can expect solutions to (7.27)–(7.28) for only a few choices of $a(y), b(x), c(y)$, and $d(x)$. Rather than attempt to determine the potentially complicated nature of such a general solution, we will content ourselves with reporting the solutions corresponding to certain tractable special choices for the conditional means and standard deviations.

Case (i): Linear regressions and conditional standard deviations.

Here we assume that (7.27) and (7.28) hold with

$$
\begin{aligned}
a(y) &= a_0 + a_1 y, \\
b(x) &= b_0 + b_1 x, \\
c(y) &= 1 + cy, \\
d(x) &= 1 + dx.
\end{aligned}
\tag{7.29}
$$

We assume that our random variables are non-negative so $x > 0$ and $y > 0$ in (7.27)–(7.29). In this case, Narumi (1923) shows that the joint density must be of the form

$$
f_{X,Y}(x, y) = (\alpha + x)^{p_1}(\beta + y)^{p_2}(\gamma + \delta_1 x + \delta_2 y)^q.
\tag{7.30}
$$

This, except for a location shift, can be identified with the Beta-2 conditionals densities discussed in Section 5.3 with the restriction that $\lambda_3 = 0$ in (5.17). Perusal of the conditional moments in (5.20) and (5.21), with $\lambda_3 = 0$, will confirm that (7.29) does hold for such densities. Note that if the support of (X, Y) is not restricted to be the positive orthant, (7.30) also includes Dirichlet distributions and other unnamed distributions.

Case (ii): Suppose that we have linear regressions and quadratic conditional variances.

Thus we assume

$$
\begin{aligned}
a(y) &= a_0 + a_1 y, \\
b(x) &= b_0 + b_1 x, \\
c(y) &= \sqrt{1 + c_1 y + c_2 y^2}, \\
d(x) &= \sqrt{1 + d_1 x + d_2 x^2}.
\end{aligned}
\tag{7.31}
$$

In this case Narumi reports that the joint density is necessarily of the form

$$
f_{X,Y}(x, y) = \left[\alpha + \beta x + \gamma y + \delta_1 x^2 + \delta_2 xy + \delta_3 y^2\right]^{-\gamma}.
\tag{7.32}
$$

This includes the nonintegrable Cauchy conditionals distribution (5.65) (when $\gamma = 1$) together with bivariate t distributions (when 2γ is an integer). It should be observed that for certain choices of γ in (7.32) conditional variances do not exist and, technically (following Narumi (1923) and Mardia (1970)), we should call $c(y)$ and $d(x)$ scedastic curves instead of conditional standard deviations.

7.7   Given One Family of Conditional Distributions and the Other Regression Function

We know that giving both families of conditional densities, of X given Y and of Y given X, is more than enough to characterize the joint density of (X, Y). It is also evident that being given both regression functions $E(X|Y = y)$ and $E(Y|X = x)$ will usually be inadequate for determining the joint density. An intermediate case would involve one family of conditional densities, say of X given Y, and the other regression function $E(Y|X = x)$.

Thus, we assume that we are given:

(i) a family of putative conditional densities

$$f_{X|Y}(x|y) = a(x, y), \quad x \in S(X), \quad y \in S(Y), \tag{7.33}$$

and

(ii) a regression function

$$E(Y|X = x) = \psi(x), \quad x \in S(X). \tag{7.34}$$

The following questions arise naturally:

(A) Are $a(x, y)$ and $\psi(x)$ compatible in the sense that there will exist a joint density function $f_{X,Y}(x, y)$ with $a(x, y)$ as its corresponding family of conditional densities and with $\psi(x)$ as its regression function of Y on X?

(B) Suppose $a(x, y)$ and $\psi(x)$ are compatible, under what conditions do they determine a unique joint density?

(C) Given $a(x, y)$, identify the class of all compatible functions ψ.

(D) Suppose that $a(x, y)$ is specified to be such that, for each $y, a(x, y)$ belongs to a given parametric family of densities, i.e., $a(x, y) = f^*(x; \underline{\theta}(y))$, and also that ψ belongs to a given parametric family of functions (e.g., linear), i.e., $\psi(x) = g^*(x; \underline{\lambda})$. Identify the class of bivariate distributions so determined.

The resolution, partial or complete, of these questions will closely parallel developments in Chapter 1, where both families of conditionals densities were assumed to be given. As in Chapter 1, there are pedagogical advantages to be gained by separating the discrete and continuous cases.

7.7.1   The Finite Discrete Case

Suppose that $S(X) = \{x_1, x_2, \ldots, x_I\}$ and $S(Y) = \{y_1, y_2, \ldots, y_J\}$ where, without loss of generality, $x_1 < x_2 < \ldots < x_I$ and $y_1 < y_2 < \ldots < y_J$. Consider any family of conditional densities

$$a_{ij} = P(X = x_i | Y = y_j) \tag{7.35}$$

and consider a candidate regression function

$$\psi_i = E(Y | X = x_i), \quad i = 1, 2, \ldots, I. \tag{7.36}$$

For $\underline{\psi}$ and A to be compatible we certainly require that $y_1 \leq \psi_i \leq y_I, \forall i = 1, 2, \ldots, I$. If there exists an appropriate marginal density for Y which will make A and $\underline{\psi}$ compatible, it, denoted by $\underline{\eta}$, will satisfy the following system of equations in restricted (nonnegative) variables:

$$\psi_i = \sum_{j=1}^{J} y_j P(Y = y_j | X = x_i) = \sum_{j=1}^{J} y_j \frac{a_{ij} \eta_j}{\sum\limits_{j=1}^{J} a_{ij} \eta_j}, \quad i = 1, \ldots, I,$$

$$\sum_{j=1}^{J} \eta_j = 1,$$

$$\eta_j \geq 0, \quad \forall j = 1, \ldots J. \tag{7.37}$$

The system (7.37) may be rewritten in the form

$$\sum_{j=1}^{J} (\psi_i - y_j) a_{ij} \eta_j = 0, \quad \forall i = 1, \ldots, I,$$

$$\sum_{j=1}^{J} \eta_j = 1,$$

$$\eta_j \geq 0, \quad \forall j = 1, \ldots J. \tag{7.38}$$

Rewritten in this form, we have a system of linear equations and we seek nonnegative solutions. Identification of the existence of a solution, or better yet identification of all possible solutions, will be the order of the day. This is, in essence, the same problem as the one we faced in Section 2.2.2 when we sought a vector $\underline{\tau}$ to make A and B compatible (cf. (2.23)).

Even if $y_1 \leq \psi_i \leq y_J, \forall i$, it is possible for the system (7.38) to fail to have a solution. When a solution exists it can be found by solving a reduced system obtained by deleting the Ith equation in the first line of (7.37). But of course this solution is not guaranteed to satisfy the deleted equation. We can illustrate this phenomenon in the simplest possible case, i.e., when $I = J = 2$.

Theorem 7.1 (Compatibility when I=J=2.) *The system of equations and inequalities obtained from (7.38) after deleting the equation associated with $i = I$, when $I = J = 2$, always has a unique solution provided that $y_1 \leq \psi_1 \leq y_2$. The pair $(\underline{\psi}, A)$ will be compatible if this solution satisfies the deleted Ith equation.* □

Proof. Note $y_1 < y_2$. Our "deleted" systems become

$$(\psi_1 - y_1)a_{11}\eta_1 + (\psi_1 - y_2)a_{12}\eta_2 = 0,$$
$$\eta_1 \qquad\qquad + \eta_2 = 1,$$
$$\eta_1 \qquad\qquad\qquad \geq 0, \qquad (7.39)$$
$$\eta_2 \geq 0.$$

The solution of the system of the first two equations is

$$\eta_1 = \frac{a_{12}(y_2 - \psi_1)}{a_{11}(\psi_1 - y_1) + a_{12}(y_2 - \psi_1)},$$
$$\eta_2 = \frac{a_{11}(\psi_1 - y_1)}{a_{11}(\psi_1 - y_1) + a_{12}(y_2 - \psi_1)}, \qquad (7.40)$$

which, according to the last two inequalities in (7.39), must be nonnegative. However, this implies that either one of the following two conditions must hold:

(I) $a_{12}(y_2-\psi_1) \geq 0, a_{11}(\psi_1-y_1) \geq 0$, and $a_{11}(\psi_1-y_1)+a_{12}(y_2-\psi_1) \geq 0$;

(II) $a_{12}(y_2-\psi_1) \leq 0, a_{11}(\psi_1-y_1) \leq 0$, and $a_{11}(\psi_1-y_1)+a_{12}(y_2-\psi_1) \leq 0$;

which, due to the non-negative character of the a_{ij}'s, lead to the equivalent conditions:

(Ia) $y_1 \leq \psi_1 \leq y_2$ and $a_{11}(\psi_1 - y_1) + a_{12}(y_2 - \psi_1) \geq 0$;

(IIa) $y_2 \leq \psi_1 \leq y_1$ and $a_{11}(\psi_1 - y_1) + a_{12}(y_2 - \psi_1) \leq 0$.

Case (Ia) always holds if $y_1 \leq \psi_1 \leq y_2$, and case (IIa) cannot occur, since $y_2 > y_1$. $\qquad\square$

If this solution does not satisfy the deleted equation, then the original full system has no solution. This occurs when $\psi_2 \neq y_1\eta_1 + y_2\eta_2$, where η_1, η_2 are as in (7.40).

It is not unusual to encounter situations in which the matrix A and partial information about ψ (namely, $\psi_1, \psi_2, \ldots, \psi_{I-1}$) will uniquely determine a compatible distribution. When $I = 2$ this is always so. When $I > 2$, it is not necessarily true, as the following examples show, even when $y_1 \leq \psi_i \leq y_J, \forall i = 1, 2, \ldots, I - 1$.

Example 7.1 (Incompatible case). Consider the case in which $I = J = 3$ and $(y_1, y_2, y_3) = (1, 2, 3)$, the following conditional probability matrix

$$A = \begin{pmatrix} 1/3 & 1/4 & 1/5 \\ 2/3 & 1/2 & 3/5 \\ 0 & 1/4 & 1/5 \end{pmatrix}, \qquad (7.41)$$

and the regression function

$$\psi = E(Y|X) = (\psi_1, \psi_2, \psi_3) = (3/2, 6/5, \psi_3). \qquad (7.42)$$

Note that $1 = y_1 \le \psi_i \le y_3 = 3$ as required.

The system (7.38) without the equation corresponding to $i = I = 3$ for this case becomes

$$
\begin{aligned}
1/6\eta_1 \; -1/8\eta_2 \; -3/10\eta_3 &= 0, \\
2/15\eta_1 \; -2/5\eta_2 \; -27/25\eta_3 &= 0, \\
\eta_1 \quad + \eta_2 \quad\quad + \eta_3 &= 1, \\
\eta_1 \quad\quad\quad\quad\quad &\ge 0, \\
\eta_2 \quad\quad &\ge 0, \\
\eta_3 &\ge 0,
\end{aligned} \tag{7.43}
$$

which is incompatible. In fact, the first three equations have as a unique solution $(\eta_1, \eta_2, \eta_3) = (1/7, 4/3, -10/21)$, which does not satisfy the last nonnegativity constraint in (7.43). □

Example 7.2 (Compatible case). Consider the case in which $I = J = 3$ and $(y_1, y_2, y_3) = (1, 2, 3)$, the following conditional probability matrix

$$
A = \begin{pmatrix} 1/3 & 1/4 & 1/5 \\ 2/3 & 1/2 & 3/5 \\ 0 & 1/4 & 1/5 \end{pmatrix}, \tag{7.44}
$$

and the regression values

$$
\psi = E(Y|X) = (\psi_1, \psi_2, \psi_3) = (86/47, 104/53, \psi_3). \tag{7.45}
$$

Note that $1 = y_1 \le \psi_i \le y_3 = 3$ as required.

The system (7.38), without the equation corresponding to $i = I = 3$, for this case becomes

$$
\begin{aligned}
13/47\eta_1 \; -2/47\eta_2 \; -11/47\eta_3 &= 0, \\
34/53\eta_1 \; -1/53\eta_2 \; -33/53\eta_3 &= 0, \\
\eta_1 \quad + \eta_2 \quad\quad + \eta_3 &= 1, \\
\eta_1 \quad\quad\quad\quad\quad &\ge 0, \\
\eta_2 \quad\quad &\ge 0, \\
\eta_3 &\ge 0.
\end{aligned} \tag{7.46}
$$

The system of inequalities is compatible and has as unique solution $(\eta_1, \eta_2, \eta_3) = (1/3, 1/3, 1/3)$. This leads to the following joint probability matrix:

$$
P = \begin{pmatrix} 1/9 & 1/12 & 1/15 \\ 2/9 & 1/6 & 3/15 \\ 0 & 1/12 & 1/15 \end{pmatrix}. \tag{7.47}
$$

It can be verified that the conditional probability matrix of X given Y corresponding to this P agrees with (7.44). From P, the other conditional probability matrix is

$$
B = P_{Y|X} = \begin{pmatrix} 20/47 & 15/47 & 12/47 \\ 20/53 & 15/53 & 18/53 \\ 0 & 15/27 & 12/27 \end{pmatrix}, \tag{7.48}
$$

and thus the regression of Y on X becomes

$$\underline{\psi} = (\psi_1, \psi_2, \psi_3) = (86/47, 104/53, 66/27), \tag{7.49}$$

that is, compatible with (7.45).   □

In general we can use Theorem 2.2 to identify the possible solutions to the system (7.38) and we can then determine whether compatibility and whether uniqueness obtains.

In direct analogy to our discussion in Chapter 2, on compatibility of matrices A and B, we can address the concept of "almost" compatibility of A and ψ (the conditional distributions of X given Y and the regression of Y on $\overline{X}$). Thus instead of seeking η so that (7.38) holds exactly we might, for a given $\epsilon > 0$ and given weights $\underline{w_{ik}}, i = 1, 2, \ldots, I, k = 1, 2$, seek $\underline{\eta}$ to satisfy the following system of inequalities:

$$\sum_{j=1}^{J} (\psi_i - y_j) a_{ij} \eta_j \leq \epsilon w_{i1}, \quad \forall i = 1, \ldots, I,$$

$$\sum_{j=1}^{J} (\psi_i - y_j) a_{ij} \eta_j \geq \epsilon w_{i2}, \quad \forall i = 1, \ldots, I, \tag{7.50}$$

$$\sum_{j=1}^{J} \eta_j = 1,$$

$$\eta_j \geq 0, \quad \forall j = 1, \ldots J,$$

where $w_{i1}, w_{i2}, i = 1, 2, \ldots, I$, are given values to reflect the relative importance of the errors in the equations in the system (7.38).

Example 7.3 (Compatible case). Consider again the case $I = J = 3$ with $(y_1, y_2, y_3) = (1, 2, 3)$, the following conditional probability matrix

$$A = \begin{pmatrix} 1/3 & 1/4 & 1/5 \\ 2/3 & 1/2 & 3/5 \\ 0 & 1/4 & 1/5 \end{pmatrix}, \tag{7.51}$$

and the regression values

$$\psi = E(Y|X) = (\psi_1, \psi_2, \psi_3) = (86/47, 104/53, \psi_3). \tag{7.52}$$

The system (7.50), without the equation corresponding to $i = I = 3$, for this case becomes using $w_{ik} = 1, \forall i, k$,

$$\begin{aligned}
+13/47\eta_1 - 2/47\eta_2 - 11/47\eta_3 - \epsilon &\leq 0, \\
-13/47\eta_1 + 2/47\eta_2 + 11/47\eta_3 - \epsilon &\leq 0, \\
+34/53\eta_1 - 1/53\eta_2 - 33/53\eta_3 - \epsilon &\leq 0, \\
-34/53\eta_1 + 1/53\eta_2 + 33/53\eta_3 - \epsilon &\leq 0, \\
\eta_1 + \eta_2 + \eta_3 &= 1, \\
-\eta_1 &\leq 0, \\
-\eta_2 &\leq 0, \\
-\eta_3 &\leq 0.
\end{aligned} \tag{7.53}$$

The general solution of the system of inequalities (7.53) using the techniques developed by Castillo, Cobo, Jubete, and Pruneda is

$$
\begin{pmatrix} \eta_1 \\ \eta_2 \\ \eta_3 \\ \epsilon \end{pmatrix} = \begin{pmatrix} 0 & 0 & \frac{1}{3} & 0 & 1 & \frac{968}{1877} & \frac{2134}{4421} & 0 & \frac{153}{2440} \\ 0 & \frac{968}{1027} & \frac{1}{3} & 0 & 0 & 0 & 0 & 1 & \frac{2287}{2440} \\ 0 & \frac{59}{1027} & \frac{1}{3} & 1 & 0 & \frac{909}{1877} & \frac{2287}{4421} & 0 & 0 \\ 1 & \frac{55}{1027} & 0 & \frac{33}{53} & \frac{34}{53} & \frac{55}{1877} & \frac{55}{4421} & \frac{2}{47} & \frac{11}{488} \end{pmatrix} \begin{pmatrix} \pi_1 \\ \lambda_1 \\ \lambda_2 \\ \lambda_3 \\ \lambda_4 \\ \lambda_5 \\ \lambda_6 \\ \lambda_7 \\ \lambda_8 \end{pmatrix}, \qquad (7.54)
$$

where $\pi_1 > 0$ and $\lambda_i \geq 0$, $\sum_{i=1}^{8} \lambda_i = 1$.

The minimum value of ϵ leading to a solution of the system (7.50) is $\epsilon = 0$, which corresponds to $\lambda_2 = 1$, $\lambda_i = 0$, $i \neq 2$. For these values of the λ's we get $(\eta_1, \eta_2, \eta_3) = (1/3, 1/3, 1/3)$, which is the compatible solution obtained in Example 7.2.

Note that the π_1 component of ϵ guarantees that a solution exist for large nonnegative values of ϵ. □

Example 7.4 (Incompatible case). Consider again the case $I = J = 3$ with $(y_1, y_2, y_3) = (1, 2, 3)$, the following conditional probability matrix

$$
A = \begin{pmatrix} 1/3 & 1/4 & 1/5 \\ 2/3 & 1/2 & 3/5 \\ 0 & 1/4 & 1/5 \end{pmatrix}, \qquad (7.55)
$$

and the regression function

$$
\psi = E(Y|X) = (\psi_1, \psi_2, \psi_3) = (3/2, 6/5, \psi_3). \qquad (7.56)
$$

The system (7.50), without the equation corresponding to $I = 3$, for this case becomes (again with $w_{ik} = 1, \forall i, k$)

$$
\begin{array}{rrrr}
1/6\eta_1 - 1/8\eta_2 & - 3/10\eta_3 - \epsilon \leq 0, \\
- 1/6\eta_1 + 1/8\eta_2 & +3/10\eta_3 - \epsilon \leq 0, \\
2/15\eta_1 - 2/5\eta_2 & - 27/25\eta_3 - \epsilon \leq 0, \\
- 2/15\eta_1 + 2/5\eta_2 & +27/25\eta_3 - \epsilon \leq 0, \\
\eta_1 + \eta_2 & +\eta_3 = 1, \\
- \eta_1 & \leq 0, \\
- \eta_2 & \leq 0, \\
- \eta_3 & \leq 0.
\end{array} \qquad (7.57)
$$

The general solution of the system of inequalities (7.57) using the techniques developed by Castillo, Cobo, Jubete, and Pruneda is:

$$
\begin{pmatrix} \eta_1 \\ \eta_2 \\ \eta_3 \\ \epsilon \end{pmatrix} = \begin{pmatrix} 0 & 0 & 23/28 & 0 & 7/11 & 1 \\ 0 & 0 & 0 & 1 & 4/11 & 0 \\ 0 & 1 & 5/28 & 0 & 0 & 0 \\ 2400 & 27/25 & 1/12 & 2/5 & 2/33 & 1/6 \end{pmatrix} \begin{pmatrix} \pi_1 \\ \lambda_1 \\ \lambda_2 \\ \lambda_3 \\ \lambda_4 \\ \lambda_5 \end{pmatrix}, \quad (7.58)
$$

where $\pi_1 > 0$ and $\lambda_i \geq 0$, $\sum_{i=1}^{5} \lambda_i = 1$.

The minimum value of ϵ leading to a solution of the system (7.57) is $\epsilon^* = 2/33$, which corresponds to $\lambda_4 = 1$, $\lambda_i = 0$, $i \neq 4$. For these values of the λ's we get $(\eta_1, \eta_2, \eta_3) = (7/11, 4/11, 0)$.

Note that the π_1 component of ϵ guarantees that the system has solution for any value of $\epsilon \geq 2/33$.   □

Note that the above development was based on the idea of finding a Y-marginal in order to analyze the compatibility of the conditional probability (a_{ij}) and the regression function ψ_i, $i = 1, 2, \ldots, I$. Alternatively, we can seek a joint probability matrix P for the same purpose. With this option, the system of inequalities analogous to (7.50) becomes

$$
\begin{aligned}
& \sum_{j=1}^{J} (\psi_i - y_j) p_{ij} \leq \epsilon w_{i1}, \quad \forall i = 1, 2, \ldots, I, \\
& \sum_{j=1}^{J} (\psi_i - y_j) p_{ij} \geq \epsilon w_{i2}, \quad \forall i = 1, 2, \ldots, I, \\
& a_{ij} \sum_{i=1}^{I} p_{ij} - p_{ij} \leq \epsilon w_{ij}, \quad \forall i = 1, 2, \ldots, I, \; j = 1, 2, \ldots, J, \\
& \sum_{j=1}^{J} p_{ij} = 1, \\
& p_{ij} \geq 0, \quad \forall i = 1, 2, \ldots, I, \; j = 1, 2, \ldots J,
\end{aligned}
\qquad (7.59)
$$

where w_{ij} are given.

This allows us to introduce two versions of the concept of ϵ-compatibility.

Definition 7.1 (ϵ-compatibility). A conditional probability matrix (a_{ij}) and a regression function ψ_i, $i = 1, 2, \ldots, I$, are said to be ϵ_1-compatible iff the system (7.50) (or (7.59)) has solution for $\epsilon \geq \epsilon_1$ and not for $\epsilon < \epsilon_1$, i.e. iff ϵ_1 is the minimum value of ϵ that allows the system (7.50) (or (7.59)) to have a solution.   □

Remark 7.1 *The extremal value of ϵ in the definition above, and the corresponding solutions η (the Y-marginal probability matrix) and P (the joint probability matrix), can be obtained by linear programming techniques, minimizing the function ϵ subject to the constraints in (7.50) (or (7.59)), respectively.*

The above definition of ϵ-compatibility suggests the following inconsistency measures:

$$IM_1 = \sum_{i=1}^{I} \left| \sum_{j=1}^{J} (\psi_i - y_j) a_{ij} \eta_j^* \right|, \tag{7.60}$$

or

$$IM_2 = \sum_{i=1}^{I} \left| \sum_{j=1}^{J} (\psi_i - y_j) p_{ij}^* \right|, \tag{7.61}$$

where η_j^*, $j = 1, 2, \ldots, J$, is one of the solutions of the system (7.50), and p_{ij}^*, $i = 1, 2, \ldots, I$, $j = 1, 2, \ldots, J$, is one of the solutions of the system (7.59), for $\epsilon = \epsilon_1$.

Instead of being given a regression function ψ, we might be given a set of conditional percentiles such as

$$\{\alpha_{r,s} : \alpha_{r,s} = P(Y \le y_s | X = x_r); (r, s) \in S \subset I \times J\}.$$

We would then wish to know if these $\alpha_{r,s}$'s are compatible with the given conditional probability matrix A.

In this case, the system of inequalities analogous to (7.50) becomes

$$\begin{aligned}
\alpha_{r,s} \sum_{j=1}^{J} a_{rj} \eta_j - \sum_{j \le s} a_{rj} \eta_j &\le \epsilon w_{rs}, \quad \forall (r, s) \in S, \\
\alpha_{r,s} \sum_{j=1}^{J} a_{rj} \eta_j - \sum_{j \le s} a_{rj} \eta_j &\ge \epsilon w_{rs}, \quad \forall (r, s) \in S, \\
\sum_{j=1}^{J} \eta_j &= 1, \\
\eta_j \ge 0, \quad \forall j &= 1, \ldots J.
\end{aligned} \tag{7.62}$$

It then becomes obvious that we could define new versions of ϵ-compatibility and related inconsistency measures.

7.7.2   The Infinite Discrete Case

Suppose $S(X) = \{0, 1, 2, \ldots\}$ and $S(Y) = \{0, 1, 2, \ldots\}$ and we are given $f_{X|Y}(x|y) = a(x, y)$ and $E(Y|X = x) = \psi(x)$. Then the density of Y, say $g(y)$, will be the solution to

$$\sum_{y=0}^{\infty} [\psi(x) - y] a(x, y) g(y) = 0, \quad \forall x = 0, 1, 2, \ldots,$$

$$\sum_{y=0}^{\infty} g(y) = 1, \tag{7.63}$$

$$g(y) \ge 0, \quad \forall y = 0, 1, 2, \ldots.$$

For certain choices of $a(x, y)$, (e.g., binomial, negative binomial) and certain ψ, this can be solved. Identifiability of mixtures of $f(x|y)$ plays a role here as it does in the continuous case discussed next.

Example 7.5 (Binomial conditionals). Suppose $S(X) = \{0, 1, 2, \ldots\}$ and $S(Y) = \{0, 1, 2, \ldots\}$. Also assume that for any $y \in S(Y)$

$$a(x, y) = \binom{y}{x} p^x (1 - p)^{y-x}, \quad x = 0, 1, 2, \ldots, y, \tag{7.64}$$

for some $p \in (0, 1)$. Assume that $f_{X|Y}(x|y) = a(x, y)$, i.e. binomial conditionals. What regression functions $\psi(x) \ (= E(Y|X = x))$ are compatible with this choice for $a(x, y)$? It is not difficult to verify that the family $a(x, y)$ is identifiable.

For this example, (7.63) assumes the form

$$\sum_{y=x}^{\infty} (\psi(x) - y) \binom{y}{x} p^x (1 - p)^{y-x} g(y) = 0, \quad \forall x = 0, 1, 2, \ldots. \tag{7.65}$$

Suppose for the moment that $\psi(x)$ is compatible with $a(x, y)$ given in (7.64). We seek a density $g(y)$ to satisfy (7.65). Divide both sides of (7.65) by $(p/(1 - p))^x$ and define

$$h(y) = \frac{(1 - p)^y g(y)}{\displaystyle\sum_{y=0}^{\infty} (1 - p)^y g(y)}. \tag{7.66}$$

We can then write (7.65) as

$$\sum_{y=x}^{\infty} (\psi(x) - y) \binom{y}{x} h(y) = 0, \quad \forall x = 0, 1, 2, \ldots. \tag{7.67}$$

Note that $h(y)$ is a discrete density with all moments finite and knowledge of $h(y)$ will suffice to determine $g(y)$. So now we need to solve (7.67). Recall that we may write $\binom{y}{x} = (y)_x / x!$ and that the generating function corresponding to the density $h(y)$, say $P_h(s)$, can be defined as

$$P_h(s) = \sum_{k=0}^{\infty} \frac{E[(Z)_k]}{k!} s^k, \tag{7.68}$$

where Z has discrete density $h(y)$. Consequently, knowledge of the factorial moments of Z will suffice to determine h. However, (7.67), after multiplying by $x!$, can be written as

$$\psi(x) \sum_{y=x}^{\infty} (y)_x h(y) = \sum_{y=x}^{\infty} y(y)_x h(y)$$

$$= \sum_{y=x}^{\infty} [(y - x) + x](y)_x h(y).$$

Consequently,

$$\psi(x)E\left[(Z)_x\right] = E\left[(Z)_{x+1}\right] + xE\left[(Z)_x\right]$$

and thus for $x = 0, 1, 2, \ldots$

$$\frac{E\left[(Z)_{x+1}\right]}{E\left[(Z)_x\right]} = (\psi(x) - x). \qquad (7.69)$$

Evidently (7.69) can be solved iteratively to obtain the sequence

$$E\left[(Z)_x\right], \quad x = 1, 2, \ldots.$$

Now we are faced with the more difficult issue of determining which sequences $\psi(x)$ can be consistent with the given family of conditional densities (7.64). The class is not empty. For example, the choice $\psi(x) = x + c$ will yield a Poisson distribution for Z in (7.69) and eventually a Poisson($c/(1-p)$) distribution for Y (from (7.66)). Evidently, a compatible ψ can be constructed, using (7.69), by beginning with any random variable Z with support $0, 1, 2, \ldots$ whose moment generating function $E(e^{tZ})$ exists for $t = -\log(1-p)$. Then (7.66) can be used to obtain the density of Y from that of Z. This certainly provides a broad class of compatible functions ψ. It is not clear whether or not other compatible choices for ψ exist. Note that Korwar (1974) discussed the special case in which $\psi(x) = ax + b$. He also treated an analogous problem involving Pascal rather than binomial conditionals. □

Papageorgiou (1983) gives some other discrete examples analogous to that discussed in Example 7.5. He discusses examples in which $a(x, y)$ is hypergeometric or negative hypergeometric. For example, if

$$a(x, y) = \frac{\dbinom{n}{x}\dbinom{N-n}{y-x}}{\dbinom{N}{y}}, \quad 0 \le x \le y, \qquad (7.70)$$

and if $E(Y|X = x) = \psi(x)$ is consistent with (7.70), then he shows that $\psi(x)$ and $a(x, y)$ uniquely determine the distribution of (X, Y). In particular, if $\psi(x) = x + (N - n)p$, then Y is necessarily a binomial(N, p) random variable (and X is binomial(n, p)). He also discusses multivariate extensions of this result. Thus if $(\underline{X}, \underline{Y})$ is such that $\underline{X}|\underline{Y} = \underline{y}$ is multivariate hypergeometric (drawing balls from urns with balls of several colors), then consistent specification of $E(Y_i|\underline{X} = \underline{x}), i = 1, 2, \ldots, \ell$, will uniquely determine the joint distribution of $(\underline{X}, \underline{Y})$.

Solutions to (7.63) (or (7.38)), for consistent functions ψ, have appeared in this literature on a case by case basis. Wesolowski (1995a, 1995b) provides a good survey up to 1995. He also provides a quite general characterization result involving conditional distributions of the power series type.

A power series distribution with support $\{0, 1, 2, \ldots\}$ and parameter θ has as its discrete density

$$f(x; \theta) = c(x)\theta^x/c^*(\theta), \quad x = 0, 1, 2, \ldots, \tag{7.71}$$

where $c(x)$ is the coefficient function and $c^*(\theta)$ (the normalizing constant) is called the series function. If X has density (7.71) we write $X \sim PS(\theta)$ (it being understood that $c(\cdot)$ and $c^*(\cdot)$ are fixed and known). Wesolowski concentrates in the case in which $X|Y = y \sim PS(y)$, $\forall y$. In such cases, the joint distribution of (X, Y) will be uniquely determined by any consistent regression function of Y on X (i.e., $\psi(x)$) provided the coefficient function $c(\cdot)$ in (7.71) is reasonably well behaved.

Theorem 7.2 (Wesolowski, 1995a.) *Let (X, Y) be a discrete random vector such that either:*

(a) *$S(X) = \{0, 1, 2, \ldots, n\}$ for some integer n and the cardinality of $S(Y)$ is $\leq n + 2$; or*

(b) *$S(X) = \{0, 1, 2, \ldots\}$ and $S(Y) \subseteq \{0, 1, 2, \ldots\}$.*

Assume that $S(X)$ and $S(Y)$ are known and that for any $x \in S(X), y \in S(Y)$ we have

$$P(X = x|Y = y) = c(x)y^x/c^*(y),$$

where $c(\cdot)$ and $c^*(\cdot)$ are known. In addition, if $S(Y)$ is not bounded assume that

$$\sum_{x \in S(X)} \sqrt[2x]{c(x)} = \infty. \tag{7.72}$$

Then the distribution of (X, Y) is uniquely determined by $E(Y|X = x) = \psi(x), x \in S(X)$. $\qquad\square$

Proof. The key to the proof lies in the introduction of a random variable Z with discrete density

$$f_Z(z) \propto P(Y = z)/c^*(z), \quad z \in S(Y). \tag{7.73}$$

It then becomes a question of determining the moments of Z using (7.63). From them since, in the finite case (or under a Carleman moment condition related to (7.72)), the moments determine the distribution of Z, we get the distribution of Z and from it (using (7.73)) the distribution of Y. $\qquad\square$

Wesolowski also provides an analogous theorem for the case in which $X|Y = y \sim PS(1/y)$ (assuming $P(Y = 0) = 0$).

Examples in which Theorem 7.2 can be successfully applied include:

(a) If $X|Y = y \sim \text{Poisson}(\lambda y), y \in S(Y)$, then $E(Y|X = x)$ determines the distribution of (X, Y) (the case in which $S(Y) = \{0, 1, 2, \ldots\}$ was discussed by Cacoullos and Papageorgiou (1995)).

(b) If

$$X|Y = y \sim \text{binomial}\left(n_1, \frac{pt^y}{1 - p + pt^y}\right), \quad y = 0, 1, 2, \ldots, n_2,$$

where $0 < p < 1$, $t \neq 1$, and $n_2 \leq n_1 + 1$, then $E(t^Y|X = x)$ uniquely determines the distribution of (X, Y). (Define $U = t^Y$ before applying the theorem to (X, U)).

(c) If $X|Y = y \sim \text{Poisson}(\lambda t^y)$, $y \in \{0, 1, 2, \ldots\}$, where $\lambda > 0, 0 < t < 1$, then $E(t^Y|X = x)$ uniquely determines the joint density of (X, Y). Here too, define $U = t^X$ before applying the theorem.

The last example is intimately related to the Poisson conditionals distribution described in Chapter 4 (see (4.99)). Recently Wesolowski has shown that the Poisson conditionals distribution can be characterized by appropriate knowledge of the conditional density of X given $Y = y$ and of $\psi(x) = E(Y|X = x)$.

Theorem 7.3 (Wesolowski, 1995a.) *If (X, Y) has $S(X) = S(Y) = 0, 1, \ldots$, and if for each $y \in S(Y)$*

$$X|Y = y \sim Poisson(\lambda_1 \lambda_3^y) \tag{7.74}$$

and for each $x \in S(X)$

$$E(Y|X = x) = \lambda_2 \lambda_3^x, \tag{7.75}$$

where $\lambda_1 > 0, \lambda_2 > 0$, and $0 < \lambda_3 < 1$, then (X, Y) has a bivariate Poisson conditionals distribution with discrete density given by (4.99).   □

Note that this theorem is somewhat unsatisfactory since it only deals with a specific form for $\psi(x)$ (i.e., (7.75),) a form which would only occur to a person who was already familiar with the Poisson conditionals distribution. A more general result was provided by Wesolowski in the Pareto conditionals case (Theorem 7.4 below).

7.7.3   The Continuous Case

In the continuous case we are given, for each $y \in S(Y)$, $f_{X|Y}(x|y) = a(x, y)$, $x \in S(X)$ and $E(Y|X = x) = \phi(x)$, $x \in S(X)$.

The corresponding density for Y, say $g(y)$, must be obtained by solving

$$\int_{S(Y)} (\psi(x) - y)\, a(x, y) g(y) d\mu_2(y) = 0, \quad \forall x \in S(X),$$

$$\int_{S(Y)} g(y) d\mu_2(y) = 1, \tag{7.76}$$

$$g(y) \geq 0, \quad \forall y \in S(Y).$$

For certain choices of $a(x, y)$ this equation can be solved.

Example 7.6 (Exponential conditionals). Suppose that the conditional densities for X given Y are exponential, i.e.,

$$f_{X|Y}(x|y) = a(x, y) = (y + \delta)e^{-(y+\delta)x}, \qquad x > 0, \qquad (7.77)$$

and suppose that for every $x > 0$ we have

$$E(Y|X = x) = \psi(x).$$

What must $f_{X,Y}(x, y)$ look like? Denote the unknown density of Y by $g(y)$. Equation (7.76) now takes the form

$$\int_0^\infty (\psi(x) - y)(y + \delta)e^{-(y+\delta)x}g(y)\, dy = 0, \quad \forall x > 0. \qquad (7.78)$$

Multiplying this by $e^{\delta x}$ yields

$$\int_0^\infty (\psi(x) - y)e^{-yx}(y + \delta)g(y)\, dy = 0, \quad \forall\, x > 0. \qquad (7.79)$$

Let $M(x)$ be the Laplace transform of the unknown nonnegative function

$$\tilde{g}(y) = (y + \delta)g(y), \qquad (7.80)$$

i.e.,

$$M(x) = \int_0^\infty e^{-xy}(y + \delta)g(y)\, dy\ . \qquad (7.81)$$

Equation (7.79) becomes

$$\psi(x)M(x) = -M'(x). \qquad (7.82)$$

Consequently, given an appropriate $\psi(x)$, we can solve (7.82) for $M(x)$. By the uniqueness of Laplace transforms this determines $\tilde{g}(y)$, from which we can determine $g(y)$, the marginal density of Y (using (7.80)). So in this example we are able to answer questions (A) and (B) on page 155. Compatible choices of $\psi(x)$, for the conditional densities (7.77), are functions such that $\exp[-\int_0^x \psi(u)\, du]$ is a Laplace transform. For example, we could take $\psi(x) = (\gamma + x)^{-1}$.   □

We can profitably consider this example from the viewpoint of identifiability.

A family of conditional densities $a(x, y)$ associates a nonnegative integrable function on $S(X)$ with any nonnegative integrable function on $S(Y)$ in a natural way as follows:

$$(Tg)(x) = \int_{S(Y)} a(x, y)g(y)\, dy. \qquad (7.83)$$

The family of conditional densities is said to be identifiable if the transformation T defined in (7.83) is invertible. The discrete examples mentioned following (7.63) and the exponential conditional example (7.77) had corresponding $a(x, y)$'s which were identifiable.

For them, it was possible to solve (7.76) (or (7.63)) for $g(y)$ for a given function $\psi(x)$. Identifiability may well be helpful in this analysis as the following discussion suggests.

Let us assume that T defined by (7.83) is invertible (i.e., $a(x, y)$ is identifiable). Equation (7.76) can be written in the form

$$\psi(x)T(g(y)) = T(yg(y)). \tag{7.84}$$

Now define a mapping S_ψ as follows:

$$S_\psi(g(y)) = T^{-1}(T(yg(y))/\psi(x)). \tag{7.85}$$

If we can show that S_ψ is a contraction mapping then it will have a unique fixed point which will be the desired solution to (7.84). Verification of the fact that S_ψ is a contraction mapping may not be easy.

The exponential conditionals example is unusual in that, for it, we were able to characterize the class of compatible $\psi(x)'s$ for the given $a(x, y)$.

Most of the results on the determination of $f(x, y)$ via $\psi(x)$ and $a(x, y)$ assume compatibility of ψ. The following theorem is representative, in it for a given $a(x, y)$ it is proved that $E(Y|X = x) = \psi(x)$ determines $f(x, y)$.

But it does not suggest how we might recognize compatible functions $\psi(x)$.

Theorem 7.4 (Wesolowski, 1994.) *If (X, Y) is an absolutely continuous random vector with $S(X) = S(Y) = (0, \infty)$ and if, for every $y > 0$,*

$$X|Y = y \sim Pareto(\frac{a + by}{1 + cy}, \alpha), \tag{7.86}$$

where $a \geq 0, b > 0, c \geq 0, \alpha > 0$, then the distribution of (X, Y) is uniquely determined by $E(Y|X = x) = \psi(x), x > 0$.   □

Thus, for example, if

$$E(Y|X = x) = \frac{a + x}{(\alpha - 1)(b + cx)}, \tag{7.87}$$

then (X, Y) must have a Pareto conditionals distribution as discussed in Section 5.2. If the c appearing in (7.86) and (7.87) is zero, then we characterize the Mardia's bivariate Pareto distribution.

Presumably functions other than (7.87) are consistent with (7.86) giving Theorem 7.4 more generality than Theorem 7.3, however no other consistent choices for $\psi(x)$ other than (7.87) come readily to mind.

7.8   Conditional Moments Only Given

What if we are just supplied with the two regression functions $\phi(y) = E(X|Y = y)$ and $\psi(x) = E(Y|X = x)$? What can we say about the joint distribution of (X, Y)? Unfortunately the answer is: "not much." As a trivial example if $\phi(y) \equiv \psi(x) \equiv 0$, an enormous variety of suitable symmetric distributions can be found to satisfy the given regression conditions. If X and Y are random variables each having only two possible values, then consistent specfications of $\phi(y)$ and $\psi(x)$ will determine the joint distribution uniquely. As soon as the cardinalities of the support sets of X and Y sum to as least 6, we will lose uniqueness; though in finite cases we could identify the structure of all solutions using the results in Castillo, Cobo, Jubete, and Pruneda (1998).

What if we add information about higher moments? Generally speaking we are still faced with difficulties. The nature of the problems that might be encountered can be glimpsed by considering a related problem: the characterization of Gaussian conditional structure. In it, first- and second-order conditional moments are used to characterize a class of distributions which include the classical normal.

Definition 7.2 (Gaussian conditional structure). A random vector (X, Y) will be said to exhibit Gaussian conditional structure if there exist constants $\alpha_1, \beta_1, \alpha_2, \beta_2, \sigma_1^2$, and σ_2^2 such that $\forall x \in S(X)$ and $\forall y \in S(Y)$

$$
\begin{aligned}
E(X|Y = y) &= \alpha_1 + \beta_1 y, \\
E(Y|X = x) &= \alpha_2 + \beta_2 x, \\
\operatorname{var}(X|Y = y) &= \sigma_1^2, \\
\operatorname{var}(Y|X = x) &= \sigma_2^2.
\end{aligned}
\tag{7.88}
$$

□

These, of course, are the first and second conditional moment expressions for classical bivariate normal distributions. But other random variables can mimic this behavior. Kagan, Linnik, and Rao (1973) provide conditions equivalent to (7.88) in terms of the joint characteristic function of (X, Y).

Examples of non-Gaussian characteristic functions satisfying (7.88) are not that easy to visualize.

The first example of this genre was provided by Kwapian sometime prior to 1985. It was first reported in Bryc and Plucinska (1985). It was in fact presented in terms of the joint characteristic function. Kwapian considers a random vector (X, Y) whose joint characteristic function is given by

$$
\phi_{X,Y}(s, t) = p \cos(s + t) + (1 - p) \cos(s - t),
\tag{7.89}
$$

where $p \in (0, 1)$ and, to avoid independence, $p \neq 1/2$. It is obvious that (7.89) does not correspond to a Gaussian random vector. Nevertheless, if

TABLE 7.1. Probability density function of (X, Y) for the Kwapian example.

$x \backslash y$	-1	1
-1	$(1-p)/2$	$p/2$
1	$p/2$	$(1-p)/2$

(X, Y) has (7.89) as its characteristic function, (7.88) will hold, and (X, Y) thus has Gaussian conditional structure.

Where did (7.89) come from? And, why does it work? The picture is clearer if we consider the joint discrete density of a random vector (X, Y) as shown in Table 7.1, where $p \in (0, 1)$ and $p \neq 1/2$. It is readily verified that this is indeed Kwapian's example (the corresponding characteristic function is given by (7.89)). But the joint distribution in Table 7.1 has marginals with only two possible values. This gives linear regression functions by default (any function with a two point domain is linear!). Constant conditional variances are a consequence of the fact that $p(1-p) = (1-p)p$.

The elegant simplicity of the Kwapian example would suggest ready extension to higher dimensions. However, only recently (Nguyen, Rempala, and Wesolowski (1996)), have any other (other than relabeled versions of the Kwapian example) non-Gaussian examples been discussed in either two or more dimensions.

Further discussion of attempts to characterize the class of distributions with Gaussian conditional structure, and of characterization of the classical normal distribution within this class, may be found in Arnold and Wesolowski (1996).

7.9   Bibliographic Notes

Narumi (1923) is the key reference for the material in Sections 7.4, 7.5, and 7.6. Section 7.7 is based in part on Arnold, Castillo, and Sarabia (1993b) and Arnold, Castillo, and Sarabia (1999a). Section 7.8 draws on Arnold and Wesolowski (1996).

Exercises

7.1 Discuss the compatibility conditions of the system (7.38) in the case $I = J = 3$. Discuss the general case.

7.2 Prove Theorem 7.3.

7.3 Let (X, Y) be a bivariate random variable. Prove that the following three sets of conditions involving conditional moments characterize the bivariate normal distribution:

(a) $E|X| < \infty$, and

$$Y|X = x \ \sim \ N(\alpha + \beta x, \sigma^2),$$
$$E(X|Y = y) \ = \ \gamma + \delta y,$$

for some real numbers $\alpha, \beta, \gamma, \delta$ with $\beta \neq 0$, $\delta \neq 0$, and $\sigma > 0$.

(b)
$$Y|X = x \sim N(\alpha + \beta x, \sigma^2),$$

with $\beta \neq 0$, $E(X) = E(Y) = 0$, $V(X) = V(Y) = 1$, and

$$E(X^2|Y = y) = 1 - \rho^2 + \rho^2 y^2,$$

where $\rho = E(XY)$, $0 \neq |\rho| < 1$,

(c)
$$Y|X = x \ \sim \ N(\mu(x), \sigma^2(x)),$$
$$E(\mu(X)|Y = y) \ = \ \alpha + \beta y, \quad \beta \neq 0,$$
$$E(\sigma^2(X)|Y - \mu(X)) \ = \ c > 0,$$

and $X \overset{d}{=} \mu(X)$ up to a change of scale and/or location.

Hint: Use characteristic functions.

(Ahsanullah and Wesolowski (1993).)

7.4 Let (X, Y) be a two-dimensional random variable, and assume that $X \overset{d}{=} Y$ and that F_X and F_Y have N as common support. Let $\{\phi_x(y) : x \in N\}$ be a family of distributions functions indexed by N.

(a) If there exists a distribution function F such that $F \equiv F_X \equiv F_Y$ and

$$\phi_x(y) = P(Y \leq y | X = x), \quad \forall y, \quad \forall x \in N,$$

then it is unique provided that $\phi_x(y)$ satisfies the following inde-composability condition: There do not exist two disjoint subsets of N, say A_1 and A_2, such that

$$\int_{A_1} d\phi_x(y) = 1, \quad \forall x \in A_1,$$

and

$$\int_{A_2} d\phi_x(y) = 1, \quad \forall x \in A_2.$$

(b) If a density exists, this can be rewritten as a homogeneous Fredholm integral equation

$$\int_R \phi'_x(y)f(x)\, dx = f(y).$$

(c) Suppose that X and Y are identically distributed and that X, given $Y = y$, is normally distributed with mean

$$E(X|Y = y) = -\frac{m_{12}y^2 + m_{11}y + m_{10}}{2(m_{22}y^2 + m_{21}y + m_{20})}$$

and variance

$$\text{var}(X|Y = y) = -\frac{1}{2(m_{22}y^2 + m_{21}y + m_{20})}.$$

Then (X, Y) is a bivariate normal conditionals distribution.

(Arnold and Pourahmadi (1988).)

7.5 If, for all $x > 0$,

$$f(x) = \int_0^\infty \frac{(\alpha + 1)(1 + y/\sigma)^{\alpha+1}}{\sigma(1 + x/\sigma + y/\sigma)^{\alpha+2}} f(y)\, dy,$$

with $\alpha > 0$, $\sigma > 0$ where f is a pdf, then

$$f(x) = \frac{\alpha}{\sigma(1 + x/\sigma)^{\alpha+1}}, \quad x > 0.$$

(Ahsanullah and Wesolowski (1993).)

7.6 Let X, Y be independent random variables. Define

$$U = aX + bY, \quad V = cX + dY,$$

where a, b, c, d are some real numbers. Assume that $ab \neq cd$. If the conditional distributions of U given V is normal, with probability 1, then X and Y are normal.

(Kagan and Wesolowski (1996).)

7.7 Suppose that $E(X|Y = y) = y, \forall y$ and $E(Y|X = x) = x, \forall x$.

(a) Show that any one of the following assumptions is sufficient to ensure that $X = Y$ with probability 1.
   (i) $E(X^2) < \infty$;
   (ii) $E|X| < \infty$; and
   (iii) $X \geq 0$ with probability 1.

(b) Construct a proper (not easy) or improper (quite easy) example in which $E(X|Y = y) = y, \forall y$ and $E(Y|X = x) = x, \forall x$ and $X \neq Y$ on a set with positive probability.

(Remark: These issues arise in discussions of the possible existence of unbiased Bayes estimates.)

(Bickel and Mallows (1988).)

7.8 Suppose that for each $y \in \mathbb{R}$, $X|Y = y \sim N\left(0, (a + by^2)^{-1}\right)$ and for each $x \in \mathbb{R}$, $E(Y|X = x) = 0$ and $E(Y^2|X = x) = (c + dx^2)^{-1}$. In addition, assume that Y has a symmetric distribution (is this crucial?). Prove that (X, Y) has a centered normal conditionals distribution (3.51).

7.9 (a) Suppose that we are given that $f_{X|Y}(x|y) = a(x, y)$, $x \in S(X)$, $y \in S(Y)$ and that, for $x \in S(X)$, the mode of the conditional density of Y given $X = x$ is given by an invertible function $\phi(x)$. Verify that the joint density of (X, Y) is completely determined by this information.

(*Hint*: Set up a differential equation that can be solved for $f_Y(y)$.)

(b) Carry through the program outlined in (a) for the case in which

$$a(x, y) = \frac{1}{\sqrt{2\pi}} \exp\left[-\frac{1}{2}(x - \alpha y)^2\right]$$

and

$$\phi(x) = \beta x.$$

8
Multivariate Extensions

8.1 Introduction

As we have seen, conditional specification in two dimensions already presents a plethora of problems and potential modeling scenarios. Much more complicated issues and problems can be expected in higher dimensions.

The issue of conditional/marginal specification in complete generality will be deferred until Chapter 10. In the present chapter we survey results and models that can be obtained readily by analogy to available two-dimensional material. In addition to developing multivariate conditionally specified models we will consider (following the lead of Bhattacharyya, who dealt with the two-dimensional normal conditionals distribution) what kind of additional conditions, in addition to having conditionals of appropriate form, are sufficient to characterize certain more classical multivariate distributions. Considerable attention will be focussed on the development of the classical multivariate normal distribution via conditional specification.

8.2 Extension by Underlining

The material in Chapters 1 and 4 was written and undoubtedly read with the assumption that the random variables X and Y were one-dimensional. However, things will continue to make sense if X and Y are of higher dimensions. If X is ℓ_1-dimensional, and better denoted by $\underline{X}$ with possible values $\underline{x} \in S_{\underline{X}}$ and Y is ℓ_2-dimensional, better denoted by $\underline{Y}$ with possible values $y \in S_{\underline{Y}}$, then all the material in Chapters 1 and 4 remains correct with $\underline{X}$'s, $\underline{Y}$'s, $\underline{x}$'s, and $\underline{y}$'s underlined and interpreted as appropriate vectors. Going further, X and Y could assume values in more abstract spaces than $\mathbb{R}^{\ell_1}$ and $\mathbb{R}^{\ell_2}$ (e.g., Hilbert spaces, etc.) and our results will, with minor editing, still remain valid.

There are of course many other ways to conditionally specify a multivariate density that do not fit into the underlining paradigm described above. For example in three dimensions, the joint density of (X, Y, Z) might be specified by giving the conditional densities of X given (Y, Z), of Y given (X, Z), and of Z given (X, Y). Clearly compatibility checks will be required here as they were in two dimensions. The scheme being used here is, evidently, to give the conditional density of each coordinate given all other coordinates. The remainder of this chapter will focus on this scheme, which is clearly a direct analogy of the scheme used in bivariate settings earlier. Discussion of other schemes is, as mentioned earlier, deferred to Chapter 10.

We could of course immediately plunge into k dimensions. However most of the ideas are already visible in three dimensions and it is helpful to start in this simplified arena.

8.3   Compatibility in Three Dimensions

For simplicity we will not only hold the dimension down to 3 but will also focus on the finite discrete case.

Consider discrete random variables X, Y, and Z with possible values x_1, $x_2, \ldots, x_I$, $y_1, y_2, \ldots, y_J$, and $z_1, z_2, \ldots, z_K$, respectively. Three candidate families of conditional densities can be denoted by A, B, and C where

$$
\begin{aligned}
a_{ijk} &= P(X = x_i | Y = y_j, Z = z_k), \\
b_{ijk} &= P(Y = y_j | X = x_i, Z = z_k), \\
c_{ijk} &= P(Z = z_k | X = x_i, Y = y_j).
\end{aligned}
\tag{8.1}
$$

Clearly we must require that $\sum_i a_{ijk} = 1$, $\forall j, k$, $\sum_j b_{ijk} = 1$, $\forall i, k$ and $\sum_k c_{ijk} = 1$, $\forall i, j$. In addition, an analog of Theorem 2.1 must hold to ensure compatibility. The densities (8.1) will be compatible iff there exist arrays $\{d_{ij}\}, \{e_{ik}\}$, and $\{f_{jk}\}$ such that

$$
\begin{aligned}
a_{ijk}/b_{ijk} &= e_{ik} f_{jk}^{-1}, \\
a_{ijk}/c_{ijk} &= d_{ij} f_{jk}^{-1}, \\
b_{ijk}/c_{ijk} &= d_{ij} e_{ik}^{-1}.
\end{aligned}
\tag{8.2}
$$

If the densities are compatible then uniqueness of the corresponding joint distribution of (X, Y, Z) is guaranteed by irreducibility of the Markov chain

(X_n, Y_n) defined on the state space $\{x_1, \ldots, x_I\} \times \{y_1, \ldots, y_J\}$ with transition probabilities defined as follows:

$$P(X_n = x_{i_2}, Y_n = y_{j_2} | X_{n-1} = x_{i_1}, Y_{n-1} = y_{j_1}) \\ = \sum_k c_{i_1 j_1 k} a_{i_2 j_1 k} b_{i_2 j_2 k}. \tag{8.3}$$

A trivial sufficient condition for uniqueness is that $a_{ijk} b_{ijk} c_{ijk} > 0$, $\forall i, j, k$. See Nerlove and Press (1986) for some discussion of alternative sufficient conditions.

8.4   Compatibility in Higher Dimensions

Now assume that $\underline{X}$ is a k-dimensional random vector with coordinates $(X_1, X_2, \ldots, X_k)$. We introduce a convenient notational convention at this point. For each coordinate random variable X_i of $\underline{X}$ we define the vector $\underline{X}_{(i)}$ to be the $(k-1)$-dimensional vector obtained from $\underline{X}$ by deleting X_i. We use the same convention for real vectors, i.e., $\underline{x}_{(i)}$ is obtained from $\underline{x}$ by deleting x_i. We concentrate on conditional specifications of the form "X_i given $\underline{X}_{(i)}$," a direct generalization to k dimensions of the material in Chapter 1 (for two dimensions) and Section 8.3 (for three dimensions).

A putative conditional specification of the joint distribution of $\underline{X}$ using the "X_i given $\underline{X}_{(i)}$" form would be as follows. For $i = 1, 2, \ldots, k$ the conditional densities should be of the form

$$f_{X_i | \underline{X}_{(i)}}(x_i | \underline{x}_{(i)}) = \phi_i(x_i; \underline{x}_{(i)}), \quad \forall \underline{x}_{(i)} \in S(\underline{X}_{(i)}), \tag{8.4}$$

where for each $\underline{x}_{(i)}$, $\int_{S(X_i)} \phi_i(x_i; \underline{x}_{(i)}) \, d\mu_i(x_i) = 1$. To ensure compatibility of the conditional specification (8.4) there must exist functions $u_i(\underline{x}_{(i)}), i = 1, 2, \ldots, k$, such that

$$\phi_1(x_1; \underline{x}_{(1)}) u_1(\underline{x}_{(1)}) = \phi_2(x_2; \underline{x}_{(2)}) u_2(\underline{x}_{(2)}) = \ldots \\ \ldots = \phi_k(x_k; \underline{x}_{(k)}) u_k(\underline{x}_{(k)}). \tag{8.5}$$

If the densities are compatible, then uniqueness of the corresponding joint density will be guaranteed by irreducibility of the Markov chain $(X_1^{(n)}, \ldots, X_{k-1}^{(n)})$ defined on the state space $S(\underline{X}_{(k)})$, in a manner analogous to that used in (8.3).

8.5   Conditionals in Prescribed Families

Consider k parametric families of densities on $\mathbb{R}$ defined by

$$\{f_i(x; \underline{\theta}_{(i)}) : \underline{\theta}_{(i)} \in \Theta_i\}, \quad i = 1, 2, \ldots, k, \qquad . \tag{8.6}$$

where $\underline{\theta}_{(i)}$ is of dimension ℓ_i and where the ith density is understood as being with respect to the measure μ_i. We are interested in k-dimensional densities that have all their conditionals in the families (8.6). Consequently, we require that for certain functions $\underline{\theta}_{(i)} : S(\underline{X}_{(i)}) \rightarrow \Theta_i$ we have, for $i = 1, 2, \ldots, k$,

$$f_{X_i | \underline{X}_{(i)}}(x_i | \underline{x}_{(i)}) = f_i(x_i; \underline{\theta}_{(i)}(\underline{x}_{(i)})). \tag{8.7}$$

If these equations are to hold, then there must exist marginal densities for the $\underline{X}_{(i)}$'s such that

$$f_{\underline{X}_{(1)}}(\underline{x}_{(1)}) f_1(x_1; \underline{\theta}_{(1)}(\underline{x}_{(1)})) = f_{\underline{X}_{(2)}}(\underline{x}_{(2)}) f_2(x_2; \underline{\theta}_{(2)}(\underline{x}_{(2)}))$$
$$\ldots = f_{\underline{X}_{(k)}}(\underline{x}_{(k)}) f_k(x_k; \underline{\theta}_{(k)}(\underline{x}_{(k)})). \tag{8.8}$$

Under certain circumstances this array of functional equations can be solved. Extended versions of the theorems in Section 1.8 will be useful in this context. The classic example in which a solution is readibly obtainable corresponds to the case in which each of the i families in (8.6) are exponential families. This will yield straightforward k-dimensional analogs of the results in Chapters 3 and 4. In addition, many of the results discussed in Chapter 5 extend readily to higher dimensions. Only the book-keeping gets worse as dimensionality increases.

8.6   Conditionals in Exponential Families

Suppose that the k families of densities $f_1, f_2, \ldots, f_k$ in (8.6) are $\ell_1, \ell_2, \ldots, \ell_k$ parameter exponential families of the form

$$f_i(t; \underline{\theta}_{(i)}) = r_i(t) \exp \left\{ \sum_{j=0}^{\ell_i} \theta_{ij} q_{ij}(t) \right\}, \quad i = 1, 2, \ldots, k, \tag{8.9}$$

(here θ_{ij} denotes the jth coordinate of $\underline{\theta}_{(i)}$ and by convention $q_{i0}(t) \equiv 1, \forall i$). We wish to identify all joint distributions for $\underline{X}$ such that (8.7) holds with the f_i's defined as in (8.9) (i.e., with conditionals in the prescribed exponential families).

By using an extended version of the Stephanos–Levi–Civita–Suto Theorem 1.3 or by taking logarithms in (8.8) and differencing with respect to $x_1, x_2, \ldots, x_k$ we may conclude that the joint density must be of the following form:

$$f_{\underline{X}}(\underline{x}) = \left[\prod_{i=1}^{k} r_i(x_i) \right] \exp \left\{ \sum_{i_1=0}^{\ell_1} \sum_{i_2=0}^{\ell_2} \cdots \sum_{i_k=0}^{\ell_k} m_{i_1, i_2, \ldots, i_k} \left[\prod_{j=1}^{k} q_{i_j}(x_j) \right] \right\}. \tag{8.10}$$

The k-dimensional array of parameters M includes one, namely $m_{00...0}$, which is a function of the others and plays the role of the normalizing constant to ensure that $f_{\underline{X}}(\underline{x})$ integrates to 1. Our experience in two dimensions warns us that the determination of appropriate constraints on the parameters in the array M, to ensure a valid (integrable) density, will be a daunting exercise. Sometimes simple sufficient conditions can be given but often a complete characterization of the admissible values of M will be unattainable. The case of independent coordinates random variables is included in (8.10). It corresponds to the choice

$$m_{i_1,i_2,\ldots,i_k} = 0 \quad \text{if} \quad \sum_{j=1}^{k} I(i_j \neq 0) > 1.$$

In the following sections we will briefly discuss some important examples of multivariate conditionals in exponential families distributions, together with some nonexponential family examples. Needless to say, the list is not exhaustive.

8.7   Multivariate Exponential Conditionals Distribution

In this case, $\underline{X}$ is assumed to be a k-dimensional random vector with $X_i > 0, i = 1, 2, \ldots, k$. For each i we require that the conditional distribution of X_i given $\underline{X}^{(i)} = \underline{x}^{(i)}$ is exponential $\mu_i(\underline{x}^{(i)})$ for some functions $\mu_i(\cdot)$. Clearly this is a conditionals in exponential families distribution and can be written in the form (8.10). However, since each exponential family has just one parameter, a slightly simpler representation is possible. The joint density must be of the following form:

$$f_{\underline{X}}(\underline{x}) = \exp\left[-\sum_{\underline{s} \in \xi_k} \lambda_{\underline{s}} \left(\prod_{i=1}^{k} x_i^{s_i}\right)\right], \quad \underline{x} > \underline{0}, \tag{8.11}$$

where ξ_k is the set of all vectors of 0's and 1's of dimension k. The parameters $\lambda_{\underline{s}}$ ($\underline{s} \neq \underline{0}$) are nonnegative, those for which $\sum_{i=1}^{k} s_i = 1$ are positive, and $\lambda_{\underline{0}}$ is such that the density integrates to 1. For $k > 2$, it is not easy to determine an analytic expression for $\lambda_{\underline{0}}$ as a function of the $\lambda_{\underline{s}}$'s. It is evident that in (8.11), X_i is stochastically decreasing in $\underline{X}^{(i)}$, so that nonpositive correlations are encountered. Submodels of (8.11) may be obtained by, for example, setting $\lambda_{\underline{s}} = 0$ for every $\underline{s}$ for which $\sum_{i=1}^{k} s_i > m$ for some integer $m < k$. Exchangeable models are associated with choices of $\lambda_{\underline{s}}$ such that $\lambda_{\underline{s}} = \lambda_{\underline{s}'}$ whenever $\sum_{i=1}^{k} s_i = \sum_{i=1}^{k} s_i'$.

Example 8.1 (Multivariate exponential conditionals). One simple exchangeable model will be discussed in some detail in Chapter 9. It is a one-parameter family of the form

$$f_{\underline{X}}(\underline{x}) = \phi(\delta) \exp\left(-\sum_{i=1}^{k} x_i - \delta \prod_{i=1}^{k} x_i\right), \quad \underline{x} > \underline{0}, \tag{8.12}$$

where $\phi(\delta)$ is the appropriate normalizing constant.   □

Finally we remark that only a modest modification of (8.11) will give the general form of all k-dimensional distributions with conditionals in one-parameter exponential families. Here, for each i, X_i given $\underline{X}^{(i)} = \underline{x}^{(i)}$ is postulated to belong to an exponential family (8.9) with $\ell_i = 1$. Instead of using the form (8.10), we can write the general form of such densities, modeled after (8.11), as follows:

$$f_{\underline{X}}(\underline{x}) = \left[\prod_{i=1}^{k} r_i(x_i)\right] \exp\left[\sum_{\underline{s} \in \xi_k} \delta_{\underline{s}} \prod_{i=1}^{k} [q_{i1}(x_i)]^{s_i}\right]. \tag{8.13}$$

Note that (8.11) and (8.13) include models with independent coordinate random variables. We merely set $\lambda_{\underline{s}}$ (respectively $\delta_{\underline{s}}$) equal to zero if $\sum_{i=1}^{k} s_i > 1$.

8.8   Multivariate Normal Conditionals Distribution

Suppose now that we require that, for each i, X_i given $\underline{X}^{(i)} = \underline{x}^{(i)}$, should be normally distributed with mean $\mu_i(\underline{x}^{(i)})$ and variance $\sigma_i^2(\underline{x}^{(i)})$. The corresponding joint density will then be of the form (8.10) with $\ell_1 = \ell_2 = \cdots = \ell_k = 2$,

$$r_i(t) = 1, \quad i = 1, 2, \ldots, k, \tag{8.14}$$

and

$$\begin{aligned} q_{i0}(t) &= 1, \quad i = 1, 2, \ldots, k, \\ q_{i1}(t) &= t, \quad i = 1, 2, \ldots, k, \\ q_{i2}(t) &= t^2, \quad i = 1, 2, \ldots, k. \end{aligned} \tag{8.15}$$

The coefficients (the m's) in (8.10) must be chosen to ensure that all the conditional variances are always positive (compare with Section 3.3) and that the density is integrable. The density can be written in the form

$$f_{\underline{X}}(\underline{x}) = \exp\left\{\sum_{\underline{i} \in T_k} m_{\underline{i}} \left[\prod_{j=1}^{k} x_j^{i_j}\right]\right\}, \tag{8.16}$$

where T_k is the set of all vectors of 0's, 1's, and 2's of dimension k, since the q_{ij}'s defined in (8.15) can be written in the form $q_{ij}(t) = t^j$.

The classical k variate normal distribution is of course a special case of (8.16). In order for (8.16) to reduce to a classical k-variate normal density we require that every $m_{\underline{i}}$, for which $\sum_{j=1}^{k} i_j > 2$, should be zero. The remaining $m_{\underline{i}}$'s must be such that the quadratic form in $\underline{x}$ that they define is negative definite, for convergence. We will investigate the role of the classical normal distribution as a special case of (8.16) in more detail in Section 8.15.

The model (8.16) reduces to one with independent normal marginals if every $m_{\underline{i}}$ for which $\sum_{j=1}^{k} I(i_j \neq 0) > 1$ is zero.

8.9   Multivariate Cauchy Conditionals Distribution

Our first nonexponential family example involves Cauchy random variables.

The k-dimensional Cauchy conditionals distribution will have the property that for each i, $X_i | \underline{X}_{(i)} = \underline{x}_{(i)}$ has a Cauchy $(\mu_i(\underline{x}_{(i)}), \sigma_i(\underline{x}_{(i)}))$ distribution for some functions μ_i and σ_i. The corresponding functional equations are readily solved (compare Section 5.5, where the bivariate case was treated). The resulting density for $\underline{X}$ is found to be of the form

$$f_{\underline{X}}(\underline{x}) = \left[\sum_{\underline{i} \in T_k} m_{\underline{i}} \left(\prod_{j=1}^{k} x_j^{i_j} \right) \right]^{-1}. \qquad (8.17)$$

To ensure that (8.17) is a valid density we must impose constraints intimately related to those required on the $m_{\underline{i}}$'s in the multivariate normal conditionals case ((8.16)). The classical elliptically symmetric k-variate Cauchy distribution is associated with the choice of $m_{\underline{i}}$'s for which $m_{\underline{i}} = 0$ whenever $\sum_{j=1}^{k} i_j > 2$.

8.10   Multivariate Uniform Conditionals Distribution

Consider a k-dimensional random variable $\underline{X}$. Suppose that for each i the conditional distribution of X_i given $\underline{X}_{(i)} = \underline{x}_{(i)}$ is uniform over some interval $(\phi_{1i}(\underline{x}_{(i)}), \phi_{2i}(\underline{x}_{(i)}))$. Evidently the joint density must be constant over some subset T of k-space of finite content constrained to be such that for each i, and each $\underline{x}_{(i)} \in \mathbb{R}^{k-1}$, the set $\{x_i : (x_1, \ldots, x_k) \in T\}$ is either empty or an interval. For example, T could be a convex set of finite content. Alternatively, it could be the region in the positive orthant under some surface (provided the region so defined is of finite content).

8.11 Multivariate Pareto Conditionals Distribution

For $\alpha > 0$ (held fixed in the present section) we say that X has a Pareto (α, σ) distribution if

$$f_X(x) = \frac{\alpha}{\sigma}\left(1 + \frac{x}{\sigma}\right)^{-(\alpha+1)}, \qquad x > 0, \qquad (8.18)$$

as in Section 5.2.

Now we seek to identify all k-dimensional distributions for which for each i the conditional density of X_i given $\underline{X}_{(i)} = \underline{x}_{(i)}$ is a member of the family (8.18) with scale parameter $\sigma_i(\underline{x}_{(i)})$, some function of $\underline{x}_{(i)}$.

This is clearly not an exponential family example. However, the arguments provided in Section 5.2, extend in a straightforward fashion to give the following general form for such multivariate Pareto(α) densities:

$$f_{\underline{X}}(\underline{x}) = \left[\sum_{\underline{s}\in\xi_k} \delta_{\underline{s}}\left(\prod_{i=1}^{k} x_i^{s_i}\right)\right]^{-(\alpha+1)}, \qquad \underline{x} > \underline{0}. \qquad (8.19)$$

Some care must be exercised to determine admissible values for the $\delta_{\underline{s}}$'s to guarantee that (8.19) is integrable (of course $\delta_{\underline{0}}$ will usually be an unattractive function of the other $\delta_{\underline{s}}$'s, chosen to make the integral over the positive orthant equal to 1). All of the $\delta_{\underline{s}}$'s must be nonnegative and some are permitted to be zero. How many and which ones depends on k and α. Refer back to the bivariate case discussed in Section 5.2 to get a flavor of the issues involved .

Correlations, when they exist, can be positive or negative depending on the choice of $\delta_{\underline{s}}$'s in (8.19). Judicious choices of $\delta_{\underline{s}}$'s will lead to models with independent marginals. Specifically, we must choose $\delta_{\underline{0}} > 0$ and for $\underline{s} \neq \underline{0}$,

$$\delta_{\underline{s}} = \delta_{\underline{0}}^{\left(1 - \sum_{i=1}^{k} s_i\right)} \prod_{i=1}^{k} (\delta_i^*)^{s_i}, \qquad (8.20)$$

where $\delta_1^* = \delta_{1000..0}, \delta_2^* = \delta_{010..0}$, etc.

Analogous multivariate extensions are possible for the generalized Pareto conditionals distributions and the Pickands–deHaan distributions discussed in Chapter 5.

For the generalized Pareto distributions we have a k-dimensional version of Model I of the form

$$f_{\underline{X}}(\underline{x}) = \left[\prod_{j=1}^{k} x_j^{\delta_j - 1}\right] \left\{\sum_{\underline{s} \in \xi_k} \lambda_{\underline{s}} \left[\prod_{j=1}^{k} x_j^{s_j \delta_j}\right]\right\}^{-(\alpha+1)}, \qquad \underline{x} > \underline{0}, \qquad (8.21)$$

where ξ_k is the set of all vectors of 0's and 1's of dimension k. The k-dimensional version of Model II is given by

$$f_{\underline{X}}(\underline{x}) = \left[\prod_{j=1}^{k} x_j^{\delta_j - 1}\right] \exp\left\{\sum_{\underline{s} \in \xi_k} \lambda_{\underline{s}} \left[\prod_{j=1}^{k} \log(\theta_j + x_j^{\delta_j})\right]\right\}, \qquad \underline{x} > \underline{0}. \quad (8.22)$$

The closely analogous k-variate forms of the Pickands–deHaan distribution are given by:

(I) $f_{\underline{X}}(\underline{x}) = [\sum_{\underline{s} \in \xi_k} \lambda_{\underline{s}} \prod_{j=1}^{k} (\sigma x_j)^{s_j}]^{1/\sigma - 1}$, $\underline{x} \in D$, where D is the set in which the expression in square brackets is positive.

(II) $f_{\underline{X}}(\underline{x}) = \exp\left[\sum_{\underline{s} \in \xi_k} \lambda_{\underline{s}} \prod_{j=1}^{k} \log(1 - \delta_i x_i)\right]^{s_i}$, $0 < x_i < \delta_i^{-1}, i = 1, 2 \ldots, k$.

8.12 Multivariate Beta Conditionals Distribution

Here we ask, for each i, that X_i given $\underline{X}_{(i)} = \underline{x}_{(i)}$ have a beta distribution with parameters $\alpha(\underline{x}_{(i)})$ and $\beta(\underline{x}_{(i)})$. The beta distribution is a two-parameter exponential family and so the beta conditionals densities will be of the form (8.10) with $\ell_1 = \ell_2 = \ldots = \ell_k = 2$,

$$r_i(t) = [t(1 - t)]^{-1} I(0 < t < 1), \qquad i = 1, 2, \ldots, k, \qquad (8.23)$$

$$\begin{aligned} q_{10}(t) &= 1, & i &= 1, 2, \ldots, k, \\ q_{i1}(t) &= \log t, & i &= 1, 2, \ldots, k, \\ q_{i2}(t) &= \log(1 - t), & i &= 1, 2, \ldots, k. \end{aligned} \qquad (8.24)$$

Most of the m's in (8.10) must be nonpositive, the exceptions being those $m_{i_1 i_2 \ldots i_k}$'s for which $\sum_{j=1}^{k} I(i_j \neq 0) = 1$. These m's must be positive. Reference back to the bivariate case in Section 4.10 will help explain these restrictions.

Life becomes more interesting if we allow scaled beta distributions as acceptable conditionals. The classical example in which all conditionals are scaled beta variables is provided by the Dirichlet distribution. We are effectively seeking a multivariate extension of the material in Section 5.9

and, again, reference to James (1975) will be instructive. We actually pose the problem in a form slightly more general than that discussed in Section 5.9 and in James' work.

Specifically, we wish to identify k-dimensional random variables such that for each i, X_i given $\underline{X}_{(i)} = \underline{x}_{(i)}$ has a scaled beta distribution with parameters $\alpha_i(\underline{x}_{(i)})$ and $\beta_i(\underline{x}_{(i)})$ and support on the interval $(0, c_i(\underline{x}_{(i)}))$. First, for compatibility, the $c_i(\underline{x}_{(i)})$'s must be such that they correspond to a joint support set of finite content for $\underline{X}$ that consists of all points in the positive orthant of k-space under some surface. Within that region, T, the joint density must be expressible as a product of marginals $f_{(i)}(\underline{x}_{(i)})$ and conditionals in k ways. Thus we must have

$$
f_{(1)}(\underline{x}_{(1)}) \frac{1}{c_1(\underline{x}_{(1)})} \left(\frac{x_1}{c_1(\underline{x}_{(1)})} \right)^{\alpha_1(\underline{x}_{(1)})-1} \left(1 - \frac{x_1}{c_1(\underline{x}_{(1)})} \right)^{\beta_1(\underline{x}_{(1)})-1}
$$

$$
= f_{(2)}(\underline{x}_{(2)}) \frac{1}{c_2(\underline{x}_{(2)})} \left(\frac{x_2}{c_2(x_{(2)})} \right)^{\alpha_2(\underline{x}_{(2)})-1} \left(1 - \frac{x_2}{c_2(\underline{x}_{(2)})} \right)^{\beta_2(\underline{x}_{(2)})-1}
$$

$$
= \cdots \text{ etc.} \tag{8.25}
$$

Only in special circumstances will (8.25) have a solution. If $\beta_i(\underline{x}_{(i)}) \equiv 1$, $i = 1, 2, \ldots, k$, then the joint density must be of the form

$$
f_{\underline{X}}(\underline{x}) = \left(\prod_{i=1}^{k} x_i \right)^{-1} \exp \left\{ \sum_{\underline{s} \in \xi_k} \delta_{\underline{s}} \left[\prod_{i=1}^{k} (\log x_i)^{s_i} \right] \right\}, \quad \underline{x} \in T. \tag{8.26}
$$

A second trivial case in which a solution is possible occurs when $c_i(\underline{x}_{(i)}) \equiv c_i$, $i = 1, 2, \ldots, k$. For in this case we merely obtain $X_i = c_i X_i^*$ where $\underline{X}^*$ has an unscaled beta conditionals distribution (given by (8.13) with r's and q's defined by (8.23) and (8.24)).

The final instance in which a solution is obtainable corresponds to the case where $\beta_i(\underline{x}_{(i)}) \equiv \beta \neq 1$, $i = 1, 2, \ldots, k$. In this case the joint density is necessarily supported on a set of the form

$$
T = \left\{ \underline{x} : x_i > 0, \sum_{i=1}^{k} c_i x_i < 1 \right\}, \tag{8.27}
$$

where the c_i's are positive. On this set, T, the joint density is of the form

$$
f_X(\underline{x}) = \left(\prod_{i=1}^{k} x_i \right)^{-1} \left(1 - \sum_{i=1}^{k} c_i x_i \right)^{\beta-1} \exp \left\{ \sum_{\underline{s} \in \xi_k} \delta_{\underline{s}} \left[\prod_{i=1}^{k} (\log x_i)^{s_i} \right] \right\}. \tag{8.28}
$$

This is essentially the form of the density introduced by James (1975).

James also discusses a variant problem as follows. Suppose for each i, $\underline{X}_{(i)}$ given $X_i = x_i$ is Dirichlet$(k - 1, \underline{\gamma}(x_i))$. It follows in this case that the k-dimensional joint density of $\underline{X}$ must be Dirichlet$(k, \underline{\gamma})$ so that no interesting new multivariate distributions will be encountered.

8.13   Multivariate Binomial Conditionals Distribution

Up till now, all our multivariate examples have involved continuous distributions. Naturally discrete examples exist. We could write appropriate choices of r's and q's in (8.13) to lead to multivariate geometric, Poisson, and binomial conditionals distributions but, by referring to the corresponding bivariate examples in Section 4.12, the interested reader can do this easily.

We will discuss an interesting variant of the binomial conditionals distribution. A close parallel with the scaled beta conditionals example will be apparent. In that example a generalization of the Dirichlet distribution was uncovered. In the present example a generalization of the multinomial distribution will be sought. Thus we seek a k-dimensional random vector $\underline{X}$ with possible values being vectors $\underline{x}$ of nonnegative integers such that $\sum_{i=1}^{k} x_i \leq n$ where n is a fixed positive integer. We wish to have, for each i, the conditional distribution of X_i given $\underline{X}_{(i)} = \underline{x}_{(i)}$ be binomial with parameters $n - \sum_{j \neq i} x_j$ and $p_i(\underline{x}_{(i)})$. By writing the joint density as a product of marginals of $\underline{X}_{(i)}$'s and conditionals of X_i given $X_{(i)}, i = 1, 2, \ldots, k$, we eventually find that the joint density must be of the form

$$f_{\underline{X}}(\underline{x}) = \left(\prod_{i=1}^{k+1} x_i!\right)^{-1} \exp\left\{\sum_{\underline{s} \in \xi_k} \lambda_{\underline{s}} \left[\prod_{i=1}^{k} x_i^{s_i}\right]\right\}, \quad x_i \geq 0, \ \sum_{i=1}^{k} x_i \leq n,$$

(8.29)

where we introduce the convenient notation $x_{k+1} = n - \sum_{i=1}^{k} x_i$. The model (8.29) includes the classical multinomial as a special case (choose $\lambda_{\underline{s}} = 0$ if $\sum_{i=1}^{k} s_i > 1$).

8.14   Further Extension by Underlining

It will be noted that almost all of the results in Sections 8.3 through 8.6 remain valid even when all the x_i's are themselves vectors of possibly different dimensions. Thus underlining (to indicate vectors) provides immediate generalization of the results.

8.15   Characterization of Multivariate Normality via Conditional Specification

Nested within the family of normal conditionals distributions (8.16) are to be found the classical k-variate normal distributions. It was remarked in Section 8.8 that they can be identified by the property that, for them, the parameters $m_{\underline{i}}$ corresponding to $\underline{i}$'s for which $\sum_{j=1}^{k} i_j > 2$ must all be zero.

These may be dubbed the non-Gaussian parameters. First Bhattacharyya (1943), and later Castillo and Galambos (1987a), addressed the issue of characterizing the classical multivariate normal distribution (or Gaussian distribution) via distributional properties in addition to normal conditionals. A survey of results of this type will be presented in this section.

It will be helpful to review certain well-known properties of the k-dimensional Gaussian (classical multivariate normal) distribution. Notational conventions used heavily in the discussion, some new, some old, are the following. If $\underline{X}$ is a k-dimensional random vector, we denote its ith coordinate by X_i and, for each i, we denote the vector $\underline{X}$ with X_i deleted by $\underline{X}_{(i)}$. For each i, j, $\underline{X}_{(i,j)}$ denotes the vector $\underline{X}$ with its ith and jth coordinates deleted. $\tilde{\underline{X}}_{(i)\ell}$ will be used to denote a subvector of $\underline{X}_{(i)}$ with ℓ coordinates. In addition, we write $\underline{X} = (\dot{\underline{X}}, \ddot{\underline{X}})$ to indicate a partitioning $\underline{X}$ into two subsets where $\dot{\underline{X}}$ includes, say, k_1 of the coordinates of $\underline{X}$ (not necessarily the first k_1 coordinates) and $\ddot{\underline{X}}$ includes the remaining coordinates.

In similar fashion, $\underline{x}$ denotes a generic point in $\mathbb{R}^k$ and $x_i, \underline{x}_{(i)}, \underline{x}_{(i,j)}, \tilde{\underline{x}}_{(i)\ell}$, $(\dot{\underline{x}}, \ddot{\underline{x}})$ are defined analogously to their random counterparts.

If a k-dimensional random variable $\underline{X}$ has a Gaussian (classical multivariate normal) distribution, then it admits a representation of the form

$$\underline{X} = \underline{\mu} + \Sigma^{1/2} \underline{Z},$$

where $Z_1, Z_2, \ldots, Z_k$ are i.i.d. standard univariate normal random variables. In such a case we write $\underline{X} \sim N^{(k)}(\underline{\mu}, \Sigma)$. Here $\underline{\mu} \in \mathbb{R}^k$ and Σ is a nonnegative definite $k \times k$ matrix. Such random variables, $\underline{X}$, have the following properties (among others):

1. All one-dimensional marginals are normal.

2. All ℓ-dimensional marginals, $\ell < k$, are ℓ-variate normal.

3. All linear combinations are normal. In fact, for any $\ell \times k$ matrix B, we have
   $$B\underline{X} \sim N^{(\ell)}(B\mu, B\Sigma B').$$

4. All conditionals are normal. Thus if we partition $\underline{X} = (\dot{\underline{X}}, \ddot{\underline{X}})$, then the conditional distribution of $\dot{\underline{X}}$, given $\ddot{\underline{X}} = \ddot{\underline{x}}$, is k_1-variate normal for every $\ddot{\underline{x}}$.

5. All regressions are linear. Thus for any i and any $j_1, j_2, \ldots, j_\ell \ (\neq i)$, $E(X_i | X_{j_1}, \ldots, X_{j_\ell})$ is a linear function of $X_{j_1}, X_{j_2}, \ldots, X_{j_\ell}$.

6. All conditional variances are constant. Thus $\text{var}(X_i | X_{j_1}, \ldots, X_{j_\ell})$ is nonrandom for any $i, j_1, \ldots, j_\ell$.

7. If Σ is positive definite, the joint density of $\underline{X}$ is elliptically contoured.

8. $\underline{X}$ has linear structures, i.e., $\underline{X}$ admits a representation of the form

$$\underline{X} = \underline{a}_0 + A\underline{Z},$$

where the Z_i's are independent random variables.

Naturally, Conditions 1–8 contain more than enough to characterize the classical k-variate normal distribution. Parsimonious selections from Conditions 1–8 undoubtedly will suffice. Most of those properties, taken individually fail to characterize the classical multivariate normal distribution. Combinations of these properties can be used to characterize the classical model. Condition 3 does characterize the classical model. Condition 4 also will characterize the classical model provided $k > 2$. None of the others alone will do it. Conditions 7 and 8, together, will characterize the classical distribution.

Condition 4, involving conditional distributions will be our major concern. As remarked above, if Condition 4 holds, then necessarily $\underline{X}$ is a classical k-variate normal random variable. But much less than the full force of Condition 4 is needed. To see this we will list three more multivariate normal properties that are subsumed by Condition 4.

If $\underline{X}$ is classical k-variate normal, then:

9. For every i, X_i given $\underline{X}_{(i)} = \underline{x}_{(i)}$ is univariate normal for each $\underline{x}_{(i)} \in \mathbb{R}^{k-1}$.

10. For each i, j, (X_i, X_j) given $\underline{X}_{(i,j)} = \underline{x}_{(i,j)}$ is classical bivariate normal for each $\underline{x}_{(i,j)} \in \mathbb{R}^{k-2}$.

11. For each i and for each subvector $\tilde{\underline{X}}_{(i)\ell}$ of $\underline{X}_{(i)}$ for each ℓ, X_i given $\tilde{\underline{X}}_{(i)\ell} = \tilde{\underline{x}}_{(i)\ell}$ is univariate normal for each $\tilde{\underline{x}}_{(i)\ell} \in \mathbb{R}^\ell$.

Of course we know that if Condition 9 holds then we merely characterize the k-variate normal conditionals distribution (see Section 8.8). More must be assumed to guarantee that the non-Gaussian parameters are forced to be zero. Bhattacharya and Castillo and Galambos addressed this goal by positing that, in addition to Condition 9, we also require that parts of one of Conditions 1, 5, or 6 hold (refer to Chapter 3 for details in the bivariate case). Our goal is to buttress Condition 9 with further conditional assumptions to achieve the end of characterizing the classical distribution.

The first result obtained in this direction was:

Theorem 8.1 *Suppose that for each i, j and for each $x_{(i,j)} \in \mathbb{R}^{k-2}$, the conditional distribution of (X_i, X_j) given $\underline{X}_{(i,j)} = \underline{x}_{(i,j)}$ is classical bivariate normal with mean vector $(\mu_i(\underline{x}_{(i,j)}), \mu_j(\underline{x}_{(i,j)}))$ and variance covariance matrix*

$$\begin{pmatrix} \sigma_{11}(\underline{x}_{i,j}) & \sigma_{12}(\underline{x}_{i,j}) \\ \sigma_{21}(\underline{x}_{i,j}) & \sigma_{22}(\underline{x}_{i,j}) \end{pmatrix}.$$

It follows that $\underline{X}$ has a classical k-variate normal distribution.   □

Proof.   Since (X_i, X_j) given $\underline{X}_{(i,j)} = \underline{x}_{(i,j)}$ is classical bivariate normal, it has univariate normal conditionals. It then follows that X_i given X_j and $\underline{X}_{(i,j)}$, i.e., given $\underline{X}_{(i)}$ is univariate normal.

It follows that $f_{\underline{X}}(\underline{x})$ has (8.16) as its density.

Now for any i and j, the conditional density of (X_i, X_j) given $\underline{X}_{(i,j)} = \underline{x}_{(i,j)}$ is postulated to be classical bivariate normal and so it will be of the form

$$f_{X_i X_j | \underline{X}_{(i,j)}}(x_i, x_j | \underline{x}_{(i,j)}) \propto \exp[Q(x_i, x_j)], \tag{8.30}$$

where $Q(x_i, x_j)$ is a quadratic form in x_i, x_j with coefficients which may depend on $\underline{x}_{(i,j)}$.

This forces many of the m_i's in (8.16) to be zero. In fact, the joint density must assume the form (using new notation for the reduced number of parameters)

$$f_{\underline{X}}(\underline{x}) = \exp\left\{ -\left[\sum_{j=1}^{k} \beta_j x_j^2 + \sum_{\underline{s} \in \xi_k} \delta_{\underline{s}} \left(\prod_{j=1}^{k} x_j^{s_j} \right) \right] \right\}, \tag{8.31}$$

where ξ_k is as defined following (8.11). However, the number of nonzero parameters can be reduced even further. By considering the conditional densities of (X_i, X_j) given $\underline{X}_{(i,j)}, \forall i, j$, which have to have positive definite quadratic forms, we can conclude that any $\delta_{\underline{s}}$ in (8.31) for which $\underline{s}$ includes more than two 1's must be zero.

It follows that the joint density is expressible in the form (again with new simplified parameters)

$$f_{\underline{X}}(\underline{x}) = \exp\left\{ -\left[\alpha + \sum_{j=1}^{k} \lambda_j x_j + \sum_{j=1}^{k} \sum_{\ell=1}^{k} \gamma_{j\ell} x_j x_\ell \right] \right\}, \tag{8.32}$$

where the matrix $(\gamma_{j\ell})_{j,\ell=1}^{k}$ is positive definite (for integrability). This of course indicates that $\underline{X}$ has a classical k-variate normal distribution.   □

Remark 8.1 Arguments similar to those used in this proof can be used to justify the following statements. Suppose that for each i and for each $\underline{x}_{(i)} \in \mathbb{R}^{k-1}$, the conditional distribution of X_i given $\underline{X}_{(i)} = \underline{x}_{(i)}$ is classical univariate normal, and in addition assume that the regression of each X_i

on $\underline{X}_{(i)}$ is linear (or assume the conditional variance of X_i given $\underline{X}_{(i)} = \underline{x}_{(i)}$ does not depend on $\underline{x}_{(i)}$). Then $\underline{X}$ has a classical k-variate normal distribution. This extends the result for $k = 2$, discussed by Bhattacharyya (1943) and Castillo and Galambos (1989).

A closely related result is the following:

Theorem 8.2   *Suppose that for each i and for each $\underline{x}_{(i)} \in \mathbb{R}^{k-1}$, the conditional distribution of X_i given $\underline{X}_{(i)} = \underline{x}_{(i)}$ is normal with mean $\mu_i(\underline{x}_{(i)})$ and variance $\sigma_i^2(\underline{x}_{(i)})$. In addition assume that for each i, j and each $\underline{x}_{(i,j)} \in \mathbb{R}^{k-2}$, the conditional distribution of X_i given $\underline{X}_{(i,j)} = \underline{x}_{(i,j)}$ is normal with mean $\mu_{ij}(\underline{x}_{(i,j)})$ and variance $\sigma_{ij}^2(\underline{x}_{(i,j)})$. It follows that $\underline{X}$ has a classical k-variate normal distribution.* □

Proof.   Since for each i, X_i given $\underline{X}_{(i)} = \underline{x}_{(i)}$ is normally distributed ($\forall \underline{x}_{(i)}$) it follows that $\underline{X}$ has a k-dimensional normal conditionals density of the form (8.16). Now fix j. Since for each i, X_i given $\underline{X}_{(i,j)} = \underline{x}_{(i,j)}$ is normally distributed for every $\underline{x}_{(i,j)}$, it follows that $\underline{X}_{(j)}$ has a $(k-1)$-dimensional normal conditionals distribution with a density that is a $(k-1)$-dimensional version of (8.16). This implies that when (8.16) is integrated with respect to x_j, the resulting marginal is again of analogous form. This can only occur if all $m_{\underline{i}}$'s with a 2 in the jth coordinate of the subscript $\underline{i}$ are zero except for $m_{00,\dots,2,\dots,0}$ (the coefficient whose subscript has a 2 in the jth coordinate and zeros elsewhere). This argument can be repeated for $j = 1, 2, \dots, k$. The joint density of $\underline{X}$ can then be expressed in the following form (using new notation for the reduced number of possibly nonzero parameters):

$$f_{\underline{X}}(\underline{x}) = \exp\left\{ - \left[\sum_{j=1}^{k} \beta_j x_j^2 + \sum_{\underline{s} \in \xi_k} m_{\underline{s}} \left(\prod_{j=1}^{k} x_j^{s_j} \right) \right] \right\}. \tag{8.33}$$

In order that (8.33) be integrable over $\mathbb{R}^k$, it is necessary that the β_j's be positive and that every $m_{\underline{s}}$ corresponding to an $\underline{s}$ with more than two nonzero coordinates must be zero.

It follows that the joint density is expressible in the form (again with new simplified parameters)

$$f_{\underline{X}}(\underline{x}) = \exp\left\{ - \left[\alpha + \sum_{j=1}^{k} \lambda_j x_j + \sum_{j=1}^{k} \sum_{\ell=1}^{k} \gamma_{j\ell} x_j x_\ell \right] \right\}. \tag{8.34}$$

For integrability the matrix $(\gamma_{j\ell})_{j=1, \ell=1}^{k \ \ k}$ must be positive definite. It is then evident that $\underline{X}$ has a classical k-variate normal distribution. □

The transition from the general normal conditional distribution to the classical k-variate normal distribution involves setting many of the parameters equal to zero. If we insist on proper (i.e., integrable) densities we must remember that setting certain of the $m_{\underline{i}}$'s equal to zero in (8.16) may require (for integrability) that others also be set equal to zero. Recall the discussion in two dimensions in Chapter 3 where it was noted that if $m_{22} = 0$ then necessarily m_{12} and m_{21} must also be zero.

Example 8.2 (Trivariate normal conditionals distribution). Consider the following normal conditionals density:

$$\begin{aligned} f(x_1, x_2, x_3) = \exp \big\{ &-[m_{000} + m_{100}x_1 + m_{010}x_2 + m_{001}x_3 \\ &+ m_{110}x_1x_2 + m_{101}x_1x_3 + m_{011}x_2x_3 \\ &+ m_{111}x_1x_2x_3 + m_{200}x_1^2 + m_{020}x_2^2 \\ &+ m_{002}x_3^2] \big\}. \end{aligned} \qquad (8.35)$$

For this density it is evident that $X_1|X_2, X_3$, $X_2|X_1, X_3$, $X_3|X_1, X_2$, $X_1|X_2$, $X_2|X_1$, $X_1|X_3$, $X_3|X_1$, $X_2|X_3$, and $X_3|X_2$ are all of the univariate normal form. The presence of a non-zero term for the parameter m_{111} in (8.35) identifies a distribution which is not classical trivariate normal.

Thus (8.35) would seem to contradict Theorem 8.2. It does not however, since close inspection of (8.35) reveals it to be "an impossible" (i.e., not integrable) model.   □

8.16   Multivariate Normality in More Abstract Settings

Bischoff (1996a) pointed out that many of the results dealing with normal conditionals and multivariate normality remain valid in abstract inner product spaces. If we let X and Y be random vectors taking values in real inner product spaces $(V, \langle \cdot, \cdot \rangle_V)$ and $(W, \langle \cdot, \cdot \rangle_W)$, respectively. Bischoff's abstract version of Bhattacharyya's theorem takes the form:

Theorem 8.3 (Bischoff (1996b).) *With respect to* $(V \oplus W, \langle \cdot, \cdot \rangle_{V \oplus W})$ *let* $f_W(\cdot|v)$ *be normal on* $(W, \langle \cdot, \cdot \rangle_W)$ *for each* $v \in V$ *with the known mean* $m_W(v) \in W$ *and the known positive definite covariance* $\Sigma_W(v) = A(v)^{-1} \in \mathcal{L}(W, W)$, *and let* $f_V(\cdot|w)$ *be normal on* $(V, \langle \cdot, \cdot \rangle_V)$ *for each* $w \in W$ *with the known mean* $m_V(0)$ *and the covariance* $\Sigma_V(0) = B(0)^{-1}$ *for* $w = 0$. *Then under the assumption*

$$f_V(v|w) \cdot f_W(w) = f(v, w) = f_W(w|v) \cdot f_V(v),$$

the Lebesgue-density f is completely determined and given with respect to
$(V \oplus W, \langle \cdot, \cdot \rangle_{V \oplus W})$ *by*

$$
\begin{aligned}
f(v,w) = {} & g_V(0) \cdot (2\pi)^{-\alpha/2} \cdot \det A(0)^{1/2} \cdot \exp\{\tfrac{1}{2}\langle m_v(0), B(0)m_V(0)\rangle_V\} \\
& \times \exp\{-\tfrac{1}{2}\langle w - m_W(0), A(0)(w - m_W(0))\rangle_W\} \\
& \times \exp\{-\tfrac{1}{2}\langle v - m_V(0), B(0)(v - m_V(0))\rangle_V\} \\
& \times \exp\{\langle w, \sum_{i=1}^{k} \langle v, C_{i,A}v + c_{i,A}\rangle_V d_i\rangle_W\} \\
& \times \exp\{-\tfrac{1}{2}\langle w, \sum_{i,j=1}^{k} \langle v_1 A_{ij}v + a_{i,j}\rangle_V (d_i \,\square\, d_j)w\rangle_W\} \\
= {} & \gamma\ \exp\left\{-\tfrac{1}{2}\langle w - m_W(v), A(0)(w - m_W(v))\rangle_W\right\} \\
& \times \exp\left\{\tfrac{1}{2}\langle m_W(v), A(v)m_W(v)\rangle_W\right\} \\
& \times \exp\left\{-\tfrac{1}{2}\langle v - m_V(0), B(0)(v - m_V(0))\rangle_V\right\} \\
& \times \exp\left\{\tfrac{1}{2}\langle m_V(0), B(0)m_V(0)\rangle_V\right\} \\
= {} & \gamma'\ \exp\left\{-\tfrac{1}{2}\langle w - m_W(0), A(0)(w - m_W(0))\rangle_W\right\} \\
& \times \exp\left\{\tfrac{1}{2}\langle m_W(0), A(0)m_W(0)\rangle_W\right\} \\
& \times \exp\left\{-\tfrac{1}{2}\langle v - m_V(w), B(w)(v - m_V(w))\rangle_V\right\} \\
& \times \exp\left\{\tfrac{1}{2}\langle m_V(w), B(w)m_V(w)\rangle_V\right\},
\end{aligned}
$$

(8.36)

where γ and γ' are constants such that the last two expressions are probability densities. □

(A convenient reference for discussion of normal distributions on inner product spaces and the linear function $\square$ is Eaton (1983).)

The step from the normal conditionals distributions, described in the above theorem, to the normal distribution on $V \oplus W$, can be made using arguments closely paralleling those used in the case where $V = W = \mathbb{R}$.

8.17   Characterizing Mardia's Multivariate Pareto Distribution via Conditional Specification

The classical normal distribution is remarkable for having all of its marginals and all of its conditionals of the same (multivariate normal) form. Few distributions share this property. One that does is Mardia's multivariate Pareto distribution. Here, then, is an opportunity to seek parallel characterizations to the normal characterizations discussed in Section 8.15. As we shall see, the parallel is remarkably close.

Recall that for $\alpha > 0$, we say that X has a Pareto(σ, α) distribution if its density is of the form

$$
f_X(x) = \frac{\alpha}{\sigma}\left(1 + \frac{x}{\sigma}\right)^{-(\alpha+1)}, \qquad x > 0. \tag{8.37}
$$

Mardia (1962) introduced an interesting k-dimensional distribution with Pareto marginals (the corresponding bivariate version of this distribution was discussed in Section 7.2). The joint survival function of this distribution is of the form

$$\bar{F}_{\underline{X}}(\underline{x}) = P(\underline{X} > \underline{x}) = \left[1 + \sum_{i=1}^{k} \left(\frac{x_i}{\sigma_i}\right)\right]^{-\alpha}, \qquad \underline{x} > 0. \qquad (8.38)$$

If $\underline{X}$ has the distribution described by (8.38) we write $\underline{X} \sim MP^{(k)}(\underline{\sigma}, \alpha)$. If we write $\underline{X} = (\dot{\underline{X}}, \ddot{\underline{X}})$, where $\dot{\underline{X}}$ is k_1-dimensional and $\ddot{\underline{X}}$ is $(k - k_1)$-dimensional, it is readily verified that $\dot{\underline{X}} \sim MP^{(k_1)}(\dot{\underline{\sigma}}, \alpha)$ (where $\dot{\underline{\sigma}}$ denotes the first k_1 coordinates of $\underline{\sigma}$). Thus the Mardia multivariate Pareto has Mardia multivariate Pareto marginals. It also has Mardia multivariate Pareto conditionals. Elementary computations confirm that $\dot{\underline{X}}|\ddot{\underline{X}} = \ddot{\underline{x}} \sim MP^{(k_1)}(c(\ddot{\underline{x}})\dot{\underline{\sigma}}, \ \alpha + k - k_1)$ where $c(\ddot{\underline{x}}) = (1 + \sum_{k_1+1}^{k} x_i/\sigma_i)$. A convenient reference for properties of the Mardia multivariate Pareto distribution is Chapter 6 of Arnold (1983).

In Section 8.11 we identified the class of all distributions for which for each i, X_i given $\underline{X}_{(i)} = \underline{x}_{(i)}$ is Pareto$(\sigma(\underline{x}_{(i)}), \alpha + k - 1)$ for every $\underline{x}_{(i)}$. This family has densities of the form

$$f_{\underline{X}}(\underline{x}) = \left[\sum_{\underline{s} \in \xi_k} \delta_{\underline{s}} \left(\prod_{i=1}^{k} x_i^{s_i}\right)\right]^{-(\alpha+k)}, \qquad \underline{x} > 0, \qquad (8.39)$$

where ξ_k is the set of all vectors of 0's and 1's of dimension k. Note that the role played by α in Section 8.11 is now being played by $\alpha + k$. This family of course includes the Mardia multivariate Pareto distributions, but it includes many other distributions with Pareto conditionals.

We must make more stringent assumptions to guarantee that the distribution is of the Mardia form. We might, for example, postulate Pareto marginals as well as conditionals. It is not difficult to verify that this will indeed characterize Mardia's distribution. It is of interest to determine whether the Mardia model can be characterized using only conditional specifications.

Paralleling the results of Section 8.15 we have the following results:

Theorem 8.4 (Characterization of Mardia's multivariate Pareto distribution). *Suppose that for each i, j and each $\underline{x}_{(i,j)} \in \mathbb{R}^{k-2}$ we have*

$$(X_i, X_j)|\underline{X}_{(i,j)} = \underline{x}_{(i,j)} \sim MP^{(2)}((\sigma_i(\underline{x}_{(i,j)}), \sigma_j(\underline{x}_{(i,j)})), \alpha + k - 2) \quad (8.40)$$

for some functions $\sigma_i(\underline{x}_{(i,j)})$ and $\sigma_j(\underline{x}_{(i,j)})$. It follows that $\underline{X}$ has a Mardia k-variate Pareto distribution.   □

Proof.   Since the $MP^{(2)}$ distribution has Pareto conditionals, (8.40) implies that X_i given $\underline{X}_{(i)} = \underline{x}_{(i)}$ is Pareto$(\sigma(\underline{x}_{(i)}), \alpha + k - 1)$. It then follows that the joint density of $\underline{X}$ is given by (8.39). However, in order for a density of the form (8.39) to have conditionals of (X_i, X_j) given $\underline{X}_{(i,j)}$ of the form (8.40), all of the $\delta_{\underline{s}}$'s in (8.39) must be zero except those corresponding to vectors $\underline{s}$ with at most one nonzero coordinate. This implies that the joint density corresponds to a Mardia k-variate Pareto distribution.   □

Theorem 8.5 (Characterization of Mardia's multivariate Pareto distribution). *Suppose that for each i and any $\underline{x}_{(i)} \in \mathbb{R}^{k-1}$ we have*

$$X_i | \underline{X}_{(i)} = \underline{x}_{(i)} \sim P(\tilde{\sigma}_i(\underline{x}_{(i)}), \alpha + k - 1) \qquad (8.41)$$

for some functions $\tilde{\sigma}_i(\underline{x}_{(i)})$ and that for every i, j and every $\underline{x}_{(i,j)} \in \mathbb{R}^{k-2}$ we have

$$X_i | \underline{X}_{(i,j)} = \underline{x}_{(i,j)} \sim P(\tilde{\tilde{\sigma}}_{i,j}(\underline{x}_{(i,j)}), \alpha + k - 1) \qquad (8.42)$$

for some functions $\tilde{\tilde{\sigma}}_{i,j}(\underline{x}_{(i,j)})$.
  It follows that $\underline{X}$ has a Mardia k-variate Pareto distribution.   □

Proof. From (8.41) we conclude that the joint density is of the form (8.39), Similarly for a fixed j, for each i we have that X_i given $\underline{X}_{(i,j)}$ is Pareto distributed so that each $\underline{X}_{(j)}$ will have a $(k-1)$-dimensional joint density the form (8.39). However it is not difficult to verify that (8.39) will have $(k-1)$-dimensional marginals of the same form iff all of the $\delta_{\underline{s}}$'s in (8.39) are zero except those corresponding to vectors $\underline{s}$ with at most one nonzero coordinate. This implies that the joint density corresponds to a Mardia multivariate Pareto distribution.   □

  In the proof of Theorem 8.5 a key idea was recognition that the hypotheses led us to a distribution for $\underline{X}$ that was conditionally specified whose $(k-1)$-dimensional marginals $\underline{X}_{(j)}, j = 1, 2, \ldots, k$, had analogous $(k-1)$-dimensional conditionally specified distributions. This places severe constraints on the parameters appearing in the conditionally specified distribution of $\underline{X}$. Many of the examples described earlier in this chapter admit analogous characterization results.
  For example, if we require a k-dimensional random variable $\underline{X}$ to have an exponential conditionals distribution (8.11) with the property that every $(k-1)$-dimensional marginal be again of the exponential conditionals form (8.11), then it follows that the coordinates of $\underline{X}$ must be independent exponential random variables.

8.18 Bibliographic Notes

Section 8.3 is based on Arnold and Press (1989b). Sections 8.5 and 8.6 draw on Arnold and Strauss (1991). Most of the examples in Sections 8.8–8.13 have not been introduced elsewhere. The key references for the characterization material in Sections 8.15 and 8.17 are Arnold, Castillo, and Sarabia (1994a, 1994c). The extension to more abstract spaces is discussed in Bischoff (1996b). The multivariate exponential conditionals distribution was described in Arnold and Strauss (1988a). James (1975) has some material relating to multivariate beta conditionals distributions.

Exercises

8.1 Let $\underline{Z} = (Z_1, \ldots, Z_n)$, $n \geq 2$, an n-dimensional random vector with continuous density function $f(\underline{z})$. Split the random vector $\underline{Z}$ into two random vectors $\underline{X} = (Z_1, \ldots, Z_k)$ and $\underline{Y} = (Z_{k+1}, \ldots, Z_n)$, with $1 \leq k < n$. It is well known that both conditional densities $f(\underline{x}|\underline{y})$ and $f(\underline{y}|\underline{x})$ are multivariate normal if the joint density $f(\underline{x}, \underline{y})$ is multivariate normal. Now, suppose that the conditional densities $f(\underline{x}|\underline{y})$ and $f(\underline{y}|\underline{x})$ are both multivariate normal. Denote the covariance matrix of the conditional density $f(\underline{y}|\underline{x})$ by $\Sigma_2(\underline{x})$. Prove that the following statements are equivalent:

(a) The probability density function $f(\underline{x}, \underline{y})$ is multivariate normal.

(b) The matrix $\Sigma_2(\underline{x})$ is constant in R^k.

(c) For the minimal eigenvalue $\lambda(\underline{x})$ of the positive definite matrix $\Sigma_2(\underline{x})$,

$$u^2 \lambda(ub_j) \to \infty, \quad as \ u \to \infty, \quad for \ j = 1, 2, \ldots, k,$$

where $b_1, \ldots, b_k$ is an arbitrary but fixed basis in R^k.

(Bischoff and Fieger (1991).)

8.2 Let (X_1, X_2, X_3) be a trivariate random variable. Assume that the random variable $(X_1, X_2)|X_3 = x_3$ is Mardia Pareto $\forall x_3$ and that $(X_2, X_3)|X_1 = x_1$ is Mardia Pareto $\forall x_1$. Do these conditions guarantee that (X_1, X_2, X_3) is a trivariate Mardia Pareto distribution?

Hint: Consider the joint density given by

$$f(x_1, x_2, x_3) \propto (1 + ax_1 + bx_2 + cx_3 + mx_1 x_3)^{-(\alpha+1)}.$$

8.3 Let $(X_1, \ldots, X_n)$ be a random variable with joint probability density function $f(x_1, \ldots, x_n)$. Denote by $f(x_i|x_{(i)})$ the conditional density of X_i given the remaining variables X_j, $j \neq i$.

(a) Prove that

$$f(x_1, \ldots, x_n) \propto \frac{\displaystyle\prod_{i=1}^{n} f(x_i|x_{(i)}^0)}{\displaystyle\prod_{i=1}^{n-1} f(x_i^0|x_{(i)}^0)}$$

for a given vector $\underline{x}^0 = (x_1^0, \ldots, x_n^0)$ for which all the conditional densities in the above expression are positive.

(b) Prove that the conditional densities $f(x_i | x_{(i)})$, $i = 1, \ldots, n$, are compatible if the expression

$$\frac{\prod\limits_{i=1}^{n} f(x_i | x_{(i)}^0)}{\prod\limits_{i=1}^{n} f(x_i^0 | x_{(i)}^0)} \times \frac{\prod\limits_{i=1}^{n} f(z_i^0 | z_{(i)}^0)}{\prod\limits_{i=1}^{n} f(x_i | z_{(i)}^0)}$$

does not depend on $(x_1, \ldots, x_n)$ for $(x_1^0, \ldots, x_n^0) \neq (z_1^0, \ldots, z_n^0)$.

(Joe (1997).)

8.4 Suppose that $X_1, X_2, \ldots, X_n$ are jointly distributed random variables, $(X_1, X_2, \ldots, X_{n-1}) \overset{d}{=} (X_2, X_3, \ldots, X_n)$, and $X_n | X_1 = x_1, X_2 = x_2, \ldots, X_{n-1} = x_{n-1}$ is normal with mean $\alpha + \sum_{j=1}^{n-1} \beta_j x_j$ and variance σ^2. Then, $(X_1, X_2, \ldots, X_n)$ are jointly multivariate normal.

(Arnold and Pourahmadi (1988).)

8.5 A nonnegative function g defined on $A \subset \mathbb{R}^2$ is called totally positive of order 2 (TP2), if for all $x_1 < y_1, x_2 < y_2$ with $x_i, y_j \in A$,

$$g(x_1, x_2) g(y_1, y_2) \geq g(x_1, y_2) g(y_1, x_2).$$

If the last inequality is reversed, then g is reversed rule of order 2 (RR2).

Consider the trivariate conditionals distributions with joint pdf:

$$f(x_1, x_2, x_3) \propto \exp[-m_{100}x_1 - m_{010}x_2 - m_{001}x_2 - q(x_1, x_2, x_3)] \times I(x_i \in \mathbb{R}^+)$$

where

$$q(x_1, x_2, x_3) = m_{110}x_1 x_2 + m_{101}x_1 x_3 + m_{011}x_2 x_3 + m_{111}x_1 x_2 x_3.$$

(a) Obtain the bivariate marginals $f(x_1, x_2)$, $f(x_1, x_3)$, and $f(x_2, x_3)$.

(b) Prove that $f(x_1, x_2)$ is reversed rule of order 2 if and only if $m_{001}m_{111} \geq m_{101}m_{011}$.

(c) If $m_{110} = 0$ and $m_{001}m_{111} < m_{101}m_{011}$, then $f(x_1, x_2)$ is totally positive of order 2.

(d) If $m_{110} = m_{001} = m_{111} = 0$, and $m_{100} = m_{010} = m_{101} = m_{011} = 1$, then $f(x_1, x_2)$ is totally positive of order 2 and $f(x_1, x_3), f(x_2, x_3)$ are reversed rule of order 2.

(Joe (1997).)

8.6 Theorems 8.1 and 8.2 require only notational modification in order to remain valid when each X_i is a vector of dimension m_i (instead of 1).

(Bischoff (1996a).)

9
Estimation in Conditionally Specified Models

9.1   Introduction

Standard estimation strategies are often difficult to implement when dealing with conditionally specified models. A variety of techniques, to some degree tailor-made for conditionally specified models, will be suggested in this chapter. The emphasis will be on bivariate models but certain multivariate cases are also discussed.

9.2   The Ubiquitous Norming Constant

Almost all of the conditionally specified distributions introduced in this book are cursed with the presence of a term $e^{m_{00}}$. All other parameters save m_{00} are constrained to belong to intervals in the real line. The parameter m_{00}, we are blithely told, is chosen to make the density integrate to 1. As a consequence m_{00} is in fact an often intractable function of the other parameters. In a few cases an explicit expression is available (e.g., the exponential conditionals density, (4.22), the Pareto conditionals density, (5.10), (5.11), and (5.12), etc.). In such cases maximum likelihood becomes less troublesome. In most cases, however, more devious means will be desirable. Pseudolikelihood and modified method of moments approaches have proved to be viable approaches. They yield consistent asymptotically normal estimates. A third approach, which involves discretization of the data in absolutely continuous cases, reduces the problem to that of estimation in a log-linear Poisson regression model.

The awkward normalizing constant hampers the likelihood enthusiast but in fact is even harder on the Bayesian investigator. Some tentative approaches to Bayesian analysis of conditionally specified densities will be sketched in Section 9.9. More work and more insight is however needed in this direction.

Throughout this chapter we focus on bivariate conditionally specified distributions. In most cases the necessary modifications to extend the discussion to higher dimensions are self-evident. Some examples are provided in Section 9.10.

9.3   Maximum Likelihood

Definition 9.1 (Maximum likelihood estimate).   Suppose that we have n observations $(X_1, Y_1), (X_2, Y_2), \ldots, (X_n, Y_n)$ from some bivariate conditionally specified density $f(x, y; \underline{\theta}), \underline{\theta} \in \Theta$. The maximum likelihood estimate of θ, say $\hat{\underline{\theta}}$, is a usually unique value of $\underline{\theta}$ for which

$$\prod_{i=1}^{n} f(X_i, Y_i; \hat{\underline{\theta}}) = \max_{\underline{\theta} \in \Theta} \prod_{i=1}^{n} f(X_i, Y_i; \underline{\theta}). \tag{9.1}$$

□

One approach to finding $\hat{\underline{\theta}}$, an approach made even more feasible as computing power increases, involves a direct search. A second, and historically more favored, approach involves solving the likelihood equations

$$\frac{\partial}{\partial \theta_j} \sum_{i=1}^{n} \log f(X_i, Y_i; \underline{\theta}) = 0, \quad j = 1, 2, \ldots, k, \tag{9.2}$$

and verifying that the solution corresponds to a true maximum. In general, the method works best when a low-dimensional sufficient statistic is available. The classical situation where maximum likelihood shines is one in which $f(x, y; \underline{\theta})$ is an exponential family of densities. Note that all of the bivariate conditionals in exponential families densities, introduced in Chapters 3 and 4, were in fact themselves exponential families of bivariate densities. We can therefore expect that maximum likelihood and perhaps necessary variations on that theme will fare well in those settings. Estimation based on samples from the "other" conditionally specified densities introduced in Chapter 5 will undoubtedly require different treatment. So, in our discussion of maximum likelihood, let us focus on conditionals in exponential families.

We will begin with a very simple example. Suppose we have observations from the conditionals in given exponential families density (4.5) with $k = 1$ and $\ell = 1$, i.e.,

$$f_{X,Y}(x, y) = r_1(x)r_2(y)\exp\{m_{00}+m_{10}q_{11}(x)+m_{01}q_{21}(y)+m_{11}q_{11}(x)q_{21}(y)\}.$$
(9.3)

Here m_{10}, m_{01} and m_{11} are constrained to make the density integrable while m_{00} is determined, as a function of the other parameters, to make the integral equal to 1. A relabeling of the parameters will help us to apply well-known estimation results. Set $m_{10} = \theta_1, m_{01} = \theta_2$, and $m_{11} = \theta_3$ and let

$$\psi(\underline{\theta}) = e^{-m_{00}}$$

$$= \int_{-\infty}^{\infty} \int_{-\infty}^{\infty} r_1(x)r_2(y)\exp\{\theta_1 q_{11}(x) + \theta_2 q_{21}(y) + \theta_3 q_{11}(x)q_{21}(y)\}\, dx\, dy.$$
(9.4)

With this notation the log-likelihood of a sample of size n from our density is

$$
\begin{aligned}
\ell(\underline{\theta}) = {} & -n\log\psi(\underline{\theta}) + \sum_{i=1}^{n}\log r_1(X_i) + \sum_{i=1}^{n}\log r_2(Y_i) \\
& +\theta_1\sum_{i=1}^{n}q_{11}(X_i) + \theta_2\sum_{i=1}^{n}q_{21}(Y_i) \\
& +\theta_3\sum_{i=1}^{n}q_{11}(X_i)q_{21}(Y_i).
\end{aligned}
$$
(9.5)

Differentiating and setting the partial derivatives equal to zero yields the likelihood equations

$$\frac{\dfrac{\partial\psi(\underline{\theta})}{\partial\theta_1}}{\psi(\underline{\theta})} = \frac{1}{n}\sum_{i=1}^{n}q_{11}(X_i),$$
(9.6)

$$\frac{\dfrac{\partial\psi(\underline{\theta})}{\partial\theta_2}}{\psi(\underline{\theta})} = \frac{1}{n}\sum_{i=1}^{n}q_{21}(Y_i),$$
(9.7)

and

$$\frac{\dfrac{\partial\psi(\underline{\theta})}{\partial\theta_3}}{\psi(\underline{\theta})} = \frac{1}{n}\sum_{i=1}^{n}q_{11}(X_i)q_{21}(Y_i).$$
(9.8)

If $\psi(\underline{\theta})$ is a simple analytic expression these equations can be easily solved (directly or iteratively). Even if $\psi(\underline{\theta})$ is ugly (which is the case for most of our conditionals in exponential families examples), there is hope. Note that

$$\frac{\frac{\partial \psi(\underline{\theta})}{\partial \theta_1}}{\psi(\underline{\theta})}$$

$$= \frac{\displaystyle\int_{-\infty}^{\infty} \int_{-\infty}^{\infty} q_{11}(x) r_1(x) r_2(y) \exp\{\theta_1 q_{11}(x) + \theta_2 q_{21}(y) + \theta_3 q_{11}(x) q_{21}(y)\} \, dx \, dy}{\displaystyle\int_{-\infty}^{\infty} \int_{-\infty}^{\infty} r_1(x) r_2(y) \exp\{\theta_1 q_{11}(x) + \theta_2 q_{21}(y) + \theta_3 q_{11}(x) q_{21}(y)\} \, dx \, dy},$$

$$(9.9)$$

which can be evaluated by numerical integration. Similar expressions are available for the other terms on the left-hand sides of (9.7) and (9.8). So a possible approach involves picking initial values of θ_1, θ_2 (maybe based on crude moment estimates) then searching for a value of θ_3 to make (9.8) hold. Take this value of θ_3 with the previous value of θ_2 and search for a value of θ_1 to make (9.6) hold. Now go to (9.7), etc. The approach is computer intensive but probably more efficient than a direct search which might involve many more numerical evaluations of $\psi(\underline{\theta})$ for various choices of $\underline{\theta}$. Having solved the likelihood equations, we can, with a little more numerical integration, write an approximation for the variance-covariance matrix of our estimate $\hat{\underline{\theta}}$. The Fisher information matrix corresponding to our model is the 3×3 matrix $I(\underline{\theta})$ with the (i,j)th element given by

$$I_{ij}(\underline{\theta}) = \frac{\psi(\underline{\theta}) \frac{\partial^2}{\partial \theta_i \partial \theta_j} \psi(\underline{\theta}) - \left(\frac{\partial}{\partial \theta_i} \psi(\underline{\theta})\right)\left(\frac{\partial}{\partial \theta_j} \psi(\underline{\theta})\right)}{(\psi(\underline{\theta}))^2}. \tag{9.10}$$

Finally, the estimated variance covariance matrix of $\hat{\underline{\theta}}$ is

$$\sum(\hat{\underline{\theta}}) = [I(\hat{\underline{\theta}})]^{-1}/n, \tag{9.11}$$

where $\hat{\underline{\theta}}$ is the solution to (9.6)–(9.8). The entries in $I(\hat{\underline{\theta}})$ may be computed by numerical integration and the resulting matrix must be inverted numerically.

Example 9.1 (Centered normal conditionals distribution). As an example of this kind of analysis consider the slow firing target data discussed by Arnold and Strauss (1991). Thirty bivariate observations are assumed to be a sample from a centered normal conditionals distribution (recall Section 3.5). The joint density in question is of the form

$$f_{X,Y}(x,y) = \exp(\theta_1 x^2 + \theta_2 y^2 + \theta_3 x^2 y^2)/\psi(\underline{\theta}), \tag{9.12}$$

where $\theta_1, \theta_2 < 0$ and $\theta_3 \leq 0$. In this example

$$q_{11}(x) = x^2, \quad q_{21}(y) = y^2. \tag{9.13}$$

The form of $\psi(\underline{\theta})$ is known. Unfortunately (referring to (3.57)), it involves confluent hypergeometric functions and the expression is not useful. The complete minimal sufficient statistics for the given data set are

$$\frac{1}{30} \sum_{i=1}^{30} X_i^2 = 8.359,$$

$$\frac{1}{30} \sum_{i=1}^{30} Y_i^2 = 5.452, \tag{9.14}$$

$$\frac{1}{30} \sum_{i=1}^{30} X_i^2 Y_i^2 = 21.310.$$

These are used on the right-hand side of (9.6), (9.7), and (9.8). Iterative solution utilizing numerical integration of expressions like (9.9) yields solutions

$$\hat{\theta}_1 = -0.0389,$$
$$\hat{\theta}_2 = -0.0597, \tag{9.15}$$
$$\hat{\theta}_3 = -0.0082.$$

$\square$

In the centered normal conditionals case, an alternative to the iterative solution of (9.14) is available. It relies on the fact that the normalizing constant, though not as simple as one might wish, is at least easy to evaluate numerically.

Example 9.2 (MLE of the centered normal conditionals model).
Suppose that a random sample $(X_1, Y_1), (X_2, Y_2), \ldots, (X_n, Y_n)$ is available from a centered normal conditional distribution with joint pdf given by (3.51). Denote the observed values by (x_i, y_i), $i = 1, 2, \ldots, n$, and define

$$\overline{x^2} = \frac{1}{n} \sum_{i=1}^{n} x_i^2, \qquad \overline{y^2} = \frac{1}{n} \sum_{i=1}^{n} y_i^2, \qquad \overline{x^2 y^2} = \frac{1}{n} \sum_{i=1}^{n} x_i^2 y_i^2.$$

The log-likelihood function, used to estimate (σ_1, σ_2, c), is given by

$$l(\sigma_1, \sigma_2, c) = n \left[-\log 2\pi + \log k(c) - \log(\sigma_1 \sigma_2) - \frac{\overline{x^2}}{2\sigma_1^2} - \frac{\overline{y^2}}{2\sigma_2^2} - \frac{c\overline{x^2 y^2}}{2\sigma_1^2 \sigma_2^2} \right].$$
$$\tag{9.16}$$

Differentiation with respect σ_1, σ_2, and c yields the following likelihood equations

$$\sigma_1^2 = \overline{x^2} + c\sigma_2^{-2}\overline{x^2 y^2}, \tag{9.17}$$
$$\sigma_2^2 = \overline{y^2} + c\sigma_1^{-2}\overline{x^2 y^2}, \tag{9.18}$$
$$\delta(c) = \overline{x^2 y^2}/(2\sigma_1^2 \sigma_2^2), \tag{9.19}$$

TABLE 9.1. Representative values of $\psi(c) = \delta^{-1/2}(c)\left[1 - 2c\delta(c)\right]$, (left side of (9.22)) in order to calculate the MLE of c.

c	$\psi(c)$	c	$\psi(c)$	c	$\psi(c)$	c	$\psi(c)$
0.0	1.41421	5.0	2.51268	10.0	2.93033	15.0	3.23418
0.5	1.73576	5.5	2.56344	10.5	2.96433	15.5	3.26098
1.0	1.89634	6.0	2.61152	11.0	2.99738	16.0	3.28726
1.5	2.01529	6.5	2.65726	11.5	3.02954	16.5	3.31306
2.0	2.11285	7.0	2.70093	12.0	3.06088	17.0	3.33838
2.5	2.19692	7.5	2.74278	12.5	3.09144	17.5	3.36327
3.0	2.27156	8.0	2.78299	13.0	3.12127	18.0	3.38773
3.5	2.33914	8.5	2.82173	13.5	3.15042	18.5	3.41179
4.0	2.40119	9.0	2.85912	14.0	3.17894	19.0	3.43546
4.5	2.45879	9.5	2.89528	14.5	3.20685	19.5	3.45876

where $\delta(c)$ is as defined in (3.64). Expressions (9.17)–(9.19) can be written in the form

$$\sigma_1^2 = \overline{x^2}\left[1 - 2c\delta(c)\right]^{-1}, \tag{9.20}$$

$$\sigma_2^2 = \overline{y^2}\left[1 - 2c\delta(c)\right]^{-1}, \tag{9.21}$$

$$\delta^{-1/2}(c)\left[1 - 2c\delta(c)\right] = \left[\frac{\overline{x^2y^2}}{2\overline{x^2} \times \overline{y^2}}\right]^{-1/2}. \tag{9.22}$$

Representative values of the defined function on the left side of (9.22) are included in Table 9.1 (taken from Sarabia (1995)). We can affirm that $\delta(c)$ is a monotone increasing function of c by direct inspection of this table. The maximum likelihood estimator of c is very easy to obtain from this table (or more extended ones) once we calculate the value of the right side of (9.22). In order to obtain the asymptotic variances of the maximum likelihood estimators, we need to calculate the Fisher information matrix. From (9.16), with $n = 1$, $i_{\sigma_1\sigma_1} = -E(\partial^2 l/\partial\sigma_1^2) = 2/\sigma_1^2$ and $i_{\sigma_2\sigma_2} = 2/\sigma_2^2$. Similar computations lead to

$$i_{\sigma_1\sigma_2} = \frac{4c\delta(c)}{\sigma_1\sigma_2}, \quad \sigma_1 i_{c\sigma_1} = \sigma_2 i_{c\sigma_2} = 2\delta(c),$$

and

$$i_{cc} = -\delta'(c) = -(1 - (2 + 8c)\delta(c) + 4c^2\delta^2(c))/4c^2.$$

□

   In principle, the same kind of analysis can be performed for samples from any density of the form (4.5). The likelihood equations will be of the form

$$E(\underline{\tilde{q}}^{(1)}(X)) = \frac{1}{n}\sum_{i=1}^{n}\underline{\tilde{q}}^{(1)}(X_i),$$

$$E(\tilde{\underline{q}}^{(2)}(Y)) = \frac{1}{n}\sum_{i=1}^{n}\tilde{\underline{q}}^{(2)}(Y_i), \tag{9.23}$$

$$E(\tilde{\underline{q}}^{(1)}(X)\tilde{\underline{q}}'^{(2)}(Y)) = \frac{1}{n}\sum_{i=1}^{n}\tilde{\underline{q}}^{(1)}(X_i)\tilde{\underline{q}}'^{(2)}(Y_i),$$

where equality between vectors denotes equality of coordinates and equality between matrices indicates elementwise equality. Thus (9.23) does indeed represent $\ell_1+\ell_2+\ell_1\ell_2$ equations in the $\ell_1+\ell_2+\ell_1\ell_2$ unknown parameters.

Castillo and Galambos (1985) investigated the performance of maximum likelihood estimates in the full (eight-parameter) normal conditionals distribution ((3.25)). Predictably, when estimating so many parameters, good results are obtainable only for relatively large sample sizes. A sample of size 200 seems adequate, a sample of size 50 is probably not.

We may, of course, use maximum likelihood estimation in any conditionally specified model, even those not involving exponential families. The price we must pay is the absence of a simple sufficient statistic, but that does not necessarily imply that the likelihood equation will be completely intractable.

Example 9.3 (Bivariate Pareto conditionals). Consider a sample from a bivariate Pareto conditionals density (5.7) with $\alpha = 1$. The normalizing constant is given in (5.13). For convenience we reparametrize by introducing

$$\delta_1 = \lambda_1/\lambda_0,$$
$$\delta_2 = \lambda_2/\lambda_0,$$

and

$$\phi = (\lambda_0\lambda_3)/(\lambda_1\lambda_2).$$

Thus our density is of the form

$$f_{X,Y}(x,y) = \frac{\delta_1\delta_2(1-\phi)}{-\log\phi}[1+\delta_1 x+\delta_2 y+\phi\delta_1\delta_2 xy]^{-2},\quad x>0,\; y>0. \tag{9.24}$$

The log-likelihood of a sample from this density is given by

$$
\begin{aligned}
\ell(\delta_1,\delta_2,\phi) \quad = \quad & n\log\delta_1 + n\log\delta_2 + n\log|1-\phi| \\
& -n\log|-\log\phi| \\
& -2\sum_{i=1}^{n}\log[1+\delta_1 X_i + \delta_2 Y_i + \phi\delta_1\delta_2 X_i Y_i]. \tag{9.25}
\end{aligned}
$$

For a fixed value of ϕ, the likelihood is maximized by solving

$$\frac{n}{\delta_1} = 2\sum_{i=1}^{n}\frac{X_i + \delta_2\phi X_i Y_i}{1+\delta_1 X_i + \delta_2 Y_i + \phi\delta_1\delta_2 X_i Y_i},$$

$$\frac{n}{\delta_2} = 2 \sum_{i=1}^{n} \frac{Y_i + \delta_1 \phi X_i Y_i}{1 + \delta_1 X_i + \delta_2 Y_i + \phi \delta_1 \delta_2 X_i Y_i}. \qquad (9.26)$$

These can be solved iteratively and then a straightforward search procedure can be used to find the optimal value of ϕ. If we are not willing or able to assume that the parameter $\alpha = 1$ in our Pareto conditionals model, then the complicated nature of the resulting normalizing constant leads us to seek alternatives to maximum likelihood estimation such as those described in the following sections.   □

9.4   Pseudolikelihood Involving Conditional Densities

The normalizing constant is the only thing that makes the above analysis nonroutine. Surely we can finesse knowledge of the normalizing constant. We know a great deal about the density without it. The perfect tool for estimation of conditionally specified distributions, especially those with conditionals in exponential families, is a particular form of pseudolikelihood (in the sense of Arnold and Strauss (1988b)) involving conditional densities. It is proposed to seek $\tilde{\underline{\theta}}$ to maximize the pseudolikelihood function.

Definition 9.2 (Pseudolikelihood estimate).   Suppose that we have n observations $(X_1, Y_1), (X_2, Y_2), \ldots, (X_n, Y_n)$ from some bivariate conditionally specified density $f(x, y; \underline{\theta}), \underline{\theta} \in \Theta$. The maximum pseudolikelihood estimate of θ, say $\tilde{\underline{\theta}}$, is a usually unique value of $\underline{\theta}$ for which

$$PL(\tilde{\underline{\theta}}) = \prod_{i=1}^{n} f_{X|Y}(X_i|Y_i; \tilde{\underline{\theta}}) f_{Y|X}(Y_i|X_i; \tilde{\underline{\theta}})$$

$$= \max_{\underline{\theta} \in \Theta} \prod_{i=1}^{n} f_{X|Y}(X_i|Y_i; \underline{\theta}) f_{Y|X}(Y_i|X_i; \underline{\theta}). \qquad (9.27)$$

□

Arnold and Strauss (1988b) show that the resulting estimate $\tilde{\underline{\theta}}$ is consistent and asymptotically normal with a potentially computable asymptotic variance. In exchange for simplicity of calculation (since the conditionals and hence the pseudolikelihood do not involve the normalizing constant) we pay a price in slightly reduced efficiency.

We will illustrate the technique in the case where the conditionals are members of one-parameter exponential families. The technique can be expected to work well in the general case of conditionals in multiparameter exponential families. The only requirement is that we should have relatively simple analytic expressions available for the β_i's appearing in (4.1) and (4.2). If not, then the conditional likelihood technique will not have any great computational advantage over ordinary maximum likelihood.

In the case $k = \ell = 1$, the joint density assumes the form (9.3) and again we reparametrize by setting $m_{10} = \theta_1, m_{01} = \theta_2$, and $m_{11} = \theta_3$. With this convention we may write the conditional densities of $X|Y = y$ as follows:

$$f_{X|Y}(x|y) = r_1(x)\beta_1(\theta_1 + \theta_3 q_{21}(y)) \exp[(\theta_1 + \theta_3 q_{21}(y))q_{11}(x)], \quad (9.28)$$

where $\beta_1(\cdot)$ is as defined in (4.2) (it represents the norming constant function for the one-parameter exponential family to which the conditional distribution of X given $Y = y$ belongs for every y). An anologous expression to (9.28) is available for $f_{X|Y}(x|y)$. Substituting these in (9.27) and taking the logarithm yields the following objective function to be maximized by suitable choices of θ_1, θ_2, and θ_3:

$$
\begin{aligned}
\log \mathrm{PL}(\underline{\theta}) = {} & \sum_{i=1}^{n} \log r_1(X_i) + \sum_{i=1}^{n} \log \beta_1(\theta_1 + \theta_3 q_{21}(Y_i)) \\
& + \theta_1 \sum_{i=1}^{n} q_{11}(X_i) + \theta_3 \sum_{i=1}^{n} q_{11}(X_i)q_{21}(Y_i) \\
& + \sum_{i=1}^{n} \log r_2(Y_i) + \sum_{i=1}^{n} \log \beta_2(\theta_2 + \theta_3 q_{11}(X_i)) \\
& + \theta_2 \sum_{i=1}^{n} q_{21}(Y_i) + \theta_3 \sum_{i=1}^{n} q_{11}(X_i)q_{21}(Y_i). \quad (9.29)
\end{aligned}
$$

Differentiating with respect to the θ_i's and equating to zero leads to the following pseudolikelihood equations:

$$- \sum_{i=1}^{n} \frac{\beta_1'(\theta_1 + \theta_3 q_{21}(Y_i))}{\beta_1(\theta_1 + \theta_3 q_{21}(Y_i))} = \sum_{i=1}^{n} q_{11}(X_i), \quad (9.30)$$

$$- \sum_{i=1}^{n} \frac{\beta_2'(\theta_2 + \theta_3 q_{11}(X_i))}{\beta_2(\theta_2 + \theta_3 q_{11}(X_i))} = \sum_{i=1}^{n} q_{21}(Y_i), \quad (9.31)$$

and

$$
\begin{aligned}
2 \sum_{i=1}^{n} q_{11}(X_i)q_{21}(Y_i) = {} & - \sum_{i=1}^{n} \frac{q_{21}(Y_i)\beta_1'(\theta_1 + \theta_3 q_{21}(Y_i))}{\beta_1(\theta_1 + \theta_3 q_{21}(Y_i))} \\
& (9.32) \\
& - \sum_{i=1}^{n} \frac{q_{11}(X_i)\beta_2'(\theta_2 + \theta_3 q_{11}(X_i))}{\beta_2(\theta_2 + \theta_3 q_{11}(X_i))}.
\end{aligned}
$$

The expression $\tau_1(\theta) = -\beta_1'(\theta)/\beta_1(\theta)$, which appears in (9.30) and (9.32), actually has a simple interpretation. Denote by Z_θ a random variable with the density (4.1) (with $k = 1$). It is readily verified that

$$\tau_1(\theta) \overset{\triangle}{=} -\beta_1'(\theta)/\beta_1(\theta) = E(q_{11}(Z_\theta)). \quad (9.33)$$

Analogously $\tau_2(\theta) = E(q_{21}(W_\theta))$, where W_θ has density (4.2).

Example 9.4 (Poisson conditionals distributions). As an example, consider the case in which all conditionals are Poisson distributions, i.e., density (4.99). In this case, $q_1(t) = q_2(t) = t$ and $\beta_i(\theta) = \exp(-e^\theta), i = 1, 2$. Consequently, $-\beta_i(\theta)/\beta_i(\theta) = e^\theta$. Thus our pseudolikelihood equations take the relatively simple form

$$e^{\theta_1} \sum_{i=1}^{n} e^{\theta_3 Y_i} = \sum_{i=1}^{n} X_i, \tag{9.34}$$

$$e^{\theta_2} \sum_{i=1}^{n} e^{\theta_3 X_i} = \sum_{i=1}^{n} Y_i, \tag{9.35}$$

and

$$e^{\theta_1} \sum_{i=1}^{n} Y_i e^{\theta_3 Y_i} + e^{\theta_2} \sum_{i=1}^{n} X_i e^{\theta_3 X_i} = \sum_{i=1}^{n} X_i Y_i. \tag{9.36}$$

Iterative solution is possible. Note that (9.34) and (9.35) give simple expressions for θ_1, θ_2 for a given value of θ_3 and the right-hand side of (9.36) is a monotone function of θ_3 for given θ_1 and θ_2. □

Example 9.5 (Centered normal conditionals distribution). Returning again to our centered normal conditionals example using the parametrization of Example 9.2, the pseudolikelihood function assumes the form

$$\text{PL}(\sigma_1, \sigma_2, c) \propto (\sigma_1 \sigma_2)^{-n} \prod_{i=1}^{n} \left[(1 + cy_i^2/\sigma_2^2)(1 + cx_i^2/\sigma_1^2) \right]^{1/2}$$
$$\exp \left[-\frac{1}{2\sigma_1^2} \sum_{i=1}^{n} (1 + cy_i^2/\sigma_2^2) x_i^2 - \frac{1}{2\sigma_2^2} \sum_{i=1}^{n} (1 + cx_i^2/\sigma_1^2) y_i^2 \right]$$

and the corresponding pseudolikelihood equations are given by

$$\sigma_1^2 + \frac{1}{n} \sum_{i=1}^{n} \frac{cx_i^2}{1 + cx_i^2/\sigma_1^2} = \overline{x^2} + \frac{2c}{\sigma_2^2} \overline{x^2 y^2},$$

$$\sigma_2^2 + \frac{1}{n} \sum_{i=1}^{n} \frac{cy_i^2}{1 + cy_i^2/\sigma_2^2} = \overline{y^2} + \frac{2c}{\sigma_1^2} \overline{x^2 y^2},$$

$$\frac{1}{n} \sum_{i=1}^{n} \frac{x_i^2/\sigma_1^2}{1 + cx_i^2/\sigma_1^2} + \frac{1}{n} \sum_{i=1}^{n} \frac{y_i^2/\sigma_2^2}{1 + cy_i^2/\sigma_2^2} = \overline{x^2 y^2}/(\sigma_1^2 \sigma_2^2).$$

These are readily solved in an iterative fashion. □

9.5   Marginal Likelihood

In the case of conditionally specified bivariate densities, the unfriendly normalizing constant is needed to specify the marginal densities, as well as the bivariate density. There will generally be little or no advantage to be gained by focussing on marginal likelihoods. Not only that, marginal likelihoods may be uninformative about some parameters in the bivariate model. For example, in the classical bivariate normal model, marginal data will tell us nothing about the correlation or covariance parameter. Of course, the motivation for using marginal data may be that it is all that is available. If that is the case, then maximization of the marginal likelihood function may well provide us with consistent asymptotic normal estimates of most and perhaps all of the θ_i's.

Definition 9.3 (Maximum marginal likelihood estimate). Suppose that we have n observations $(X_1, Y_1), (X_2, Y_2), \ldots, (X_n, Y_n)$ from some bivariate conditionally specified density $f(x, y; \underline{\theta}), \underline{\theta} \in \Theta$. The maximum marginal likelihood estimate of θ, say $\tilde{\underline{\theta}}$, is a usually unique value of $\underline{\theta}$ for which

$$\text{ML}(\tilde{\underline{\theta}}) = \prod_{i=1}^{n} f_X(X_i; \tilde{\underline{\theta}}) \prod_{i=1}^{n} f_Y(Y_i; \tilde{\underline{\theta}}) = \max_{\underline{\theta} \in \Theta} \prod_{i=1}^{n} f_X(X_i; \underline{\theta}) \prod_{i=1}^{n} f_Y(Y_i; \underline{\theta}).$$
(9.37)

$\square$

For example, Castillo and Galambos (1985) report on the successful use of such an approach for estimating the eight parameters of the normal conditionals model ((3.25)).

9.6   An Efficiency Comparison

Evidently the marginal and conditional likelihood approaches will lead to loss of efficiency when compared to the maximum likelihood solution. There are some indications that the loss may not be as great as one might fear; at least in the case of conditional likelihood. We should be prepared for potentially catastrophic drops in efficiency when we resort to marginal inference. For example, some parameters may not even be marginally estimable. The classical bivariate normal example reminds us that this can occur in models that would definitely not be considered pathological. We can contrive situations in which conditional inference will be uninformative. For example, if X is Poisson(θ) and $Y = X$ with probability 1. In general, we expect that conditional inference will be effective but somewhat inefficient (since the estimates obtained are usually not functions of the minimal sufficient statistics). Conditional likelihood estimates can be fully efficient. If we consider a sample from a classical five parameter bivariate normal distribution, we find that the maximum likelihood and maximum conditional likelihood estimates are identical.

The example we study in detail is one in which all three methods (full likelihood, marginal likelihood, and conditional likelihood) are relatively straightforward and all are informative about the unknown parameter. Details regarding the necessary computations are to be found in Arnold and Strauss (1988b).

Example 9.6 (Bivariate exponential conditionals model). We consider n observations from a bivariate exponential conditionals model with unit scale parameters. Thus the density from (4.14) can be written in the form

$$f_{X,Y}(x,y) = k(\theta) \exp\{-(x + y + \theta xy)\}, \qquad x, y > 0, \tag{9.38}$$

where

$$k(\theta) = \left[\int_0^\infty e^{-u}(1 + \theta u)^{-1} du\right]^{-1}. \tag{9.39}$$

The parameter θ reflects dependence and is necessarily nonnegative ($\theta = 0$ corresponds to independent standard exponential marginals). The likelihood equation corresponding to a sample $(X_1, Y_1), \ldots, (X_n, Y_n)$ from (9.38) reduces to

$$[1 + \theta - k(\theta)]/\theta^2 = \frac{1}{n}\sum_{i=1}^n X_i Y_i. \tag{9.40}$$

It is relatively easy to solve (9.40) numerically, although the implied use of (9.39) means that a considerable number of numerical integrations may be required. Denote the solution to (9.40) by $\hat{\theta}$. The asymptotic distribution of this maximum likelihood estimate is then

$$\hat{\theta} \overset{.}{\sim} N(\theta, [nI(\theta)]^{-1}), \tag{9.41}$$

where
$$I(\theta) = [\theta^2 + 2\theta - k(\theta)(k(\theta) + \theta - 1)]/\theta^4, \qquad \theta \neq 0. \tag{9.42}$$

Next we turn to marginal inference. The marginal density of X derived from (9.38) is

$$f_X(x) = k(\theta)(1 + \theta x)^{-1}e^{-x}, \qquad x > 0. \tag{9.43}$$

The marginal density of Y is identical to that of X. Our marginal likelihood equation

$$\frac{d}{d\theta}\sum_{i=1}^n[\log f_X(X_i; \theta) + \log f_Y(Y_i; \theta)] = 0 \tag{9.44}$$

simplifies to

$$[1 + \theta - k(\theta)]/\theta^2 = \frac{1}{2n} \sum_{i=1}^{n} \left[\frac{X_i}{1 + \theta X_i} + \frac{Y_i}{1 + \theta Y_i} \right]. \tag{9.45}$$

This can be solved by a straightforward search procedure. Denoting the resulting marginal likelihood estimate by $\hat{\theta}_M$ we may verify that

$$\hat{\theta}_M \overset{\cdot}{\sim} N(\theta, [nI_M(\theta)]^{-1}), \tag{9.46}$$

where

$$I_M(\theta) = \frac{[1 + 2k^2(\theta) - (\theta + 3)k(\theta)]^2}{\theta^4[(\theta + 7)k(\theta) - 4k^2(\theta) - 3 + 2\theta^2 M(\theta)]}, \tag{9.47}$$

in which

$$M(\theta) = e^{1/\theta}k(\theta)\theta^{-2} \int_0^{\theta} e^{-1/u}[k(u)]^{-1}\, du.$$

We observe that the marginal likelihood equation (9.45), involving $k(\theta)$ as it does, is actually more troublesome to solve than is the full likelihood equation (9.40). Presumably we would only use $\hat{\theta}_M$ if only marginal data were available.

Now consider conditional or pseudolikelihood estimation. Our objective function, to be maximized, is

$$\log PL(\theta) = \sum_{i=1}^{n} [\log(1 + \theta Y_i) - (1 + \theta Y_i)X_i$$
$$+ \log(1 + \theta X_i) - (1 + \theta X_i)Y_i]. \tag{9.48}$$

The equation $\dfrac{\partial \log PL(\theta)}{\partial \theta} = 0$ simplifies to the form

$$\sum_{i=1}^{n} \frac{X_i}{1 + \theta X_i} + \sum_{i=1}^{n} \frac{Y_i}{1 + \theta Y_i} = 2 \sum_{i=1}^{n} X_i Y_i. \tag{9.49}$$

The left-hand side is a decreasing function of θ so that a simple search procedure can be used to solve (9.49). If we denote the solution to (9.49) by $\hat{\theta}_C$, we may verify (see Arnold and Strauss (1988b)) that

$$\hat{\theta}_C \overset{\cdot}{\sim} N(\theta, [nI_C(\theta)]^{-1}), \tag{9.50}$$

where

$$I_C(\theta) = \frac{[1 + 4\theta + 2\theta^2 - (1 + 3\theta)k(\theta)]^2}{\theta^4[1 + 8\theta + 4\theta^2 - k(\theta)(1 + 7\theta) + 2\theta^2 M(\theta)]}, \tag{9.51}$$

in which $M(\theta)$ is as defined following (9.47). By referring to (9.42), (9.47), and (9.51) it is possible to graphically compare the asymptotic variances of

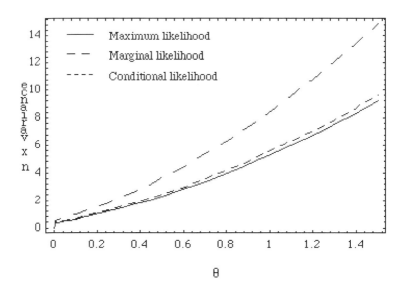

FIGURE 9.1. Asymptotic variances of competing estimates.

the three competing estimates. Reference to Figure 9.1 confirms the assertion that the conditional estimate is remarkably efficient when compared to the maximum likelihood estimate, no matter what the true value of θ may be. The marginal estimate rapidly becomes seriously inefficient as θ increases, underlining our earlier conclusion that we would only use such an estimate in situations where only marginal information were available.   □

9.7   Method of Moments Estimates

Long before Fisher presented and popularized likelihood estimation techniques, consistent asymptotically normal parametric estimates were routinely identified using a variety of schemes fitting loosely into the category of "method of moments" techniques. Roughly speaking, the technique could be quite routinely employed when we had available n vector observations $\underline{X}_1, \ldots, \underline{X}_n$ from some parametric family of densities $f(\underline{x}; \underline{\theta})$ where $\underline{\theta} = (\theta_1, \ldots, \theta_k)$ was k-dimensional.

Definition 9.4 (Method of moments estimate).   Suppose that we have n observations $(X_1, Y_1), (X_2, Y_2), \ldots, (X_n, Y_n)$ from some bivariate conditionally specified density $f(x, y; \underline{\theta}), \underline{\theta} \in \Theta$. Choose judiciously k functions $\phi_1, \ldots, \phi_k$ such that the expressions

$$E_{\underline{\theta}}(\phi_i(\underline{X})) = g_i(\underline{\theta}), \quad i = 1, 2, \ldots, k, \tag{9.52}$$

are analytically computable, involve collectively each $\theta_i, i = 1, 2, \ldots, k$, and are functionally independent. Set up the k equations

$$g_i(\underline{\theta}) = \frac{1}{n} \sum_{j=1}^{n} \phi_i(\underline{X}_j), \qquad i = 1, 2, \ldots, k, \tag{9.53}$$

and solve for $\underline{\theta}$. The resulting solution $\tilde{\underline{\theta}}$ is, under regularity conditions, a consistent asymptotically normal estimate of $\underline{\theta}$, which is called a method of moments estimate. □

In the case of conditionally specified densities, the method of moments technique works quite well but may well involve repeated recomputation of the awkward density norming constant in our efforts to solve the equations of the form (9.53). In the case of densities involving conditionals in exponential families, a judicious choice of ϕ_i's in (9.53) actually gives method of moment equations which are exactly the likelihood equations displayed in (9.23). The solution of these equations, as is the case for almost any other set of moment equations, will repeatedly involve the normalizing constant; feasible but tiresome.

A clever swindle due to Strauss (introduced in Arnold and Strauss (1988a)) helps us avoid the problems with the normalizing constant. Of course a small price must be paid in terms of reduced efficiency but, when the normalizing constant is particularly fearsome, that may be judged to be a very small price to pay. The technique will work well in most conditionally specified models. It relies on a simple device of treating the awkward normalizing constant, say $k(\underline{\theta}) = \theta_0$, as an additional parameter in addition to the k θ_i's and setting up one additional moment equation, $k + 1$ in all, and solving for $\theta_0, \theta_1, \ldots, \theta_k$. The estimated values obtained for $\theta_1, \ldots, \theta_k$ in this way are clearly again consistent asymptotically normal estimates. A simple example, once again the exponential conditionals case, will illustrate the technique.

Example 9.7 (Exponential conditionals distribution). Suppose we have n observations from the density

$$f_{X,Y}(x, y) = k(\underline{\theta}) \exp[-(\theta_1 x + \theta_2 y + \theta_3 xy)], \quad x, y > 0, \tag{9.54}$$

where $k(\underline{\theta})$ is such that the integral over the positive quadrant equals 1. Denote $k(\underline{\theta})$ by θ_0 and treat (9.54) as a density involving four parameters $\theta_0, \theta_1, \theta_2, \theta_3$ (ignoring for convenience the fact that θ_0 is a function of $\theta_1, \theta_2, \theta_3$).

Now consider the following four functions of (X, Y):

$$
\begin{aligned}
\phi_1(X, Y) &= X, \\
\phi_2(X, Y) &= Y, \\
\phi_3(X, Y) &= XY, \\
\phi_4(X, Y) &= (X + Y)^2.
\end{aligned}
\tag{9.55}
$$

Note that the number of ϕ_i's is one more than the true dimension of the parameter space. We propose to equate sample averages of the ϕ_i's in (9.55) to their expectations written as functions of $\theta_1, \theta_2, \theta_3$, and θ_0 and solve for the parameters. It is not difficult to verify that

$$
\begin{aligned}
E(X) &= (\theta_0 - 1)\theta_2/\theta_3, \\
E(Y) &= (\theta_0 - 1)\theta_1/\theta_3, \\
E(XY) &= \frac{\theta_1\theta_2(\theta_0 - \theta_0^2) + \theta_3(1 - \theta_1 - \theta_2) + \theta_0\theta_3(\theta_1 + \theta_2)}{\theta_3^2},
\end{aligned}
\tag{9.56}
$$

and

$$
E[(X + Y)^2] = \frac{1}{\theta_3^2}\left[(1 - \theta_0)(\theta_1^2 + \theta_2^2) + \theta_0\theta_1\theta_2\theta_3\left[\frac{1}{\theta_1} + \frac{1}{\theta_2}\right]\right] + 2E(XY).
$$

Consistent estimates are then obtainable by solving the system

$$
\frac{1}{n}\sum_{i=1}^{n} X_i = E(X),
$$

$$
\frac{1}{n}\sum_{i=1}^{n} Y_i = E(Y),
$$

$$
\frac{1}{n}\sum_{i=1}^{n} X_iY_i = E(XY),
\tag{9.57}
$$

$$
\frac{1}{n}\sum_{i=1}^{n} (X_i + Y_i)^2 = E[(X + Y)^2],
$$

for $\theta_0, \theta_1, \theta_2$, and θ_3.

A slight variation on this scheme actually permits us to write explicitly a set of consistent asymptotically normal estimates. Define, following Arnold and Strauss (1988a),

$$\bar{X} = \frac{1}{n}\sum_{i=1}^{n} X_i,$$

$$\bar{Y} = \frac{1}{n}\sum_{i=1}^{n} Y_i,$$

$$S_X^2 = \frac{1}{n}\sum_{i=1}^{n}(X_i - \bar{X})^2, \qquad (9.58)$$

$$S_Y^2 = \frac{1}{n}\sum_{i=1}^{n}(Y_i - \bar{Y})^2,$$

$$S_{XY} = \frac{1}{n}\sum_{i=1}^{n}(X_i - \bar{X})(Y_i - \bar{Y}),$$

and

$$R_{XY} = S_{XY}/\sqrt{S_X^2 S_Y^2}.$$

By the strong law of large numbers we know that, as $n \to \infty$,

$$\bar{X} \xrightarrow{a.s.} (\theta_0 - 1)\theta_2/\theta_3,$$

$$\bar{Y} \xrightarrow{a.s.} (\theta_0 - 1)\theta_1/\theta_3,$$

$$R_{XY} \xrightarrow{a.s.} \rho(X,Y) = \frac{\left(\dfrac{\theta_3}{\theta_1\theta_2} + \theta_0 - \theta_0^2\right)}{\theta_0\left(1 + \dfrac{\theta_3}{\theta_1\theta_2} - \theta_0\right)}, \qquad (9.59)$$

$$T = \sqrt{S_X^2 S_Y^2}/\bar{X}\bar{Y} \xrightarrow{a.s.} \frac{\theta_0\left(1 + \dfrac{\theta_3}{\theta_1\theta_2} - \theta_0\right)}{(\theta_0 - 1)^2}.$$

Now we equate the left- and right-hand sides of the expressions in (9.59) and solve. We obtain the following relatively simple consistent asymptotically normal estimates:

$$\begin{aligned}
\tilde{\theta}_0 &= T/(1 + R_{XY}T), \\
\tilde{\theta}_1 &= \tilde{\theta}_0/[\bar{X}(\tilde{\theta}_0 + T(\tilde{\theta}_0 - 1))], \\
\tilde{\theta}_2 &= \tilde{\theta}_0/[\bar{Y}(\tilde{\theta}_0 + T(\tilde{\theta}_0 - 1))], \qquad (9.60) \\
\tilde{\theta}_3 &= \tilde{\theta}_0(\tilde{\theta}_0 - 1)/[\bar{X}\bar{Y}(\tilde{\theta}_0 + T(\tilde{\theta}_0 - 1))].
\end{aligned}$$

Note that in these expressions the true value of θ_0 is always ≥ 1. The true correlation is always nonpositive so that the estimated value of θ_0 given in (9.60) will usually be also ≥ 1. If $\tilde{\theta}_0 < 1$, then this can be taken as an indication that the data do not fit the exponential conditionals model.   □

Example 9.8 (Centered normal conditionals distribution). Recall that we have data $(X_1, Y_1), (X_2, Y_2), \ldots, (X_n, Y_n)$ with common density (3.51). Treat the normalizing constant $\delta(c) = \delta$ as a fourth parameter and define

$$\overline{X^2} = \frac{1}{n} \sum_{i=1}^{n} X_i^2,$$

$$\overline{Y^2} = \frac{1}{n} \sum_{i=1}^{n} Y_i^2,$$

$$S_{X^2}^2 = \frac{1}{n} \sum_{i=1}^{n} (X_i^2 - \overline{X^2})^2,$$

$$S_{Y^2}^2 = \frac{1}{n} \sum_{i=1}^{n} (Y_i^2 - \overline{Y^2})^2,$$

$$S_{X^2 Y^2} = \frac{1}{n} \sum_{i=1}^{n} (X_i^2 - \overline{X^2})(Y_i^2 - \overline{Y^2}),$$

then by the strong law of large numbers, we may conclude that as $n \to \infty$:

$$\overline{X^2} \xrightarrow{a.s.} \sigma_1^2 (1 - 2c\delta),$$

$$\overline{Y^2} \xrightarrow{a.s.} \sigma_2^2 (1 - 2c\delta),$$

$$T = \frac{S_{X^2} S_{Y^2}}{\overline{X^2} \times \overline{Y^2}} \xrightarrow{a.s.} \frac{1 + 2\delta - 4c^2\delta^2}{(1 - 2c\delta)^2},$$

$$R = \frac{S_{X^2 Y^2}}{S_{X^2} S_{Y^2}} \xrightarrow{a.s.} \frac{1 - 2\delta - 4c\delta + 4c^2\delta^2}{-1 - 2\delta + 4c^2\delta^2}.$$

By equating these statistics to their a.s. limits and solving, the following strongly consistent estimators are obtained:

$$\hat{c} = \frac{1}{4}\left[1 + RT - 2T - \frac{1 - T^2}{1 + RT}\right], \tag{9.61}$$

$$\hat{\delta} = \frac{2(1 + RT)}{(1 - R)^2 T^2}, \tag{9.62}$$

$$\hat{\sigma_1^2} = \frac{\overline{X^2}}{1 - 2\hat{c}\hat{\delta}}, \tag{9.63}$$

$$\hat{\sigma_2^2} = \frac{\overline{Y^2}}{1 - 2\hat{c}\hat{\delta}}, \tag{9.64}$$

and, therefore, $\hat{k} = \exp(\hat{c}\hat{\delta})$ will be the estimator of the normalizing constant. $\qquad\square$

Yet another manner to obtain suitable moment equations exists in the exponential conditionals case. Note that if (X, Y) have joint density given by (9.54) it follows that:

$$(\theta_1 + \theta_3 Y)X \sim \Gamma(1,1) \tag{9.65}$$

and

$$(\theta_2 + \theta_3 X)Y \sim \Gamma(1,1). \tag{9.66}$$

Equating sample moments of the variables on the left-hand sides of (9.65) and (9.66), with their corresponding known theoretical expectations, will give us a plethora of moment equations to choose from (and, incidentally, no guidance on how to choose among them). For example we could set up the following three equations:

$$\theta_1 \frac{1}{n} \sum_{i=1}^{n} X_i + \theta_3 \frac{1}{n} \sum_{i=1}^{n} X_i Y_i = 1,$$

$$\theta_2 \frac{1}{n} \sum_{i=1}^{n} Y_i + \theta_3 \frac{1}{n} \sum_{i=1}^{n} X_i Y_i = 1, \tag{9.67}$$

and

$$\theta_1^2 \frac{1}{n} \sum_{i=1}^{n} X_i^2 + \theta_3^2 \frac{1}{n} \sum_{i=1}^{n} X_i^2 Y_i^2 + 2\theta_1 \theta_3 \frac{1}{n} \sum_{i=1}^{n} X_i^2 Y_i = 2,$$

and solve for θ_1, θ_2, and θ_3.

Another possibility would involve choosing θ_1, θ_2, and θ_3 to minimize some weighted sum of squared deviations between sample and theoretical moments corresponding to (9.65) and (9.66). Thus our objective function might be

$$\sum_{j=1}^{M} e^{-j^2} \left\{ \left[\frac{1}{n} \sum_{\ell=1}^{j} \binom{j}{\ell} \theta_1^\ell \theta_3^{j-\ell} \sum_{i=1}^{n} X_i^j Y_i^{j-\ell} - (j!) \right]^2 \right.$$

$$\left. + \left[\frac{1}{n} \sum_{\ell=1}^{j} \binom{j}{\ell} \theta_2^\ell \theta_3^{j-\ell} \sum_{i=1}^{n} X_i^{j-\ell} Y_i^j - (j!) \right]^2 \right\}, \tag{9.68}$$

where M is not too large. Other possible schemes will occur to the reader. Note that techniques of this type can also be used for the general normal conditionals distribution. In that case, we can write

$$\phi_1(Y)(X - \phi_2(Y)) \sim N(0,1) \tag{9.69}$$

and

$$\phi_3(X)(Y - \phi_4(X)) \sim N(0,1) \tag{9.70}$$

for suitably chosen ϕ_1, ϕ_2, ϕ_3 and ϕ_4. Then we can equate several sample and theoretical moments of (9.69) and (9.70) and solve to estimate the parameters. Another situation where the technique works is the case where the conditionals are gamma distributed with unknown but constant shape parameters. In that case, $X(\theta_1 + \theta_3 Y)$ and $Y(\theta_2 + \theta_3 X)$ have gamma distributions with unit scale and constant shapes, so equations analogous to (9.67) are readily set up.

Example 9.9 (Gamma conditionals distribution Model II). In the above situation, if we assume in addition that $\theta_1 = \theta_2 = 1$, not only can we set up equations analogous to (9.67) but we can even obtain closed-form solutions. The common density of the (X_i, Y_i)'s in this case can be written as

$$f_{X,Y}(x, y) = \frac{k_{r,s}(c)}{\Gamma(r)\Gamma(s)} x^{r-1} y^{s-1} e^{-x-y-cxy}.$$

Then we have

$$X(1 + cY) \sim \Gamma(r, 1) \text{ independent of } Y, \tag{9.71}$$

$$Y(1 + cX) \sim \Gamma(s, 1) \text{ independent of } X, \tag{9.72}$$

and, consequently,

$$E\left[XY(1 + cY)\right] = E\left[X(1 + cY)\right] E\left[Y\right], \tag{9.73}$$

$$E\left[XY(1 + cX)\right] = E\left[Y(1 + cX)\right] E\left[X\right]. \tag{9.74}$$

Addition of these equations, and some algebraic manipulation, leads to

$$c = -2 \frac{\text{cov}(X, Y)}{\text{cov}(XY, X + Y)}. \tag{9.75}$$

From (9.71) and (9.72) we get

$$E\left[X(1 + cY)\right] = r,$$
$$E\left[Y(1 + cX)\right] = s. \tag{9.76}$$

Now we define the sample moments

$$\overline{X} = \frac{1}{n} \sum_{i=1}^{n} X_i,$$
$$\overline{Y} = \frac{1}{n} \sum_{i=1}^{n} Y_i,$$
$$\overline{XY} = \frac{1}{n} \sum_{i=1}^{n} X_i Y_i, \tag{9.77}$$
$$S_{X,Y} = \frac{1}{n} \sum_{i=1}^{n} (X_i - \overline{X})(Y_i - \overline{Y}),$$

and

$$S_{XY,X+Y} = \frac{1}{n} \sum_{i=1}^{n} \left(X_i Y_i - \overline{XY} \right) \left(X_i + Y_i - \overline{X} - \overline{Y} \right). \tag{9.78}$$

Finally, using (9.75) and (9.76), equating the statistics in (9.77) and (9.78) to their population values, we obtain the following consistent estimates of the parameters:

$$\hat{c} = -2\frac{S_{X,Y}}{S_{XY,X+Y}}, \tag{9.79}$$

$$\hat{r} = \overline{X} - \hat{c}\overline{XY}, \tag{9.80}$$

$$\hat{s} = \overline{Y} - \hat{c}\overline{XY}. \tag{9.81}$$

$\square$

9.8   Log-Linear Poisson Regression Estimates

The curse of the normalizing constant has been shown to be avoidable by small swindles such as treating conditional rather than full likelihood or by introducing an extra parameter of convenience. Consistent, but generally not fully efficient estimators, result from these manoeuvres. In this section another simple swindle is introduced. The motivation and explanations are simplest in finite discrete cases; so we begin with such an example.

Suppose $(X_1, Y_1), (X_2, Y_2), \ldots, (X_n, Y_n)$ are i.i.d. random variables with a common binomial-conditionals distribution (recall Section 4.12). Here n_1, n_2 are fixed and known and the joint density for (X, Y) can be written in the reparametrized form

$$f_{X,Y}(x, y) = \exp\left[\theta_0 + c(x, y) + \theta_1 x + \theta_2 y + \theta_3 xy \right]$$

$$\times I(x \in \{0, 1, 2, \ldots, n_1\}) I(y \in \{0, 1, 2, \ldots, n_2\}), \tag{9.82}$$

where $c(x, y) = \log \binom{n_1}{x} + \log \binom{n_2}{y}$, $\theta_1 \in \mathbb{R}, \theta_2 \in \mathbb{R}, \theta_3 \in \mathbb{R}$. Of course θ_0 is the awkward normalizing constant. We wish to estimate $(\theta_1, \theta_2, \theta_3)$. For each possible value (i, j) of (X, Y) let N_{ij} denote the number of observations for which $X = i$ and $Y = j$. Let $\underline{N}$ denote the two-way contingency table of N_{ij}'s. The random variable $\underline{N}$ has a multinomial distribution, i.e. $\underline{N} \sim \text{multinomial}(n, \underline{p})$, where

$$\log p_{ij} = \theta_0 + c(i, j) + \theta_1 i + \theta_2 j + \theta_3 ij. \tag{9.83}$$

But instead, if $\underline{N}^*$ were thought of as having coordinates that are independent Poisson(p_{ij}) random variables, then the conditional distribution of $\underline{N}^*$ given $\sum_{i=1}^{n_1} \sum_{j=1}^{n_2} N_{ij}^* = n$, would be, again, multinomial$(n, \underline{p})$. Consequently, the likelihood associated with (9.83) is identical to the likelihood associated with an array of independent Poisson random variables N_{ij} with means $\mu_{ij} = p_{ij}$.

Thus, we are in a Poisson regression situation, in which the logarithms of the means of the N_{ij}'s are linear functions of covariables $c(i,j)$ (with known coefficient 1), i, j, and ij. Standard Poisson regression algorithms will then yield maximum likelihood estimates of θ_1, θ_2, and θ_3 (the "intercept" θ_0 is of no interest to us).

Such estimates being maximum likelihood estimates will be consistent and efficient asymptotically normal estimates.

Lindsey (1974) recommends use of this approach in many nondiscrete cases via grouping of the data. Moschopoulous and Staniswalis (1994) pointed out its potential utility for conditionally specified models. Grouping will undoubtedly reduce the efficiency of the resulting estimates, nevertheless, good consistent asymtotically normal estimates can be obtained in this manner for all our models involving conditionals in exponential families. And, as Moschopoulous and Staniswalis (1994) show, they can be remarkably efficient. We briefly describe the technique here. More details can be found in Moschopoulous and Staniswalis (1994), in Lindsey and Mersch (1992), and in Lindsey (1974).

Suppose that $(X_1, Y_1), (X_2, Y_2), \ldots, (X_n, Y_n)$ is a random sample from a general bivariate conditionals in exponential families distribution with density (4.5). Thus

$$\log f_{X,Y}(x,y) = c(x,y) + \left[\underline{q}^{(1)}(x)' M \underline{q}^{(2)}(y)\right], \quad x \in S(X), \quad y \in S(Y).$$
(9.84)

We will group the data to transform our problem into a multinomial problem. Following Moschopoulous and Staniswalis (1994), we let $\mathcal{R}(\Delta_1, \Delta_2) = \{R_{ij}\}_{i=1,\infty}^{j=1,\infty}$ denote a partition of the support of $f_{X,Y}$ where each $R_{ij} = \{(x,y), x \in ((i-1)\Delta_1, i\Delta_1]$ and $y \in ((j-1)\Delta_2, j\Delta_2]\}$. Hence each element R_{ij} in the partition $\mathcal{R}$ is a rectangle with area $\Delta_1\Delta_2$ and is centered at the point $(x_i, y_j) = ((i - 0.5)\Delta_1, (j - 0.5)\Delta_2)$. Let N_{ij} denote the number of observations in the sample $(X_1, Y_1), (X_2, Y_2), \ldots, (X_n, Y_n)$ falling in R_{ij}.

Instead of maximizing the likelihood of the original data (the (X_i, Y_i)'s) we deal with the multinomial likelihood of the N_{ij}'s. Observe that

$$P((X,Y) \in R_{ij}) \doteq \Delta_1\Delta_2 f_{X,Y}(x_i, y_j).$$
(9.85)

So the likelihood based on the N_{ij}'s is

$$L(M) \doteq \prod_{i=-\infty}^{\infty} \prod_{j=-\infty}^{\infty} (\Delta_1\Delta_2 f_{X,Y}(x_i, y_j))^{N_{ij}}.$$
(9.86)

Of course, in practice, only a finite number of the N_{ij}'s will be nonzero (they sum to n), so only a finite number of factors in the right-hand side of (9.86) need be considered. But from (9.84) it is obvious that p_{ij}, the probability associated with cell (i, j) (i.e., with R_{ij}), will satisfy

$$\log p_{ij} = -\log \Delta_1 - \log \Delta_2 + c(x_i, y_j) + \underline{q}^{(1)}(x_i)' M \underline{q}^{(2)}(y_j). \qquad (9.87)$$

So, as in our simple binomial conditionals example, we can estimate the elements of M by Poisson regression.

9.9 Bayesian Estimates

The general paradigm for Bayesian inference involves specification of an acceptable likelihood function $f(\underline{x}; \underline{\theta})$ for the data, elicitation of an appropriate prior density for the parameter $\underline{\theta}$, say $g(\underline{\theta})$, and then, via Bayes theorem, determination of the resulting posterior density of $\underline{\theta}$ given $\underline{X} = \underline{x}$, i.e.,

$$f(\underline{\theta}|\underline{x}) \propto f(\underline{x}; \underline{\theta}) g(\underline{\theta}), \qquad (9.88)$$

to be used for subsequent inference regarding $\underline{\theta}$. Naturally, if the likelihood $f(\underline{x}; \underline{\theta})$ is only known up to a constant of proportionality, as is the case in many conditionally specified models, the Bayesian technique will flounder.

Definition 9.5 (Bayes estimate). The mean or the mode of the posterior density is used as an estimate of $\underline{\theta}$ and is known as the Bayes estimate. □

The computation of the posterior mean or mode will generally be impossible if $f(\underline{x}; \underline{\theta})$ is only known to be of the form

$$c(\underline{\theta}) g(\underline{x}; \underline{\theta}),$$

where $c(\underline{\theta})$, to be chosen so that $\int f(\underline{x}; \underline{\theta})\, d\underline{x} = 1$, can only be numerically evaluated for each choice of $\underline{\theta}$.

A computer intensive solution is of course possible. For a conditionals in exponential families data set our likelihood can be written in the form

$$\prod_{i=1}^{n} f(x_i, y_i; \underline{\theta}) = [\psi(\underline{\theta})]^{-n} \exp\left[-\sum_{j=1}^{M} \theta_j T_j(\underline{x}, \underline{y}) \right], \qquad (9.89)$$

where $\underline{\theta} \in \Theta$, a subset of $\mathbb{R}^M$ (here M, the dimension of the parameter space, can be quite large). Any integral of a function of $\underline{\theta}$ over Θ can be approximated to a desired level of accuracy by averaging the values of the integrand evaluated at a large finite number of equally spaced points in Θ. Denote this grid of points in Θ by G.

For illustrative purposes we will focus on the case in which $\underline{\theta} \geq \underline{0}$.

Now assume a joint prior for $\underline{\theta}$ with independent gamma marginals, i.e.,

$$\theta_i \sim \Gamma(\alpha_i, \lambda_i)$$

with α_i, λ_i chosen to match our elicited prior mean and variance of the parameter θ_i. It follows that the posterior expectation of any parametric function $h(\underline{\theta})$ given $X = x$ will be well approximated by

$$E(h(\underline{\theta})|\underline{X} = \underline{x}) \doteq \frac{\sum\limits_{\underline{\theta} \in G} h(\underline{\theta})[\psi(\underline{\theta})]^{-n} \left(\prod\limits_{i=1}^{M} \theta_i^{\alpha_i - 1} \right) \exp \left[-\sum\limits_{j=1}^{M} \theta_j [T_j(\underline{x}, \underline{y}) + \lambda_j] \right]}{\sum\limits_{\underline{\theta} \in G} [\psi(\underline{\theta})]^{-n} \left(\prod\limits_{i=1}^{M} \theta_i^{\alpha_i - 1} \right) \exp \left[-\sum\limits_{j=1}^{M} \theta_j [T_j(\underline{x}, \underline{y}) + \lambda_j] \right]}.$$

$$(9.90)$$

The choice $h(\underline{\theta}) = \theta_j$ will then give us the approximate squared error loss Bayes estimate of θ_j. Implementation of this approach will require evaluation of $\psi(\underline{\theta})$ at each of the many points in G.

A second alternative is to use independent gamma priors and then use the mode of the posterior as our estimate of $\underline{\theta}$. The computations required for this approach are exactly the same as those required for maximum likelihood estimation.

A third avenue involves incorporation of the normalizing constant in the prior. The general technique will be clear after working through an example. Consider a sample of size n from a conditionals in one-parameter exponential families distribution with unit scale parameters and one unknown interaction parameter θ. Thus our likelihood is

$$\prod_{i=1}^{n} f_{X,Y}(x_i, y_i; \theta) = [\psi(\theta)]^{-n} \exp \left\{ \sum_{i=1}^{n} [q_1(x_i) + q_2(y_i) - \theta q_1(x_i) q_2(y_i)] \right\},$$

$$(9.91)$$

where $\theta > 0$.

Consider the family of prior densities

$$g_{\alpha,\lambda}(\theta) \propto [\psi(\theta)]^n \theta^{\alpha - 1} e^{-\lambda \theta}. \qquad (9.92)$$

Now elicit from the researcher an appropriate prior mean and variance for θ and choose α and λ so that the prior $g_{\alpha,\lambda}(\theta)$ defined in (9.92) has that mean and that variance. This will require a small amount of numerical integration but can be accomplished. With the appropriate choice, say α^* and λ^*, of hyper-parameters in (9.92) the posterior density becomes

$$\begin{aligned} f(\theta|\underline{x}, \underline{y}) &= f(\underline{x}, \underline{y}; \theta) g_{\alpha^*, \lambda^*}(\theta) \\ &\propto \theta^{\alpha^* - 1} e^{-\theta[\lambda^* + \sum_{i=1}^{n} q_1(x_i) q_2(y_i)]}. \end{aligned} \qquad (9.93)$$

Evidently the posterior is a gamma density and, for example, the squared error loss Bayes estimate of θ will be the posterior mean, namely,

$$\hat{\theta}_B = \alpha^* \left[\lambda^* + \sum_{i=1}^{n} q_1(X_i)q_2(Y_i) \right]^{-1} . \tag{9.94}$$

In the case of a conditionals in exponential families model involving several parameters $\theta_1, \theta_2, \ldots, \theta_M$ we merely take a convenient joint prior from the family

$$g_{\underline{\alpha}, \underline{\lambda}}(\underline{\theta}) = [\psi(\underline{\theta})]^n \prod_{j=1}^{M} \theta_j^{\alpha_j - 1} e^{-\lambda_j \theta_j} . \tag{9.95}$$

We might reasonably choose $\underline{\alpha}$ and $\underline{\lambda}$ to match prior means and variances of the θ_i's. Having done that (a nontrivial job involving considerable numerical integration, see below) the posterior density will factor into independent gamma marginals, and Bayes estimates of the coordinates of $\underline{\theta}$ are immediately available. A Bayesian purist might object to the appearance of n in the prior. Surely changing the sample size shouldn't affect the prior. The response to this criticism is that the change in n has not affected the prior. It has, however, changed the class of convenience priors that we select from to approximate our prior. From this viewpoint, the approach is philosophically no more objectionable than is the customary use of conjugate prior families in routine Bayesian analysis.

The fly in the ointment associated with the use of priors like (9.95) is the job of matching up prior means and variances. It will probably involve evaluating $\psi(\underline{\theta})$ at all points in some grid to evaluate numerically the moments of (9.95). If this is done then the work is essentially equivalent to that needed for an approximate standard Bayesian analysis using independent gamma priors. The technique, involving the use of (9.95), would have value if we envisioned analyzing several data sets using the same prior.

9.9.1   Pseudo-Bayes Approach

A final entry in the list of Bayesian and Bayesian motivated methodologies may be dubbed the pseudo-Bayes approach.

Definition 9.6 (Pseudo-Bayes estimate). If we replace the likelihood

$$f(\underline{x}, \underline{y}; \underline{\theta}) = \prod_{i=1}^{n} f(x_i; y_i; \underline{\theta})$$

by the pseudolikelihood

$$\prod_{i=1}^{n} f(x_i|y_i; \underline{\theta})f(y_i|x_i; \underline{\theta})$$

and then seek the mode or mean of the resulting pseudo-posterior density of $\underline{\theta}$, i.e.,

$$\tilde{f}(\underline{\theta}|\underline{x},\underline{y}) \propto g(\underline{\theta}) \prod_{i=1}^{n} f(x|y_i;\underline{\theta})f(y_i|x_i;\underline{\theta}), \qquad (9.96)$$

we obtain a new estimate, which is called the pseudo-Bayes estimate.   □

This ad hoc approach performs remarkably well. Naturally, for large n, the pseudo-Bayes estimate will be very close to the pseudolikelihood estimate (or conditional likelihood estimate if you wish) which is consistent and asymptotically normal. It is not clear precisely how the use of (9.96) accommodates the prior information implicit in the assumption of $g(\underline{\theta})$ as a prior. It is evident that (9.96) is not a true posterior density for any non-data-dependent prior. Some preliminary work on properties of such estimates has been reported in Arnold and Press (1990).

One disturbing feature of this approach is that the pseudo-posterior density appears often to have smaller variance than the true posterior density (when it can be calculated). In a sense, (9.96) is behaving as if there are $2n$, instead of n, observations. A quick fix would be to replace the pseudolikelihood by its square root, i.e., use as a pseudo-posterior the function

$$\tilde{\tilde{f}}(\underline{\theta}|\underline{x},\underline{y}) \propto g(\underline{\theta})\sqrt{\prod_{i=1}^{n} f(x_i|y_i;\underline{\theta})f(y_i|x_i;\underline{\theta})}. \qquad (9.97)$$

However, reliable rules for recommending which of (9.96) or (9.97) (or indeed a form involving some other power of the pseudolikelihood) to use are lacking at this time.

9.10 Multivariate Examples

Although the concepts discussed in bivariate settings in this chapter usually extend readily to higher dimensions, the associated book-keeping can be depressing. We will limit ourselves to a brief discussion of two quite simple k-variate models.

Example 9.10 (**k-variate exponential conditionals**). Suppose that we have a sample $(\underline{X}^{(1)},\ldots,\underline{X}^{(n)})$ from the joint density

$$f_{\underline{X}}(\underline{x}) = \phi(\delta)\exp\left[-\sum_{i=1}^{k} x_i - \delta\prod_{i=1}^{k} x_i\right], \quad \underline{x} \geq 0, \qquad (9.98)$$

where $\phi(\delta)$ is the appropriate normalizing constant. This is a simple example with exponential conditionals. In this case, a method of moments approach works well for estimating δ. Note that for each i,

$$X_i | \underline{X}_{(i)} = \underline{x}_{(i)} \sim \Gamma \left(1, 1 + \delta \prod_{j \neq i} x_j \right)$$

and so

$$X_i \left(1 + \delta \prod_{j \neq i} X_j \right) \sim \Gamma(1, 1).$$

Consequently

$$E(X_i) + \delta E \left(\prod_{j=1}^{k} X_j \right) = 1$$

and

$$\sum_{i=1}^{k} E(X_i) + k \delta E \left(\prod_{j=1}^{k} X_j \right) = k. \tag{9.99}$$

From (9.99), equating sample and population moments, we find a simple consistent estimate of δ of the form

$$\hat{\delta} = \frac{1 - \dfrac{1}{nk} \sum_{i=1}^{n} \sum_{j=1}^{k} X_j^{(i)}}{\dfrac{1}{n} \sum_{i=1}^{n} \prod_{j=1}^{k} X_j^{(i)}}. \tag{9.100}$$

□

Example 9.11 (Simplified centered k-variate normal conditionals). Suppose that we have a sample $(\underline{X}^{(1)}, \ldots, \underline{X}^{(n)})$ from the joint density

$$f_{\underline{X}}(\underline{x}) = \psi_k(c)(2\pi)^{-k/2} (\sigma_1 \ldots \sigma_k)^{-1} \exp \left\{ \frac{1}{2} \left[\sum_{i=1}^{k} \left(\frac{x_i}{\sigma_i} \right)^2 + c \prod_{i=1}^{k} \left(\frac{x_i}{\sigma_i} \right)^2 \right] \right\}, \tag{9.101}$$

where $\sigma_i > 0, i = 1, 2, \ldots, k, c \geq 0$, and $\psi_k(c)$ is an appropriate normalizing constant. Clearly (9.101) is a particularly simple k-dimensional density with normal conditionals.

The moment generating function of $(X_1^2, \ldots, X_k^2)$ is given by

$$\begin{aligned}
M(s_1, \ldots, s_k) &= E(e^{s_1 X_1^2 + \ldots + s_k X_k^2}) \\
&= \frac{(1 - 2s_1 \sigma_1^2)^{-1/2} \cdots (1 - 2s_k \sigma_k^2)^{-1/2} \psi_k(c)}{\psi_k \left[c(1 - 2s_1 \sigma_1^2)^{-1} \cdots (1 - 2s_k \sigma_k^2)^{-1} \right]}. \tag{9.102}
\end{aligned}$$

Unfortunately, for dimensions higher than two, the normalizing constant $\psi_k(c)$ does not satisfy a simple differential equation, and this fact makes the estimation process somewhat more difficult. First, we verify that for $i, j \in \{1, 2, ..., k\}$,

$$E(X_i^2) = \sigma_i^2(1 - 2c\delta_k(c)), \tag{9.103}$$

$$V(X_i^2) = \sigma_i^4(2 - 8c\delta_k(c) - 4c^2\delta_k'(c)), \tag{9.104}$$

$$E(X_i^2 X_j^2) = (\sigma_i\sigma_j)^2(1 - 8c\delta_k(c) + 4c^2\delta_k^2(c) - 4c^2\delta_k'(c)), \text{ if } i \neq j, \tag{9.105}$$

$$\rho(X_i^2, X_j^2) = \frac{2c\delta_k(c) + 2c^2\delta_k'(c)}{-1 + 4c\delta_k(c) + 2c^2\delta_k'(c)}, \tag{9.106}$$

$$E(\prod_{i=1}^{k} X_i^2) = 2\delta_k(c)\prod_{i=1}^{k}\sigma_i^2, \tag{9.107}$$

where $\delta_k(c) = \psi_k'(c)/\psi_k(c)$ and $\delta_k'(c) = d\delta_k(c)/dc$. Formula (9.107) can be easily deduced taking into account the fact that $E(\partial \log f_{\underline{X}}(\underline{x})/\partial c) = 0$. For convenience, for estimation by the method of moments, we consider δ_k and δ_k' to be new parameters. By the strong law of large numbers, as $n \to \infty$),

$$\overline{X_i^2} \xrightarrow{a.s.} E(X_i^2) = \sigma_i^2(1 - 2c\delta_k), \qquad i = 1, 2, ..., k, \tag{9.108}$$

$$\left[\prod_{i=1}^{k} \frac{S_{X_i^2}}{\overline{X_i^2}}\right]^{2/k} \xrightarrow{a.s.} \left[\prod_{i=1}^{k} \frac{\sqrt{V(X_i^2)}}{E(X_i^2)}\right]^{2/k} = \frac{2 - 8c\delta_k - 4c^2\delta_k'}{(1 - 2c\delta_k)^2}, \tag{9.109}$$

$$\frac{2}{k(k-1)}\sum_{i<j} r(X_i^2, X_j^2) \xrightarrow{a.s.} \frac{2c\delta_k + 2c^2\delta_k'}{-1 + 4c\delta_k + 2c^2\delta_k'}, \tag{9.110}$$

$$\frac{\overline{\prod_{i=1}^{k} X_i^2}}{\prod_{i=1}^{k} \overline{X_i^2}} \xrightarrow{a.s.} \frac{E(\prod_{i=1}^{k} X_i^2)}{\prod_{i=1}^{k} E(X_i^2)} = \frac{2\delta_k}{(1 - 2c\delta_k)^k}. \tag{9.111}$$

Observe that, when $k = 2$, only (9.108) to (9.110) need to be used, and (9.111) is unnecessary. We denote by u, v, and w the statistics defined on the left sides of (9.109)–(9.111). If we equate the left and right sides of (9.108)–(9.111) and solve for c, δ_k, and δ_k' we obtain the strongly consistent parameter estimators

$$\hat{c} = 2g(U, V)/W\left[1 - 2g(U, V)\right]^k, \tag{9.112}$$

$$\hat{\delta}_k = g(U, V)/\hat{c}, \tag{9.113}$$

$$\hat{\sigma}_i^2 = \overline{X_i^2}/(1 - 2\hat{c}\hat{\delta}_k), \quad i = 1, 2, ..., k, \tag{9.114}$$

where

$$g(U, V) = V(U - 1)/2U(V - 1). \tag{9.115}$$

$\square$

9.11   Bibliographic Notes

Castillo and Galambos (1985) studied maximum likelihood for the normal conditionals model. They also considered marginal likelihood. Arnold (1987) discussed maximum likelihood for the Pareto conditionals model. Conditional likelihood techniques were discussed in Arnold (?) and Arnold and Strauss (1988a, 1988b, 1991). Estimates obtained using the method of moments were introduced in Arnold and Strauss (1988a). The Poisson regression estimation approach (Section 9.8) was introduced by Moschopoulous and Staniswalis (1994). Most of the Bayesian estimation material has not appeared elsewhere. The exception is the pseudo-Bayesian material which is based on preliminary research contained in a working paper by Arnold and Press (1990).

Exercises

9.1 Consider a bivariate binary distribution (X, Y) such that if $p_{ij} = P(X = i, Y = j)$, for $i, j = 0, 1$,

$$p_{00} = p_{10} = p_{01} = \theta, \quad p_{11} = 1 - 3\theta, \quad 0 \le \theta \le 1/3.$$

Suppose we have a random sample of size n, and let N_{ij}, $i, j = 0, 1$, be the corresponding frequencies.

(a) Prove that the maximum likelihood estimator of θ is given by

$$\hat{\theta} = \frac{n - n_{11}}{3n}$$

with variance

$$\frac{\theta(1 - 3\theta)}{3n}.$$

(b) Prove that the estimator which maximizes the marginal likelihood is

$$\hat{\theta}_m = \frac{2n_{00} + n_{01} + n_{10}}{4n}$$

with variance

$$\frac{\theta(3 - 8\theta)}{8n}.$$

(c) Prove that the pseudolikelihood estimator of θ is

$$\hat{\theta}_c = \frac{n_{01} + n_{10}}{3(n_{01} + n_{10}) + 2n_{11}}$$

with asymptotic variance

$$\frac{\theta(1 - 3\theta)(1 - \theta)}{2n}.$$

(d) Prove that the pseudolikelihood estimator and the marginal likelihood estimates have less than full efficiency for $0 < \theta < 1/3$.

(e) Prove that the variance functions of $\hat{\theta}_m$ and $\hat{\theta}_c$ cross at $\theta = 1/6$, showing that in general we cannot expect $\hat{\theta}_m$ to be a uniformly better or worse estimator than $\hat{\theta}_c$.

(Arnold and Strauss (1991).)

9.2 Let $(X_1, \ldots, X_n)$ be a random variable with joint pdf,

$$f(x_1, \ldots, x_n) = \phi_n(c) \prod_{i=1}^n \frac{1}{\sigma_i} \exp\left(-\sum_{i=1}^n \frac{x_i}{\sigma_i} - c \prod_{i=1}^n \frac{x_i}{\sigma_i}\right),$$
$$x_i \geq 0, \quad i = 1, \ldots, n,$$

where $\sigma_i > 0$, $i = 1, \ldots, n$, $c \geq 0$, and $\phi_n(c)$ is the normalizing constant.

(a) Describe a method for obtaining $\phi_n(c)$ using simulations from independent exponential random variables with mean σ_i, $i = 1, \ldots, n$.

(b) Show that

$$X_i | X_{(i)} = x_{(i)} \sim \mathrm{Exp}\left[\frac{1}{\sigma_i}\left(1 + c\prod_{j\neq i}\frac{x_i}{\sigma_j}\right)\right]$$

with $i = 1, \ldots, n$.

(c) If $\delta_n(c) = d\log\phi_n(c)/dc$, show that

$$\begin{aligned}
E(X_i) &= \sigma_i\left[1 - \delta_n(c)\right], \\
\mathrm{var}(X_i) &= \sigma_i^2\left[1 - 2c\delta_n(c) - c^2\delta_n'(c)\right], \\
E(X_iX_j) &= \sigma_i\sigma_j\left[1 - 3c\delta_n(c) + c^2\delta_n^2(c) - c^2\delta_n'(c)\right], \quad i \neq j, \\
\rho(X_i, X_j) &= \frac{c\delta_n(c) + c^2\delta_n'(c)}{-1 + 2c\delta_n(c) + c^2\delta_n'(c)}, \\
E(X_1\cdots X_n) &= \sigma_1\cdots\sigma_n\delta_n(c).
\end{aligned}$$

(d) From a multivariate random sample of size m, and using (c), obtain consistent estimators of the parameters $\sigma_1, \ldots, \sigma_n$ and c.

9.3 Consider the bivariate joint pdf given by

$$f(x, y; \theta) = k(\theta)xy\exp(\theta\log x\log y), \quad 0 < x, \; y < 1,$$

with $\theta \leq 0$.

(a) Obtain $k(\theta)$.

(b) What is the sufficient statistic for θ?

(c) Obtain estimators for θ using maximum likelihood, marginal likelihood and pseudolikelihood. Obtain the asymptotic variances and compare them.

9.4 Consider the Model II with gamma conditionals given by (4.45) and with joint pdf

$$f(x,y) = \frac{k_{r,s}(c)}{\sigma_1^r \sigma_2^s \Gamma(r)\Gamma(s)} x^{r-1} y^{s-1} \exp\left(-x/\sigma_1 - y/\sigma_2 - cxy/\sigma_1\sigma_2\right).$$

Assume that r and s are known parameters, and that we wish to estimate σ_1, σ_2, and c from a bivariate random sample of size n.

(a) If $\delta_{r,s}(c) = d \log k_{r,s}(c)/dc$, prove that the maximum likelihood estimators of the parameters satisfy the equations

$$\sigma_1 = \frac{\overline{X}}{r - c\delta_{r,s}(c)},$$

$$\sigma_2 = \frac{\overline{Y}}{s - c\delta_{r,s}(c)},$$

$$\frac{\overline{XY}}{\overline{X} \times \overline{Y}} = \frac{\delta_{r,s}(c)}{[r - c\delta_{r,s}(c)][s - c\delta_{r,s}(c)]}.$$

(b) For a particular value of r and s, obtain a table for evaluating the maximum likelihood estimator of the parameter c.

(c) Obtain the Fisher information matrix and the asymptotic variances.

9.5 Suppose that a sample of size n is available from a centered Cauchy conditionals distribution (5.75). Outline a suitable algorithm for obtaining pseudolikelihood estimates of the three parameters in the model.

9.6 Suppose that a sample of size n is available from a Pareto conditionals distribution (5.7). Derive the pseudolikelihood equations for estimating the parameters of the model and outline an iterative algorithm for solving them.

(Arnold (1991).)

9.7 Suppose that a sample of size n is available from a bivariate distribution with conditionals in exponential families (4.5). Verify that the maximum likelihood estimates of the parameters of the model are solutions to the equations

$$E[\tilde{\underline{q}}^{(1)}(X)] = \frac{1}{n} \sum_{i=1}^{n} \tilde{\underline{q}}^{(1)}(X_i),$$

$$E[\tilde{\underline{q}}^{(2)}(Y)] = \frac{1}{n} \sum_{i=1}^{n} \tilde{\underline{q}}^{(2)}(Y_i),$$

$$E[(\tilde{\underline{q}}^{(1)}(X))(\tilde{\underline{q}}^{(2)}(Y))'] = \frac{1}{n} \sum_{i=1}^{n} (\tilde{\underline{q}}^{(1)}(X_i))(\tilde{\underline{q}}^{(2)}(Y_i))'.$$

(Arnold and Strauss (1991).)

10
Marginal and Conditional Specification in General

10.1 Introduction

A k-dimensional density function is determined by certain combinations of marginal and conditional densities. It would be desirable to identify all possible such specifications. Considerable progress can be made in this direction. A key result is a uniqueness theorem due to Gelman and Speed (1993). However, the issue is clouded by the existence of certain non-standard examples dating back at least to Seshadri and Patil (1964). We will begin by surveying marginal and conditional specification in the bivariate case and then enumerate carefully the available results in higher dimensions. Throughout we will assume absolute continuity with respect to some convenient dominating measure.

10.2 Specifying Bivariate Densities

For a two-dimensional random variable (X, Y) it is common to use some combinations of marginal and/or conditional densities to describe the joint density of (X, Y). Thus, we might specify the marginal density of X, $f_X(x)$, and for each possible value x of X, specify the conditional density of Y given $X = x$, i.e., $f_{Y|X}(y|x)$. Clearly, this yields enough information to characterize the joint density $f_{X,Y}(x, y)$ uniquely. This is a trivial characterization result. But others are of course possible. If we are given both families of conditional densities, i.e., $f_{X|Y}(x|y), \forall y \in S_Y$, and $f_{Y|X}(y|x), \forall x \in S_X$, then, as outlined in Chapter 1, we can identify suitable consistency conditions and identify situations in which such conditional specifications uniquely determine the joint density $f_{X,Y}(x,y)$. But, what if we are given one marginal density, say $f_X(x)$ and the "wrong" family of conditional densities, i.e., $f_{X|Y}(x|y), y \in S_Y$? Can we characterize the joint density in this case? The perhaps surprising answer is: "Sometimes."

It is convenient to first consider the finite discrete case and then to extend to more general settings (just as we did in Chapter 1). Thus we assume that X and Y are discrete variables with possible values $x_1, x_2, \ldots, x_I$ and $y_1, y_2, \ldots, y_J$, respectively. Suppose that we are given a matrix A and a vector $\underline{\tau}$ and we ask whether there exists a compatible joint distribution for (X, Y), i.e. one such that

$$P(X = x_i | Y = y_j) = a_{ij}, \quad \forall i, j, \tag{10.1}$$

and

$$P(X = x_i) = \tau_i, \quad \forall i. \tag{10.2}$$

If a compatible distribution exists, it is natural to then ask if it is unique.

The problem is actually quite easy to solve. What we seek is a compatible marginal distribution for Y say $\underline{\eta}$ (where $\eta_j = P(Y = y_j)$) which can be combined with A to completely specify the joint distribution of (X, Y). Such a vector $\underline{\eta}$ must clearly satisfy

$$\tau_i = \sum_{j=1}^{J} a_{ij} \eta_j, \quad \forall i. \tag{10.3}$$

This is equivalent to the statement that $\underline{\tau}$ belongs to the convex hull of the columns of A. A necessary and sufficient condition for this to be true is that

$$\underline{\tau} \cdot \underline{c} \leq \max_{1 \leq j \leq J} \left\{ A^{(j)} \cdot \underline{c} \right\}, \quad \forall \underline{c} \in \mathbb{R}^I. \tag{10.4}$$

An algorithm (due to Vardi and Lee (1993)) for determining $\underline{\tau}$, given a compatible pair A and $\underline{\tau}$, will be described after we introduce the more general case.

Of course, a member of the convex hull of the columns of A is not usually a unique convex combination of those columns. Even if all the columns of A are extreme points of the convex hull they determine, it is possible that points in the convex hull may not be represented as a unique convex combination of the columns of A.

Example 10.1 (Uniqueness example). Consider

$$A = \begin{pmatrix} 1/4 & 3/5 \\ 3/4 & 2/5 \end{pmatrix}$$

and
$$\underline{\tau} = (1/3 \quad 2/3).$$

Here A and $\underline{\tau}$ are compatible and $\underline{\tau}$ is a unique convex combination of the columns of A. The unique compatible joint distribution $P = (p_{ij})$ where $p_{ij} = P(X = x_i, Y = y_j)$ is given by

$$P = \begin{pmatrix} 4/21 & 3/21 \\ 12/21 & 2/21 \end{pmatrix}.$$

□

Example 10.2 (Multiple solutions). Consider

$$A = \begin{pmatrix} 1/4 & 3/5 & 1/8 \\ 3/4 & 2/5 & 7/8 \end{pmatrix}$$

and
$$\underline{\tau} = (1/3 \quad 2/3).$$

Here A and $\underline{\tau}$ are compatible but there is not a unique compatible P but a continuum of compatible P's. □

Example 10.3 (Incompatible example). Consider

$$A = \begin{pmatrix} 1/4 & 3/5 \\ 3/4 & 2/5 \end{pmatrix}$$

and
$$\underline{\tau} = (1/16 \quad 15/16).$$

These are not compatible since $\underline{\tau}$ is not a convex combination of the columns of A. □

In a more general setting, suppose that (X, Y) is absolutely continuous with respect to $\mu_1 \times \mu_2$ on $S_X \times S_Y$. Suppose that we are given two functions $u(x)$ and $a(x, y)$ and we ask whether there exists a compatible joint distribution for (X, Y), i.e., one such that

$$f_X(x) = u(x), \quad \forall x \in S_X, \tag{10.5}$$

and, for each $y \in S_Y$,

$$f_{X|Y}(x|y) = a(x, y), \quad \forall x \in S_X. \tag{10.6}$$

In addition, we may ask when there is a unique such compatible distribution.

It is clear that u and a will be compatible if there exists a suitable density for Y, say $w(y)$, such that

$$u(x) = \int_{S_Y} a(x, y) w(y) \; dy, \quad \forall x \in S_X. \tag{10.7}$$

Thus u and a are compatible if and only if u can be expressed as a mixture of the given conditional densities $\{a(x, y) : y \in S_Y\}$. Uniqueness of the compatible distribution $f_{X,Y}(x, y) = w(y)a(x, y)$ will be encountered if and only if the family of conditional densities is identifiable.

Example 10.4 (Uniqueness). Suppose

$$a(x, y) = ye^{-xy}I(x > 0)$$

and

$$u(x) = (1 + x)^{-2}I(x > 0).$$

It may be verified that these are compatible. The corresponding density for Y is

$$w(y) = e^{-y}I(y > 0).$$

Identifiability of the family $(ye^{-xy}I(x > 0, y > 0))$ may be verified using the uniqueness property of Laplace transforms, consequently there is a unique joint density corresponding to the given a and u, namely

$$f_{X,Y}(x, y) = ye^{-(x+1)y}I(x > 0)I(y > 0).$$

$\square$

Suppose now that we are given u and a. How can we identify the corresponding mixing density $w(y)$? A clever iterative solution is available from Vardi and Lee (1993). Let $w_0(y)$ be an arbitrary strictly positive density on S_Y. Now, for $n = 0, 1, \ldots$, define

$$w_{n+1}(y) = w_n(y) \int_{S_X} \frac{a(x, y)u(x)}{\int_{S_Y} w_n(y')a(x, y')\ dy'}\ dx. \tag{10.8}$$

Theorem 10.1 (Convergence of the Vardi and Lee scheme). *The iterative scheme (10.8) will always converge. If a and u are compatible it will converge to an appropriate mixing distribution $w(y)$.* $\square$

If a and u are incompatible, (10.8) will converge to a function $\hat{w}(y)$, where $\hat{w}$ minimizes the Kullback-Leibler information distance between $u(x)$ and

$$\int_{S_Y} a(x, y)w(y)\ dy$$

for nonnegative functions $w(y)$.

In the finite discrete case, (10.8) assumes the form

$$\underline{\eta}_0 > 0 \text{ is arbitrary}$$

and for each j, for $n = 0, 1, 2, \ldots$,

$$\eta_j^{(n+1)} = \eta_j^{(n)} \sum_{i=1}^{I} \left[a_{ij}\tau_i \Big/ \left(\sum_{j'=1}^{J} \eta_{j'}^{(n)}a_{ij'} \right) \right]. \tag{10.9}$$

TABLE 10.1. η marginals obtained after different iterations using the Vardi and Lee method in Example 10.5.

Iteration (n)	$\underline{\eta}_n$	Iteration (n)	$\underline{\eta}_n$
1	(0.5328, 0.4672)	20	(0.7394, 0.2606)
2	(0.5616, 0.4384)	30	(0.7543, 0.2457)
3	(0.5867, 0.4133)	40	(0.7593, 0.2407)
4	(0.6085, 0.3915)	50	(0.7610, 0.2390)
5	(0.6274, 0.3726)	60	(0.7616, 0.2384)
6	(0.6437, 0.3563)	70	(0.7618, 0.2382)
7	(0.6579, 0.3421)	80	(0.7619, 0.2381)
8	(0.6702, 0.3298)	90	(0.7619, 0.2381)
9	(0.6809, 0.3191)	100	(0.7619, 0.2381)
10	(0.6902, 0.3098)	∞	$(16/21, 5/21)$

Example 10.5 (Uniqueness example). Consider again

$$A = \begin{pmatrix} 1/4 & 3/5 \\ 3/4 & 2/5 \end{pmatrix}$$

and

$$\underline{\tau} = (\, 1/3 \quad 2/3 \,).$$

Using the Vardi and Lee iterative method in (10.9) we get the results in Table 10.1. Note that the convergence is quite slow.

Using the asymptotic value of $\underline{\eta} = (16/21, 5/21)$ we get a joint probability

$$P = \begin{pmatrix} 4/21 & 12/21 \\ 3/21 & 2/21 \end{pmatrix} \simeq \begin{pmatrix} 0.190476 & 0.571429 \\ 0.142857 & 0.0952381 \end{pmatrix}.$$

Since its X-marginal coincides with $\underline{\tau}$, it follows that A and τ are compatible.   □

Example 10.6 (Multiple solutions). Consider once more

$$A = \begin{pmatrix} 1/4 & 3/5 & 1/8 \\ 3/4 & 2/5 & 7/8 \end{pmatrix}$$

and

$$\underline{\tau} = (\, 1/3 \quad 2/3 \,).$$

Using the Vardi and Lee iterative method in (10.9) we get

$$\underline{\eta} = (0.328009, 0.352278, 0.319712)$$

and the corresponding P becomes

$$P = \begin{pmatrix} 0.0820023 & 0.211367 & 0.0399641 \\ 0.246007 & 0.140911 & 0.279748 \end{pmatrix}.$$

Since the resulting X-marginal

$$P_X = (\,0.333333 \quad 0.666667\,)$$

coincides with $\underline{\tau}$, it follows that A and τ are compatible.

Note that we are not able to detect that there is more than one compatible P matrix, though we do obtain one of them. □

Example 10.7 (Incompatible example). Consider

$$A = \begin{pmatrix} 1/4 & 3/5 \\ 3/4 & 2/5 \end{pmatrix}$$

and

$$\underline{\tau} = (\,1/16 \quad 15/16\,).$$

Using the Vardi and Lee iterative method in (10.9) we get

$$\underline{\eta} = (1,0)$$

and the corresponding P becomes

$$P = \begin{pmatrix} 0.25 & 0 \\ 0.75 & 0 \end{pmatrix}.$$

Since the resulting X-marginal

$$P_X = (\,0.25 \quad 0.75\,)$$

does not coincide with $\underline{\tau}$, then the given A and $\underline{\tau}$ are incompatible. □

Where did (10.8) come from? It can be visualized as follows.

Act as if the density of Y is $w_n(y)$. Combine this with $a(x,y)$ to get a joint density for (X,Y). Compute the family of conditional densities of Y given X from this density. Combine this with the marginal $u(x)$ to get a new joint density for (X,Y) whose Y marginal will be denoted by $w_{n+1}(y)$.

Algorithm 10.1 (Obtaining a compatible joint distribution of (X,Y) given the conditional $X|Y$ and the marginal of X).

Input. *A conditional probability density function $a(x,y) = f_{X|Y}(x|y)$, an X-marginal probability density function $f_X(x)$, and an Error.*

Output. *The corresponding compatible Y-marginal density function $f_Y(y)$ and the joint probability density function $f_{X,Y}(x,y)$ or a close alternative.*

Step 1. *Make Error1 = 1 and choose an arbitrary Y-marginal probability density function $f_0(y)$.*

Step 2. *Make $f_Y(y) = f_0(y)$.*

Step 3. *Calculate the joint density $f_{X,Y}(x,y)$ using*

$$f_{X,Y}(x,y) = f_Y(y)a(x,y).$$

Step 4. *Calculate the conditional density $f_{Y|X}(y|x)$ using*

$$f_{Y|X}(y|x) = \frac{f_{X,Y}(x,y)}{\int_{S(Y)} f_{X,Y}(x,y)\,dy}.$$

Step 5. *Calculate the updated Y-marginal probability density function $f_0(y)$ using*

$$f_0(y) = \int_{S(X)} f_{Y|X}(y,x)f_X(x)\,dx.$$

Step 6. *Calculate the error by*

$$Error1 = \int_{S(Y)} |f_Y(y) - f_0(y)|\,dy.$$

Step 7. *If Error1 > Error, go to Step 2; otherwise return the marginal probability density $f_0(y)$ and the joint probability density function $f_{X,Y}(x,y)$ and exit.*

□

10.3   To Higher Dimensions with Care

In k-dimensions we can envision a wide variety of combinations of marginal and/or conditional densities which might be used to characterize, perhaps uniquely, a k-dimensional density. Gelman and Speed (1993) addressed the question of identifying all such consistent specifications of multivariate distributions. We will review and develop their theorem in the following section. The theorem was presented in the form of necessary and sufficient conditions. Reduced to two dimensions, the theorem asserts that only consistent specifications of one of the following forms will uniquely determine $f_{X,Y}(x,y)$:

(i) $f_{X|Y}(x|y)$ and $f_{Y|X}(y|x)$;

(ii) $f_{X|Y}(x|y)$ and $f_Y(y)$; or

(iii) $f_{Y|X}(y|x)$ and $f_X(x)$.

But what about Example 10.4 above? In it, $f_X(x)$ and $f_{X|Y}(x|y)$ determine (uniquely) $f_{X,Y}(x,y)$. A careful review of Gelman and Speed's proof suggests an appropriate amendment to make their theorem true. They assumed a positivity condition, i.e., that $f_{X,Y}(x,y) > 0$, $\forall x \in S_X, y \in S_Y$. Under this assumption, the above list (i)–(iii) includes all situations in which <u>any</u> consistent specification will uniquely determine $f_{X,Y}(x,y)$.

10.4   Conditional/Marginal Specification in k Dimensions

We are concerned with k-dimensional random variables $(X_1, \ldots, X_k)$ with density $f(x_1, x_2, \ldots, x_k)$. For any vector $\underline{x}$ we define, for $i = 1, 2, \ldots, k$,

$\underline{x}_{(i)}$ to be the vector $\underline{x}$ with the ith coordinate deleted;

$\underline{x}_{(i,j)}$ to be the vector $\underline{x}$ with the ith and jth coordinates deleted;

$\dot{\underline{x}}_i$ to be the vector including the first i coordinates of $\underline{x}$; and

$\ddot{\underline{x}}_i$ to be vector $\underline{x}$ after deleting the first i coordinates.

Thus always $\underline{x} = (\dot{\underline{x}}_i, \ddot{\underline{x}}_i)$. Analogously, we define the associated random vectors $\underline{X}_{(i)}, \underline{X}_{(i,j)}, \dot{\underline{X}}_i$, and $\ddot{\underline{X}}_i$. In k dimensions our standard unambiguous notation for conditional densities, for example, $f_{X_1,X_2|X_3,X_4}(x_1, x_2|x_3, x_4)$, will prove cumbersome. Throughout the rest of this chapter we will, hopefully without loss of clarity and/or specificity, sometimes delete the subscripts. The expressions are cleaner and the missing subscripts could always be reintroduced by the reader if desired. Thus we will sometimes write $f(x_1, x_2|x_3, x_4)$ instead of $f_{X_1,X_2|X_3,X_4}(x_1, x_2|x_3, x_4)$.

We focus on situations in which a finite number of conditional and/or marginal densities is given and we wish to determine whether the given set of densities is consistent and whether they uniquely specify the joint density $f(x_1, \ldots, x_k)$. Following Gelman and Speed (1993), we can envision that the given densities are represented by functions of the form $f_{A|B}(a|b)$ where A and B are disjoint subsets of $(X_1, X_2, \ldots, X_k)$ and A is nonempty while B might be empty (if that particular given density were a marginal rather than a conditional).

Models specified in this way are known as *conditionally specified probability models*.

Definition 10.1 (Conditionally specified probability models). Consider a set of variables $\underline{X} = \{X_1, \ldots, X_k\}$. A conditionally specified probability model is a set of conditional and/or marginal probability distributions of $\underline{X}$ of the form

$$P = \{f_{A_i|B_i}(a_i|b_i); \ i = 1, \ldots, m\}, \tag{10.10}$$

which uniquely defines the joint probability density of $\underline{X}$, where A_i and B_i are disjoint subsets of $(X_1, \ldots, X_k)$ and $A_i \neq \emptyset$, for each $i = 1, 2, \ldots, m$. $\square$

Next, we discuss compatibility and other problems associated with conditionally specified probability models. In particular, we address the following questions:

- **Question 10.1. Uniqueness:**
  Does the given set of conditional probability densities define a unique joint probability density? In other words, does the set of conditional densities provide enough constraints for the existence of *at most* one joint probability density?

- **Question 10.2. Consistency or compatibility:**
  Is the given set of conditional probability densities compatible with a joint probability density for $\underline{X}$?

- **Question 10.3. Parsimony:**
  If the answer to Question 10.1 is yes, can any of the given conditional probability densities be ignored or deleted without loss of information?

- **Question 10.4. Reduction:**
  If the answer to Question 10.2 is yes, can any set of conditional probability densities be reduced to a minimum (e.g., by removing some of the conditioning variables)?

Of special interest are the conditional probabilities in canonical form.

Definition 10.2 (Conditional densities in canonical form). Any function of the form $f_{A|B}(a|b) = f_{A|B}(x_{i_1}, \ldots, x_{i_\ell}|b)$ can be replaced by ℓ functions of the form

$$f_{X_{i_j}|B,\tilde{X}}(x_{i_j}|b, \{x_{i_j} : j > m\}), \quad m = 1, 2, \ldots, \ell,$$

and in this way any given collection of densities can be replaced by a list in canonical form

$$f_{X_i|C_{ij}}(x_i|c_{ij}), \quad i = 1, 2, \ldots, k, \ \ j = 1, 2, \ldots, n_i, \quad (10.11)$$

where each C_{ij} is a subvector of $\underline{X}_{(i)}$. $\square$

An example will help clarify this notation.

Example 10.8 (Conditional densities in canonical form). Suppose $k = 4$ and we are given the following set of marginal and conditional densities:

$$f_{X_1, X_2|X_3}(x_1, x_2|x_3), \quad f_{X_3|X_4}(x_3|x_4),$$

$$f_{X_2}(x_2), \quad f_{X_3|X_1,X_4}(x_3|x_1,x_4), \tag{10.12}$$

$$f_{X_2,X_3,X_4|X_1}(x_2,x_3,x_4|x_1),$$

$$f_{X_1,X_2}(x_1,x_2).$$

In canonical form this is equivalent to the following list of given univariate conditionals and marginals:

$$f_{X_1|X_2,X_3}(x_1|x_2,x_3), \quad f_{X_2|X_3}(x_2|x_3), \quad f_{X_3|X_4}(x_3|x_4)$$

$$f_{X_2}(x_2), \quad f_{X_3|X_1,X_4}(x_3|x_1,x_4), \quad f_{X_2|X_1,X_3,X_4}(x_2|x_1,x_3,x_4),$$

$$f_{X_3|X_1,X_4}(x_3|x_1,x_4), \quad f_{X_4|X_1}(x_4|x_1), \quad f_{X_1|X_2}(x_1|x_2). \tag{10.13}$$

This is a canonical form. It is not claimed to be, nor is it, a unique canonical form. As Gelman and Speed (1993) point out, it is necessary to check for consistency. For example, in the above list (10.13), the conditional density $f_{X_2|X_1}(x_2|x_1)$ must agree with the form derivable from the given densities for $f_{X_2}(x_2)$ and $f_{X_1|X_2}(x_1|x_2)$ (which together determine $f_{X_1,X_2}(x_1,x_2)$ and hence $f_{X_2|X_1}(x_2|x_1)$). □

10.5   Checking Uniqueness

The uniqueness theorem of Gelman and Speed (1993) may be stated as follows:

Theorem 10.2 (Uniqueness). *Suppose that a collection of given marginals and conditionals in canonical form (10.11) is consistent in the sense that there exists a joint density $f(x_1, \ldots, x_k)$ with the given marginals and conditionals. The density $f(x_1, x_2, \ldots, x_k)$ will be essentially unique provided that, after possibly permuting subscripts, the list in canonical form contains a set of the form*

$$f_{X_i|A_i,\underline{\ddot{X}}_i}(x_i|A_i,\underline{\ddot{x}}_i), \quad i = 1, 2, \ldots, k, \tag{10.14}$$

where each A_i is a possibly empty subset of $\underline{\dot{X}}_{i-1}$. If any A_j's are nonempty the list (10.14) must be checked for consistency. □

In conditional specification problems, life is simplest when the range of each X_i is a finite set. In this case we can express the joint density in the form of a k-dimensional contingency table. Gelman and Speed actually provide a proof of their theorem in this setting and, in addition, they assume no empty cells, their positivity condition. With a little care, the theorem remains valid in more general cases. The set of $\underline{x}$'s on which $f_{\underline{X}}(\underline{x}) > 0$, say $D(\underline{X})$, does not have to be a Cartesian product. We do, in our consistency check, have to make sure that only vectors or subvectors of elements of

$D(\underline{X})$ are referred to (10.14). In the discrete case this corresponds to the natural requirement that the list (10.14) never involves conditioning on an event of zero probability. There are two other potential problems to be faced in the more general setting. First, the assumption of positivity of $f_X(\underline{x})$ on a Cartesian product was important in the Gelman and Speed uniqueness proof. Our experience in Chapter 1 suggests that some limitations on the support set of $f_X(\underline{x})$ will be necessary (to avoid examples such as that provided by Gourieroux and Montfort (1979)).

The resolution provided by Arnold and Press (1989b) in the bivariate case may also be used in the present situation. We merely envision a Markov process which generates X's by cycling through the list of conditional densities (10.14). If this process is indecomposable we have a unique marginal distribution of X_k which together with the list (10.14) uniquely determines $f_{X_1,\ldots,X_k}(x_1,\ldots,x_k)$.

The other crucial issue to be faced in more general settings involves integrability. If the range of the X_i's is unbounded (above or below or both), then it is possible that the unique solution $f_X(x_1, x_2 \ldots, x_k)$ obtained from (10.14) is not integrable and so does not represent a proper distribution. Several examples of this kind are catalogued in Chapter 6. It should be remarked that the assumption of a positivity condition (i.e., $f_X(x_1,\ldots,x_k) > 0$ on some Cartesian product set) does not obviate the need to check for integrability of the solution if the set on which $f_X(x_1,\ldots,x_k) > 0$ is unbounded.

Two simple examples will illustrate the need for the integrability check.

Example 10.9 (Uniform conditionals). In the first example we postulate that our random variables are positive and, for each j, X_j given $\underline{X}_{(j)} = \underline{x}_{(j)}$ is uniform on the interval $(0, (\prod_{i\neq j} x_i)^{-1})$. Everything is in order except that any putative joint density for this model would have to be constant on the set $\{\underline{x} : \underline{x} \in \mathbb{R}^k, x_i > 0, \ i = 1, 2, \ldots, k, \prod_{i=1}^{k} x_i < 1\}$. Such a function fails to be integrable and we have an "impossible" model.   □

Example 10.10 (Nonintegrable even with positivity condition). In our second example, the positivity condition will be satisfied on the positive orthant (a Cartesian product of $\mathbb{R}^+$ with itself k times). We assume that, for each j, and each $\underline{x}_{(j)} > \underline{0}$,

$$f_{X_j|\underline{X}_{(j)}}(x_j|\underline{x}_{(j)}) = \left(\prod_{i\neq j} x_i\right) \exp\left[-\left(\prod_{i=1}^{k} x_i\right)\right], \quad x_j > 0.$$

If there were to be a joint density compatible with these conditionals, it would have to be proportional to $\exp[-(\prod_{i=1}^{k} x_i)]$ over the positive orthant and such a function would not be integrable. The earliest reference to this example is in Besag (1974) in his discussion of auto-exponential distributions.   □

Note that, as observed at the end of Section 10.3, if we impose a positivity condition (to avoid Gourieroux and Montfort (1979) examples), then only collections of marginals and conditionals that include a set of the form (10.14) will uniquely determine $f(\underline{x})$ for any consistent specification.

10.6   Checking Compatibility

In two dimensions, a simple condition for consistency was available. If we were given $f(x_1|x_2)$ and $f(x_2)$, consistency was automatic. If we were given $f(x_1|x_2)$ and $f(x_2|x_1)$, the necessary condition was the existence of two functions $g_1(x_2)$ and $g_2(x_1)$ at least one of which was integrable such that

$$\frac{f(x_1|x_2)}{f(x_2|x_1)} = \frac{g_2(x_1)}{g_1(x_2)}.$$

When we consider a set of conditionals and marginals in canonical form (10.14), consistency will be guaranteed iff there exists a function $g(\underline{x}_{(1)})$ which is integrable such that for $i = 2, \ldots, k$

$$f(x_i|A_i, \ddot{\underline{x}}_i) = g(x_i|A_i, \ddot{\underline{x}}_i),$$

where each of the conditional densities $g(x_i|A_i, \ddot{\underline{x}}_i)$ is computed using the joint density

$$f(x_1|\underline{x}_{(1)})g(\underline{x}_{(1)})/\left[\int g(\underline{x}_{(1)})\, d\underline{x}_{(1)}\right].$$

The compatibility can be checked one step at a time. After we have checked compatibility of the first m densities in the canonical list, consider A_{m+1}. If it is empty then the $(m+1)$st density is automatically compatible. If it is not empty then compute $f(A_{m+1}|\ddot{\underline{x}}_m)$ from the already determined conditional density $f(\dot{\underline{x}}_m|\ddot{\underline{x}}_m)$. Note that $f(A_{m+1}|\ddot{\underline{x}}_m) = f(A_{m+1}|x_{m+1}, \ddot{\underline{x}}_{m+1})$ and we can check consistency of the $(m+1)$st density by considering the ratio

$$\frac{f(x_{m+1}|A_{m+1}, \ddot{\underline{x}}_{m+1})}{f(A_{m+1}|x_{m+1}, \ddot{\underline{x}}_{m+1})}.$$

If this equals

$$\frac{g(x_{m+1}, \ddot{\underline{x}}_{m+1})}{h(A_{m+1}, \ddot{\underline{x}}_{m+1})},$$

where the numerator is an integrable function of x_{m+1}, then the $(m+1)$st density in the list of conditionals is compatible. It is, in fact, possible to identify the form of the most general expression for $f(x_{m+1}|A_{m+1}, \ddot{\underline{x}}_{m+1})$ which will be compatible with the already determined density $f(\dot{\underline{x}}_m|\ddot{\underline{x}}_m)$. The expression in question is

$$f(x_{m+1}|A_{m+1}, \ddot{\underline{x}}_{m+1}) = \frac{f(x_{m+1}|\ddot{\underline{x}}_{m+1})\int f(\dot{\underline{x}}_m|\ddot{\underline{x}}_m)\, dA^*_{m+1}}{\int f(x_{m+1}|\ddot{\underline{x}}_{m+1})\int f(\dot{\underline{x}}_m|\ddot{\underline{x}}_m)\, dA^*_{m+1} dx_{m+1}},$$

$$(10.15)$$

where $A^*_{m+1} = \underaccent{\dot}{x}_m \cap A^c_{m+1}$. Note that (10.15) depends on the already determined $f(\underaccent{\dot}{x}_m|\underaccent{\ddot}{x}_m)$ and the arbitrary compatible $f(x_{m+1}|\underaccent{\ddot}{x}_{m+1})$.

There is an alternative way of phrasing the requirement that the ratio

$$\tau(x_{m+1}, A_{m+1}, \underaccent{\ddot}{x}_{m+1}) = \frac{f(x_{m+1}|A_{m+1}, \underaccent{\ddot}{x}_{m+1})}{f(A_{m+1}|x_{m+1}, \underaccent{\ddot}{x}_{m+1})} \tag{10.16}$$

factors into a form

$$\frac{g(x_{m+1}, \underaccent{\ddot}{x}_{m+1})}{h(A_{m+1}, \underaccent{\ddot}{x}_{m+1})},$$

where the numerator is integrable as a function of x_{m+1}. Instead we merely ask that the function (10.16) be an integrable function of x_{m+1} and that

$$\frac{\tau(x_{m+1}, A_{m+1}, \underaccent{\ddot}{x}_{m+1})}{\int \tau(x_{m+1}, A_{m+1}, \underaccent{\ddot}{x}_{m+1}) \, dx_{m+1}}$$

be independent of A_{m+1} for every $\underaccent{\ddot}{x}_{m+1}$.

We may summarize this in the form of a theorem:

Theorem 10.3 (Compatibility). *Suppose that we are given an array of conditional densities in canonical form as follows:*

$$\{f(x_i|A_i, B_i) : i = 1, 2, \ldots, k\}, \tag{10.17}$$

where for each i, $A_i \subset \underaccent{\dot}{x}_{i-1}$ and $B_i \subset \underaccent{\ddot}{x}_i$.

Then a necessary and sufficient condition for the set (10.17) to be compatible with at least one joint density $f(x_1, x_2, \ldots, x_k)$ is that for each $i = 1, 2, \ldots, k$ either $A_i = \emptyset$ or

$$R_i = \frac{f(x_i|A_i, \underaccent{\ddot}{x}_i)/f(A_i|x_i, \underaccent{\ddot}{x}_i)}{\int_{x_i} f(x_i|A_i, \underaccent{\ddot}{x}_i)/f(A_i|x_i, \underaccent{\ddot}{x}_i) \, d\mu_i(x_i)} \tag{10.18}$$

is independent of A_i for any $\underaccent{\ddot}{x}_i$. □

Theorem 10.3 also suggests an algorithm for checking the consistency of a given set of conditional probability densities, one conditional probability density at a time, and for constructing a canonical form with $A_i = \emptyset$, $i = 1, \ldots, n$ (see Figure 10.1).

Definition 10.3 (Standard canonical form). When $A_i = \emptyset$ or $B_i = \emptyset$ for all i, we say that the canonical form is a *standard canonical form* and we call the term $f_{X_i|\underaccent{\ddot}{X}_i}(x_i|\underaccent{\ddot}{x}_i)$, or $f_{X_i|\underaccent{\dot}{X}_{i-1}}(x_i|\underaccent{\dot}{x}_{i-1})$, a *standard canonical component*. □

If the nested set of conditonal probabilities in Theorem 10.3 are given in standard canonical form, the consistency is guaranteed. Otherwise, the set of conditional probability densities must be checked for consistency.

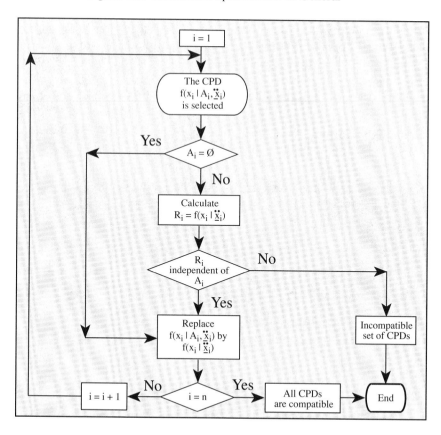

FIGURE 10.1. A flow chart for checking consistency of a canonical set of conditional probability densities that satisfies the uniqueness conditions

Algorithm 10.2 (Checking compatibility of a set of conditional probability densities).

- **Input.** *A set*

$$P = \{f(x_i|a_i, \ddot{\underline{x}}_i); i = 1, 2, \ldots, n\}$$

  of conditional probability densities in canonical form that satisfies the uniqueness condition.

- **Output.** *True or False, depending on whether or not the set of conditional probability densities in* P *is consistent.*

1. *The first conditional probability* $f(x_1|\ddot{\underline{x}}_1)$ *must be given in any case. Otherwise, the given set of conditional probability densities does not satisfy the uniqueness condition. Thus, at the initial step, we start with* $f(x_1|\ddot{\underline{x}}_1)$.

2. *At the ith step we are given either $f(x_i|\ddot{\underline{x}}_i)$ or $f(x_i|a_i, \ddot{\underline{x}}_i)$, where the set $A_i \subset \dot{\underline{x}}_{i-1}$. If $A_i = \emptyset$ proceed to Step 5; otherwise, we calculate $f(a_i|x_i, \ddot{\underline{x}}_i)$ by marginalizing $f(\dot{\underline{x}}_{i-1}|x_i, \ddot{\underline{x}}_i)$ over all variables in $\underline{\dot{X}}_{i-1}$ other than those in A_i, that is, using*

$$f(a_i|x_i, \ddot{\underline{x}}_i) = \int_{\underline{\dot{X}}_{i-1} \setminus A_i} f(\dot{\underline{x}}_{i-1}|x_i, \ddot{\underline{x}}_i) \ d\dot{\underline{x}}_{i-1} \setminus a_i. \qquad (10.19)$$

3. *Calculate the standard canonical component $R_i = f(x_i|\ddot{\underline{x}}_i)$ based on the previous and the new information using (10.18).*

4. *If R_i is independent of the variables in the set A_i, then go to Step 5; otherwise the given $f(x_i|a_i, \ddot{\underline{x}}_i)$ is not compatible with the previous conditional probabilities.*

5. *Calculate $f(\dot{\underline{x}}_{i-1}, x_i|\ddot{\underline{x}}_i) = f(\dot{\underline{x}}_{i-1}|x_i, \ddot{\underline{x}}_i)f(x_i|\ddot{\underline{x}}_i)$.*

6. *Repeat Steps 2 to 5 until all conditional probability densities have been analyzed.*

□

Therefore, given the set P of conditional probability densities we can determine whether or not it is consistent using Algorithm 10.2 or Theorem 10.3. This provides the answer to Question 6.6.

Example 10.11 (Possibly inconsistent set of conditional probability densities). Consider the following set P of conditional probability densities, which was given by a human expert to describe the joint probability density of five variables in X:

$$\{f(x_1|x_2, x_3, x_4, x_5), f(x_2|x_3, x_4, x_5), f(x_3|x_1, x_4, x_5), f(x_4|x_2, x_5), f(x_5)\}.$$

The set P can be shown to satisfy the uniqueness conditions. Thus, we can use Algorithm 10.2 to determine whether this set is compatible with a joint probability density:

For $i = 1$, the first conditional density, $f(x_1|x_2, x_3, x_4, x_5)$, is always compatible because $\ddot{\underline{X}}_1 = \{X_2, X_3, X_4, X_5\}$ and $A_1 = \emptyset$.

For $i = 2$, the second conditional density, $f(x_2|x_3, x_4, x_5)$, is compatible because $\ddot{\underline{X}}_2 = \{X_3, X_4, X_5\}$ and $A_2 = \emptyset$. Here, we have $\underline{\dot{X}}_1 = \{X_1\}$; hence

$$f(\dot{\underline{x}}_1, x_2|\ddot{\underline{x}}_2) = f(x_1, x_2|x_3, x_4, x_5) = f(x_1|x_2, x_3, x_4, x_5)f(x_2|x_3, x_4, x_5).$$

For $i = 3$, the third conditional density, $f(x_3|x_1, x_4, x_5)$, must be checked for compatibility because we have $\underline{\dot{X}}_3 = \{X_4, X_5\}$ and $A_3 = \{X_1\} \neq \emptyset$. Here, we have $\underline{\dot{X}}_2 = \{X_1, X_2\}$, and we need to compute $f(a_3|x_3, \ddot{x}_3)$ $= f(x_1|x_3, x_4, x_5)$ using (10.19). We obtain

$$f(x_1|x_3, x_4, x_5) = \int_{S(X_2)} f(x_1, x_2|x_3, x_4, x_5)\, dx_2.$$

We also need to compute R_3 using (10.18). We obtain

$$R_3 = f(x_3|x_4, x_5) = \frac{f(x_3|x_1, x_4, x_5)/f(x_1|x_3, x_4, x_5)}{\int_{S(X_3)} (f(x_3|x_1, x_4, x_5)/f(x_1|x_3, x_4, x_5))\, dx_3}.$$

Then, if R_3 does not depend on X_1, $f(x_3|x_1, x_4, x_5)$ is compatible with the previous two conditional probability densities. Otherwise the set is incompatible. For the sake of illustration, suppose that it is compatible. In this case, we replace $f(x_3|x_1, x_4, x_5)$ by R_3. We also calculate

$$f(\dot{x}_2, x_3|\ddot{x}_3) = f(x_1, x_2, x_3|x_4, x_5) = f(x_1, x_2|x_3, x_4, x_5)f(x_3|x_4, x_5).$$

For $i = 4$, the fourth conditional density, $f(x_4|x_2, x_5)$, must be checked for compatibility because we have $\underline{\dot{X}}_4 = \{X_5\}$ and $A_4 = \{X_2\} \neq \emptyset$. Here, we have $\underline{\dot{X}}_3 = \{X_1, X_2, X_3\}$, and we need to compute $f(a_4|x_4, \ddot{x}_4) = f(x_2|x_4, x_5)$ using (10.19). We obtain

$$f(x_2|x_4, x_5) = \int_{S(X_3) \times S(X_1)} f(x_1, x_2, x_3|x_4, x_5)\, dx_3\, dx_1.$$

We also need to compute R_4 using (10.18). We obtain

$$R_4 = f(x_4|x_5) = \frac{f(x_4|x_2, x_5)/f(x_2|x_4, x_5)}{\int_{S(X_4)} (f(x_4|x_2, x_5)/f(x_2|x_4, x_5))\, dx_4}.$$

Then, if R_4 does not depend on X_2, $f(x_4|x_2, x_5)$ is compatible with the previous three conditional probability densities. Otherwise the set is incompatible. Again, for the sake of illustration, suppose that it is compatible. In this case, we replace $f(x_4|x_2, x_5)$ by R_4. We also calculate

$$f(\dot{x}_3, x_4|\ddot{x}_4) = f(x_1, x_2, x_3, x_4|x_5) = f(x_1, x_2, x_3|x_4, x_5)f(x_4|x_5).$$

For $i = 5$, the fifth conditional density $f(x_5)$ is compatible because $\underline{\dot{X}}_5 = \emptyset$ and $A_5 = \emptyset$.

Therefore, if R_3 does not depend on X_1 and R_4 does not depend on X_2, then P is consistent; otherwise P is inconsistent. □

From the above example, we can see that Theorem 10.3 has the following important practical implications:

1. Every well-behaved joint probability density for the set of variables $(X_1, X_2, \ldots, X_k)$ can be represented by a conditionally specified probabilistic model.

2. Increasing the standard canonical form by superfluous information (extra conditional probability density) leads to the need for checking its compatibility.

3. If the set of conditional probability densities is given in standard canonical form, the conditional probability densities can be completely arbitrary, that is, they are not restricted by conditions other than those implied by the probability axioms.

4. Any conditional probability density of the form $f(x_i | a_i, \ddot{x}_i)$, with $A_i \neq \emptyset$, can be replaced by the conditional probability in standard canonical form $f(x_i | \ddot{x}_i)$ without affecting the joint probability density of the variables. The standard canonical form can be obtained using (10.18).

Theorems 10.2 and 10.3 answer Questions 10.1 and 10.2, posed in Section 10.4. Thus, when defining a conditionally specified probabilistic model, it is preferable to specify only the minimal set of conditional probabilities needed to define the joint probability density uniquely. Any extra information will require unnecessary additional computational effort, not only for its assessment but also for checking that the extra information is indeed consistent with the previously given conditional probability densities. This answers Question 10.3.

Furthermore, given a set P of conditional probability densities that is consistent and leads to a unique joint probability density of $\underline{X}$, we can replace P by another P' in standard canonical form leading to the same joint probability density. Further reduction of P' violates the uniqueness conditions. Also, redundant information hurts because it requires consistency checking. This answers Question 10.4.

10.6.1 Assessment of Conditionally Specified Probabilistic Models

Note that Theorem 10.3 assumes a set of conditional probability densities that already satisfies uniqueness. Therefore, uniqueness has to be checked first before compatibility. Thus, initially, we convert the specified set of conditional probability density to a canonical form, then we check uniqueness using Theorem 10.2, and then we check compatibility by means of Theorem 10.3. This topic in the context of Bayesian networks has been studied by Castillo, Gutiérrez, and Hadi (1997).

10.7   Overspecification

In the two-dimensional case, instead of being given $f(x|y)$ and $f(y|x)$ for every x, y, it is clearly adequate to be given $f(x|y)$ for every y and $f(y|x_0)$ for one fixed value of x_0 provided that $f(y|x_0) > 0$, $\forall y$. We can obtain the marginal density of y by normalizing the ratio

$$\frac{f(y|x_0)}{f(x_0|y)}$$

provided it is integrable. Compatibility is guaranteed. This observation can be carried to higher dimensions. In any one of a canonical list of conditional and/or marginal densities, if we are given

$$f(x_{m+1}|A_{m+1}^{(0)}, \ddot{\underline{x}}_{m+1}) \tag{10.20}$$

for particular values of $A_{m+1}^{(0)}$ of the conditioning variables A_{m+1}, our job is actually simplified (provided (10.20) is positive for every x_{m+1}). We need only obtain $f(A_{m+1}|x_{m+1}, \ddot{\underline{x}}_{m+1})$ from previously derived densities and normalize the assumed integrable ratio

$$\frac{f(x_{m+1}|A_{m+1}^{(0)}, \ddot{\underline{x}}_{m+1})}{f(A_{m+1}^{(0)}|x_{m+1}, \ddot{\underline{x}}_{m+1})}$$

to obtain

$$f(x_{m+1}|\ddot{\underline{x}}_{m+1}).$$

10.8   Marginals and Conditionals in Specified Families

Recall the bivariate conditional specification scenario discussed in detail in Section 1.7. Let $f_1(x; \underline{\theta})$, $\underline{\theta} \in \Theta$, denote a k-parameter family of densities and $f_2(y; \underline{\tau})$, $\underline{\tau} \in T$, a possibly different ℓ-parameter family of densities. We seek to identify all possible bivariate densities $f(x, y)$ for which both sets of conditional densities are in the specified families. Thus we require that

$$f(x|y) = f_1(x; \underline{\theta}(y)), \quad \forall y,$$

and

$$f(y|x) = f_2(y; \underline{\tau}(x)), \quad \forall x, \tag{10.21}$$

for some functions $\underline{\theta}(y)$ and $\underline{\tau}(x)$. The problem was resolved by assuming that a density with the required properties did exist with marginals

$g(x), h(y)$, and then setting up and hopefully solving the resulting functional equation of the following form:

$$h(y)f_1(x; \underline{\theta}(y)) = g(x)f_2(y; \underline{\tau}(x)). \tag{10.22}$$

A multivariate extension of this program was also described in Chapter 8. In such a setting we envision a k-dimensional density $f(x_1, \ldots, x_k)$ and k parametric families of densities $f_i(x; \underline{\theta}^{(i)})$, $\underline{\theta}^{(i)} \in \Theta_i$, $i = 1, 2, \ldots, k$. We then ask that the conditional densities belong to the given parametric families, i.e., that for $i = 1, 2, \ldots, k$,

$$f(x_i | \underline{x}_{(i)}) = f_i(x_i; \underline{\theta}^{(i)}(\underline{x}_{(i)})), \quad \forall \underline{x}_{(i)} . \tag{10.23}$$

If a density satisfying (10.23) is to exist it must have marginals which can be denoted by $h_i(\underline{x}_{(i)})$, $i = 1, 2, \ldots, k$. The joint density can then be written as a product of a marginal and a conditional in k different ways and we need to solve the following system of functional equations:

$$h_1(\underline{x}_{(1)})f_1(x_1; \underline{\theta}^{(1)}(\underline{x}_{(1)})) = h_2(\underline{x}_{(2)})f_2(x_2; \underline{\theta}^{(2)}(\underline{x}_{(2)}))$$
$$\ldots = h_k(\underline{x}_{(k)})f_k(x_k; \underline{\theta}^{(k)}(\underline{x}_{(k)})). \tag{10.24}$$

Examples in which this program can be carried out successfully are documented in Chapter 8.

In principle, there is nothing to prevent us from formulating an analogous problem in which arbitrary marginal and/or conditional densities are posited to belong to specified parametric families of densities. The notational bookkeeping is somewhat troubling but the concept is clear. An abbreviated form of the general problem may be stated as follows using the notation used in Definition 10.1. Let $f_1(\cdot; \underline{\theta}^{(1)}), f_2(\cdot; \underline{\theta}^{(2)}), \ldots, f_m(\cdot; \underline{\theta}^{(m)})$ denote m parametric families of distributions. We seek to identify all possible densities $f(x_1, \ldots, x_k)$ for which for each $j = 1, 2, \ldots, m$,

$$f(A_j | B_j) = f_j(A_j; \underline{\theta}^{(j)}(B_j)), \quad \forall B_j. \tag{10.25}$$

If a solution exists, it will have marginals including $h_j(B_j)$, $j = 1, 2, \ldots, m$, and we will seek to solve the following system of functional equations:

$$h_1(B_1)f_1(A_1; \underline{\theta}^{(1)}(B_1)) = h_2(B_2)f_2(A_2; \underline{\theta}^{(2)}(B_2))$$
$$\ldots = h_m(B_m)f_m(A_m; \underline{\theta}^{(m)}(B_m)). \tag{10.26}$$

Note that m could be less than, equal to, or greater than k, the dimension of $\underline{X}$. Generally speaking, increasing m will reduce the solution set. In extreme cases, the solution class might be very broad or, at the other extreme, empty. The case in which $A_i = x_i$ and $B_i = \underline{x}_{(i)}$, $i = 1, 2, \ldots, k$ (with $m = k$), has already been mentioned in Chapter 8. The case in which $A_i = x_i$ and $B_i = \emptyset$, $i = 1, 2, \ldots, k$, is a case of marginal specification. An enormous number of possible solutions exist. For example, we might try to identify the class of all possible distributions with univariate normal marginals. Random vectors with such distributions admit the representation

$$(X_1, \ldots, X_k) = (\mu_1 + \sigma_1 \Phi^{-1}(U_1), \ldots, \mu_k + \sigma_k \Phi^{-1}(U_k)),$$

where $\underline{\mu} \in \mathbb{R}^k, \underline{\sigma} \in \mathbb{R}^k_+, \Phi^{-1}$ is the standard normal quantile function and $\underline{U}$ is a completely arbitrary random vector with uniform $(0, 1)$ marginals.

We will focus attention only on examples in which the B_j's are non-empty, that is to say, on cases which really do involve conditional specification. No complete catalog can possibly be given but the examples in the following sections should give the flavor of the kinds of models and characterizations that such specifications can lead to.

10.9 X_i Given $\underline{X}_{(i)}$

Reference should be made to the material in Chapter 8 where a spectrum of such conditionally specified models is described.

10.10 $\underline{X}_{(i)}$ Given X_i

Suppose that we require that, for every i, the conditional density of $\underline{X}_{(i)}$ given $X_i = x_i$ should belong to some ℓ_i parameter family of densities

$$\{f_i(\cdot; \underline{\theta}^{(i)}); \underline{\theta}^{(i)} \in \Theta^{(i)}\}.$$

The resulting system of functional equations to solve, involving the marginals of X_i's, is given by

$$
\begin{aligned}
h_1(x_1)f_1(\underline{x}_{(1)}; \underline{\theta}^{(1)}(x_1)) &= h_2(x_2)f_2(\underline{x}_{(2)}; \underline{\theta}^{(2)}(x_2)) \\
\cdots &= h_k(x_k)f_k(\underline{x}_{(k)}; \underline{\theta}^{(k)}(x_k)). \quad (10.27)
\end{aligned}
$$

The exponential family case is relatively easy to resolve. In this setting we assume that

$$f_i(\underline{x}_{(i)}; \underline{\theta}^{(i)}(x_i)) = h_i(x_i) \exp\left\{\sum_{j=0}^{\ell_i} \theta_j^{(i)}(x_i) T_j^{(i)}(\underline{x}_{(i)})\right\}, \qquad \underline{x}_{(i)} \in \mathcal{X}_{(i)},$$

(10.28)

$i = 1, 2, \ldots, k$, where by convention $\theta_0^{(i)}(x_i) \equiv 1$, for each i. If we assume that (10.27) holds with conditionals of the form (10.28) then we will have a large number of consistency checks to perform. Actually just two of the conditional densities, e.g., $\underline{X}_{(1)}|X_1 = x_1$ and $\underline{X}_{(2)}|X_2 = x_2$, will be enough to specify the joint distribution and the remaining conditions must be checked for consistency.

To begin with, let us assume that we are given that (10.27) and (10.28) hold for $i = 1, 2, \ldots, k$. If we introduce the notation

$$\theta^{(i)}_{\ell_i+1}(x_i) = \log h_i(x_i), \tag{10.29}$$

we can write the joint density of $(X_1, \ldots, X_k)$ in k apparently different but equal ways:

$$\exp\left\{\sum_{j=0}^{\ell_i+1} \theta^{(1)}_j(x_1)T^{(1)}_j(\underline{x}_{(1)})\right\} = \exp\left\{\sum_{j=0}^{\ell_2+1} \theta^{(2)}_j(x_2)T^{(2)}_j(\underline{x}_{(2)})\right\}$$

$$= \ldots \tag{10.30}$$

$$= \exp\left\{\sum_{j=0}^{\ell_k+1} \theta^{(k)}_j(x_k)T^{(k)}_j(\underline{x}_{(k)})\right\},$$

where

$$T^{(i)}_{\ell_i+1}(\underline{x}_{(i)}) \equiv 1, \quad \forall i.$$

Taking logarithms in (10.30) we arrive at a system of functional equations amenable to solution using the Stephanos (1904) theorem (Theorem 1.3). For example, holding $\underline{x}_{(1,2)}$ fixed in the first equation of (10.30), we may conclude that

$$T^{(1)}_j(\underline{x}_{(1)}) = \sum_{j'=0}^{\ell_2+1} C_{j,j'}(\underline{x}_{(1,2)})\theta^{(2)}_{j'}(x_2).$$

Continued application of the Stephanos result leads to the conclusion that the joint density is necessarily of the form

$$f(\underline{x}) = \sum_{j_1=0}^{\ell_1+1}\sum_{j_2=0}^{\ell_2+1}\cdots\sum_{j_k=0}^{\ell_k+1} c_{j_1 j_2 \ldots j_k}\left[\prod_{i=1}^{k} \theta^{(i)}_{j_i}(x_i)\right]. \tag{10.31}$$

Comparing (10.30) and (10.31) we realize that compatibility will be encountered only if the given functions $T^{(i)}_j(\underline{x}_{(i)})$ have representations of the form

$$T^{(i)}_j(\underline{x}_{(i)}) = \sum_{\{j'_\ell s \text{ with } \ell \neq i\}} \cdots \sum c_{j_1 \cdots j \cdots j_k}\left[\prod_{i' \neq i} \theta^{(i')}_{j_{i'}}(x_{i'})\right]. \tag{10.32}$$

If a representation such as (10.32) does not hold then no density exists with the given conditionals. If the representation does exist, we can read off the forms of the functions $\{\theta^{(i)}_j(x_i)\}$.

Suppose now that we are given all but the last one of the conditional densities (10.30). What must the joint density look like? Repeated application of the Stephanos theorem leads to the following expression for the joint density:

$$f(\underline{x}) = \sum_{j_1=0}^{\ell_1+1} \sum_{j_2=0}^{\ell_2+1} \cdots \sum_{j_{k-1}=0}^{\ell_{k-1}+1} C_{j_1,\ldots,j_{k-1}}(x_k) \prod_{i=1}^{k-1} \theta_{j_i}^{(i)}(x_i) \qquad (10.33)$$

for appropriate functions $C_{j_1,\ldots,j_{k-1}}(x_k)$. Referring back to the conditional densities, we have

$$T_j^{(i)}(\underline{x}_{(i)}) = \sum_{\substack{j_\ell's \text{ with } \ell \leq k-1 \\ \text{and } \ell \neq i}} \cdots \sum C_{j_1,\ldots,j,\ldots,j_{k-1}}(x_k) \prod_{\substack{i \neq i' \\ i \leq k-1}} \theta_{j_i}^{(i)}(x_i). \qquad (10.34)$$

The given $T_j^{(i)}$'s must admit a representation (10.34) in order to be consistent and, when they do have such a representation, we can read off the corresponding $C_{j_1,\ldots,j_{k-1}}(x_k)$'s and $\theta_j^{(i)}(x_i)$'s. Naturally, the family (10.33) is more general than (10.31). The model (10.31) arises only when

$$C_{j_1,\ldots,j_{k-1}}(x_k) = \sum_{j_k=0}^{\ell_k+1} c_{j_1,\ldots,j_k} \theta_{j_k}^k(x_k) \qquad (10.35)$$

for some functions $\{\theta_j^k(x_k)\}$.

Next notice that nothing in the above development precludes the x_i's from being vectors rather than scalars. This allows us to immediately discern the consequences of only assuming that the first m of the equations in (10.28) hold. We merely write a new $\underline{x}$ vector as $(x_1, x_2, \ldots, x_m, \ddot{\underline{x}}_m)$ and arrive at a joint density of the form (cf. (10.33))

$$f(\underline{x}) = \sum_{j_1=0}^{\ell_1+1} \cdots \sum_{j_m=0}^{\ell_m+1} C_{j_1,\ldots,j_m}(\ddot{\underline{x}}_m) \prod_{i=1}^{m} \theta_{j_i}^{(i)}(x_i). \qquad (10.36)$$

It may be observed that models obtained by assuming all conditionals of $X_{(i)}$ given $X_i = x_i$ $(i = 1, 2, \ldots, k)$ are in exponential families as in (10.28), are identical to the class of all densities for which X_i given $\underline{X}_{(i)} = \underline{x}_{(i)}$ $(i = 1, 2, \ldots, k)$ are in appropriate exponential families. We get more general models only when we assume that (10.28) holds only for some but not all i.

Finally, we remark that analogous arguments work in certain nonexponential family situations (the multivariate Pareto is a prize example). The key issue is that equality among the available versions of the joint density (the expressions analogous to (10.27)) should lead to solvable systems of functional equations.

10.11 The Case of X Given Y, Y Given Z, and Z Given X

Suppose that we assume that, for every y, the conditional distribution of X given $Y = y$ is a member of a given k_1-parameter family, i.e., $f(x|y) = f_1(x; \underline{\theta}^{(1)}(y))$. In addition, we make analogous assumptions regarding Y given Z and Z given X; i.e., $f(y|z) = f_2(y; \underline{\theta}^{(2)}(z))$ and $f(z|x) = f_3(z; \underline{\theta}^{(3)}(x))$ for specified k_2- and k_3-parameter families f_2 and f_3. What can be said of the joint distribution of (X, Y, Z)? The disappointing answer to this question is: Not very much. The problem is that $f(x|y)$, $f(y|z)$, and $f(z|x)$ do <u>not</u> uniquely determine $f(x, y, z)$ since they are completely determined by the bivariate marginals. Alternatively, it is clear that the list $f(x|y)$, $f(y|z)$, and $f(z|x)$ does not contain, nor is it equivalent to, a set of densities of the form (10.14), needed in the Gelman and Speed theorem.

For example, if we assume that

$$X|Y = y \sim N(\alpha_1 + \beta_1 y, \sigma_1^2),$$

$$Y|Z = z \sim N(\alpha_2 + \beta_2 z, \sigma_2^2),$$

and

$$Z|X = x \sim N(\alpha_3 + \beta_3 x, \sigma_3^2),$$

then there is a unique classical trivariate normal distribution with these conditionals. However, there are many other trivariate distributions with the same bivariate marginals which necessarily have the same conditionals for $X|Y$, $Y|Z$, and $Z|X$.

It is appropriate at this point to emphasize that the negative features of this example are <u>not</u> in conflict with the many "nearest neighbor" specifications of Markovian spatial processes. For simplicity, consider a spatial process with only four locations which for convenience can be visualized as corners of a square. Let X, Y, Z, W denote the values of the process at the four locations. The process will often be defined in terms of the conditional distribution of the value at a point given the values at its (two) neighboring points. So we are given the distribution of X given W and Y, of Y given X and Z, of Z given Y and W, and of W given Z and X. Stated in that fashion, our discussion, in the last paragraphs, clearly indicates that a broad spectrum of possible models exist, since in the given conditional specification only trivariate marginals of (X, Y, Z, W) are involved. The nearest neighbor specification (if consistent) does yield a unique distribution if explicitly, or (often) implicitly, what we assume is that the conditional distribution of X given Y, Z, and W depends <u>only</u> on W and Y, the conditional distribution of Y given the rest depends <u>only</u> on X and Z, etc. Some authors are regretably not precise in their formulations of such processes.

10.12   Logistic Regression Models

Let Y be a binary random variable such that

$$P(Y = y) = \frac{e^{\sigma y}}{1 + e^{\sigma}}, \quad y = 0, 1, \tag{10.37}$$

where $\sigma > 0$ is a scale parameter. Note that (10.37) represents a one-parameter exponential family.

Definition 10.4 (Logit transformation of a binary random variable). Let Y be a random variable with pdf given by (10.37). The logit transformation of $P(Y = y)$ is defined as

$$\text{logit}[P(Y = y)] = \log\left[\frac{P(Y = 1)}{P(Y = 0)}\right] = \sigma. \tag{10.38}$$

$\square$

First we consider the case of two binary random variables Y_1 and Y_2. We are interested in the most general bivariate random variable (Y_1, Y_2) such that its conditional logit transformations satisfy

$$\text{logit}[P(Y_1 = y_1 | Y_2 = y_2)] = \sigma_1(y_2) \tag{10.39}$$

and

$$\text{logit}[P(Y_2 = y_2 | Y_1 = y_1)] = \sigma_2(y_1). \tag{10.40}$$

Since (10.37) is an exponential family, the most general distribution with conditionals of the form (10.39) and (10.40) is given by (4.5). Thus the joint pdf of the bivariate random variable (Y_1, Y_2) becomes

$$P(Y_1 = y_1, Y_2 = y_2) = \frac{e^{m_{10}y_1 + m_{01}y_2 + m_{11}y_1 y_2}}{1 + e^{m_{10}} + e^{m_{01}} + e^{m_{10} + m_{01} + m_{11}}}. \tag{10.41}$$

Note that, in this case, the normalizing constant has been explicitly obtained. From (10.41) it can be easily shown that

$$\text{logit}[P(Y_1 = y_1 | Y_2 = y_2)] = \sigma_1(y_2) = m_{10} + m_{11}y_2, \tag{10.42}$$

and

$$\text{logit}[P(Y_2 = y_2 | Y_1 = y_1)] = \sigma_2(y_1) = m_{01} + m_{11}y_1. \tag{10.43}$$

If we assume that the constants m_{10} and m_{01} depend on a covariate vector $\underline{x}$, such that

$$m_{10} = \alpha_1 + \underline{\beta}'_1 \underline{x}$$

and

$$m_{01} = \alpha_2 + \underline{\beta}'_2 \underline{x},$$

an extension of (10.41) is possible.

In this case, model (10.41) becomes

$$P(Y_1 = y_1, Y_2 = y_2 | \underline{x}) \propto \exp[(\alpha_1 + \underline{\beta}_1' \underline{x}) y_1 + (\alpha_2 + \underline{\beta}_2' \underline{x}) y_2 + m_{11} y_1 y_2], \tag{10.44}$$

which has been considered by Joe (1997).

The bivariate model (10.41) can also be generalized to the k-dimensional case, as has been done with other models in Chapter 8. Joe (1997) considers a particular multivariate extension of (10.44) of the following form. Assume that we are interested in the k-dimensional variable $(Y_1, \ldots, Y_k)$ such that, for each $i = 1, 2, \ldots, k$, the following holds:

$$\text{logit}[P(Y_i = y_i | Y_j = y_j, j \neq i, \underline{x})] = \alpha_i + \underline{\beta}_i' \underline{x} + \sum_{j \neq i} m_{ij} y_j. \tag{10.45}$$

Then the necessary and sufficient conditions for compatibility of these conditional distributions are that $m_{ij} = m_{ji}, i \neq j$ (compare with (10.42) and (10.43)). The corresponding joint pdf is

$$P(Y_1 = y_1, \ldots, Y_k = y_k | \underline{x}) = k \exp\left[\sum_{i=1}^{k} (\alpha_i + \underline{\beta}_i' \underline{x}) y_i + \sum_{i<j} m_{ij} y_i y_j\right], \tag{10.46}$$

where the normalizing constant is given by

$$k^{-1} = \sum_{y_1=0}^{1} \cdots \sum_{y_k=0}^{1} \exp\left[\sum_{i=1}^{k} (\alpha_i + \underline{\beta}_i' \underline{x}) y_i + \sum_{1 \leq i < j \leq k} m_{ij} y_i y_j\right]. \tag{10.47}$$

The parameters m_{ij} in (10.46) can be interpreted as conditional log-odds ratios.

10.13 Bibliographic Notes

Some of the ideas in Section 10.2 were discussed in Arnold, Athreya, and Sethuraman (1998). Gelman and Speed (1993) is a key reference for general discussion of consistent conditional and marginal specification of multivariate distributions. Further elaboration of the problems involved may be found in Arnold, Castillo, and Sarabia (1992, 1993a, 1993b, 1995, 1996b) and Castillo (1988, 1992).

Exercises

10.1 Let (X_1, X_2, X_3) be a trivariate random variable.

(a) Assuming compatibility, are $f(x_1, x_2 | x_3)$ and $f(x_2, x_3 | x_1)$ enough to determine $f(x_1, x_2, x_3)$?

(b) If (a) is true, obtain the most general trivariate distributions such that $f(x_1, x_2 | x_3)$ and $f(x_2, x_3 | x_1)$ are in exponential families.

10.2 Generalize the model (10.46) in the sense of including terms with higher interactions, i.e., including terms of the form $y_i y_j y_k$, $y_i y_j y_k y_r$, etc.

10.3 Check the compatibility and uniqueness of the set of conditional distributions

$$f(x_1 | x_2, x_3) = \sqrt{\frac{3}{22\pi}} \exp[-(3x_1 - x_2 + x_3)^2 / 66],$$

$$f(x_2 | x_1, x_3) = \sqrt{\frac{2}{11\pi}} \exp[-(x_1 - 4x_2 + 4x_3)^2 / 88],$$

$$f(x_3 | x_1, x_1) = \sqrt{\frac{15}{22\pi}} \exp[-(x_1 - 4x_2 + 15x_3)^2 / 330].$$

10.4 Given $f_X(x)$ and $f_{X|Y}(x|y)$, show that a sufficient condition for $f_Y(y)$, and hence for $f_{X,Y}(x, y)$, to be unique is that the conditional density of X given Y is of the exponential form

$$f_{X|Y}(x|y) = \exp[yA(x) + B(x) + C(y)],$$

where an interval is contained in the range of $A(x)$.
(Seshadri and Patil (1964).)

10.5 Let (X, Y) be a bivariate distribution, and assume that the marginal distribution of X is exponential.

(a) If the conditional density of X given Y is

$$f(x|y) = e^{-x(1+\delta y)}[(1 + \delta x)(1 + \delta y) - \delta],$$

with $0 \le \delta \le 1$, $x, y \ge 0$, then show that the marginal distribution of Y is unique and is the exponential distribution.

(b) If the conditional density of X given Y is,

$$f(x|y) = e^{-x}(1 + \alpha - 2\alpha e^{-y}) - 2\alpha e^{-2x}(1 - 2e^{-y}),$$

with $-1 \le \alpha \le 1$, $x, y \ge 0$, show that the marginal density of Y is not unique.

(Seshadri and Patil (1964).)

11
Conditional Survival Models

11.1 Introduction

In this chapter we will deal with k-dimensional random variables usually with positive coordinates, which can be visualized as representing times to failures of k distinct types. It is envisioned however that an individual will inevitably eventually suffer all k types of failures so that the coordinates of the random vectors are finite with probability 1. In actual practice, multivariate data sets of this type are relatively rare. What this means is that many multivariate survival models may well find their domain of application in the study of data sets involving k-dimensional data which realistically have little to do with survival or times to failure. In a sense they may be best considered to be k-variate distributions motivated by mathematically natural multivariate extensions of univariate survival models. In this sense they deserve the name "multivariate survival distributions," though the same should not be construed as implying limitations in their potential fields of applications.

The most commonly used univariate survival models are exponential, Weibull, Pareto, gamma, log-normal, and generalizations of these (often including powers). It is to be expected then that many popular multivariate survival models will have marginals and/or conditionals in these families.

We have already encountered multivariate conditionally specified distributions related to many of these basic survival models. The emphasis in this chapter is on models derived via alternative conditional specification methods perhaps better suited to the survival context. Most of our presentation will be for the bivariate case. Multivariate extensions are often readily envisioned and will be briefly noted.

Up until now, when we have spoken of conditionally specified bivariate distributions, we have referred to joint densities $f_{X,Y}(x,y)$ with all conditionals of X given $Y = y$ belonging to a particular parametric family and all conditionals of Y given $X = x$ belonging to a second, possibly different, parametric family. In the context of bivariate survival models, it is more natural to condition on component survivals, i.e., on events such as $\{X > x\}$ and $\{Y > y\}$ rather than conditioning on particular values of X and Y.

Most of the questions discussed in Chapter 1 can be asked anew in this new context; i.e., conditioning on $\{X > x\}$ and $\{Y > y\}$ instead of $X = x$ and $Y = y$. The answers will lead to distinct models from those discussed in the earlier chapters of this book.

Several related conditional specification paradigms (motivated by survival considerations) will also be discussed in this chapter.

11.2   Conditional Survival in the Bivariate Case

Consider bivariate random variables (X, Y) (discrete or absolutely continuous) with the set of possible values for X (respectively Y) denoted by $S(X)$ (respectively $S(Y)$). Suppose that for each $(x, y) \in S(X) \times S(Y)$ we are given $f_X(x|Y > y)$ and $f_Y(y|X > x)$. Reasonable questions to ask at this juncture include:

(i) Are the given families of conditional densities compatible?

(ii) If they are, do they determine a unique joint density for (X, Y)?

Since conditional survival functions are uniquely determined by conditional densities, an equivalent more convenient formulation of the problem is available. Thus we ask about compatibility of putative families of conditional survival functions of the forms

$$P(X > x|Y > y), \quad (x, y) \in S(X) \times S(Y),$$

and

$$P(Y > y|X > x), \quad (x, y) \in S(X) \times S(Y).$$

The compatibility issue is readily resolved as follows:

Theorem 11.1 (Compatibility of conditional survival functions).
Two families of conditional survival functions

$$P(X > x|Y > y) = a(x, y), \ (x, y) \in S(X) \times S(Y),$$
$$and \quad P(Y > y|X > x) = b(x, y), \ (x, y) \in S(X) \times S(Y), \quad (11.1)$$

are compatible if and only if their ratio factors, i.e., if and only if there exist functions $u(x), x \in S(X)$ and $v(y), y \in S(Y)$ such that

$$\frac{a(x,y)}{b(x,y)} = \frac{u(x)}{v(y)}, \quad (x,y) \in S(X) \times S(Y), \qquad (11.2)$$

where $u(x)$ is a one-dimensional survival function (nonincreasing, right continuous, and of total variation 1).   □

Proof.   If $a(x,y)$ and $b(x,y)$ are to be compatible there must exist corresponding marginal survival functions $P(X > x) = u(x)$ and $P(Y > y) = v(y)$. Writing the event $P(X > x, Y > y) = a(x,y)v(y) = b(x,y)u(x)$ yields (11.2).   □

In Theorem 1.1 an analogous result was presented involving $f_{X|Y}(x|y)$ and $f_{Y|X}(y|x)$. However after proving existence of a solution, using Theorem 1.1, additional assumptions were required to guarantee uniqueness. In the present setting, life is simpler. If there is any pair $u(x), v(y)$ for which (11.2) holds, it is readily verified that they are unique. Thus two families of survival functions (11.1) will uniquely determine a joint distribution (via a joint survival function) if their ratio factors are as in (11.2).

Remark 11.1 In many reliability contexts, it is not practical to consider the event $\{X > x\}$ conditioned on the event $\{Y > y\}$. In those cases, it might be easier to envision conditioning on the event $\{\min(X,Y) > y\}$ (cf. the "dynamic construction" discussed in Section 1.2).

11.3   Conditional Survival Functions in Parametric Families

Rather than specify the precise form of $P(X > x | Y > y)$ we might only require that for, each $y \in S(Y)$, it be a member of a specified parametric family of survival functions. An analogous requirement, that for each $x \in S(X)$, $P(Y > y | X > x)$ should be a member of a possibly different parametric family of survival functions, would also be imposed. What kind of joint survival functions will be determined by such constraints?

Consider a k-parameter family of survival functions denoted by

$$\{\overline{F}_1(x;\underline{\theta}) : \underline{\theta} \in \Theta_1\},$$

where $\Theta_1 \subset \mathbb{R}^{k_1}$, and a possibly different k_2-parameter family of survival functions denoted by

$$\{\overline{F}_2(x;\underline{\tau}) : \underline{\tau} \in T\},$$

where $T \subset \mathbb{R}^{k_2}$. We are interested in all possible joint distributions for (X, Y) such that for each $y \in S(Y)$

$$P(X > x | Y > y) = \overline{F}_1(x; \underline{\theta}(y)), \quad \forall x \in S(X), \tag{11.3}$$

and for each $x \in S(X)$

$$P(Y > y | X > x) = \overline{F}_2(y; \underline{\tau}(x)), \quad \forall y \in S(Y), \tag{11.4}$$

for some functions $\underline{\theta}(y)$ and $\underline{\tau}(x)$.

If (11.3) and (11.4) are both to hold, there must exist marginal survival functions for X and Y denoted by $\overline{F}_X(x)$ and $\overline{F}_Y(y)$ such that

$$\overline{F}_Y(y)\overline{F}_1(x; \underline{\theta}(y)) = \overline{F}_X(x)\overline{F}_2(y; \underline{\tau}(x)), \quad \forall x, y. \tag{11.5}$$

This is true since both the left- and right-hand sides of (11.5) must equal $P(X > x, Y > y)$. Whether we can solve this functional equation depends, of course, on the structure of the functions $\overline{F}_1$ and $\overline{F}_2$. If we can solve the functional equation (in $\underline{\theta}(y)$, $\underline{\tau}(x)$, $\overline{F}_X(x)$, and $\overline{F}_Y(y)$) we do need to check that the expressions obtained for $\overline{F}_X(x)$ and $\overline{F}_Y(y)$ are valid survival functions. In Chapter 1, exponential families provided particularly convenient examples in which the functional equation analogous to (11.5) (i.e., (1.26)) could be solved. In the survival context, exponential families do not typically play such a prominent role. Nevertheless, it is sometimes possible to solve (11.5). Let us begin with an example which will itself suggest possible generalizations.

Suppose that for each $y > 0$, the conditional survival function for X given $Y > y$ is exponential with intensity parameter $\theta(y)$. Analogously, suppose that Y given $X > x$ is also always exponential. Thus for some functions $\theta(y)$ and $\tau(x)$ we have

$$P(X > x | Y > y) = \exp[-\theta(y)x] \tag{11.6}$$

and

$$P(Y > y | X > x) = \exp[-\tau(x)y]. \tag{11.7}$$

Equations (11.6) and (11.7) are to hold for every $x > 0, y > 0$. If there is to be a joint survival function for (X, Y) consistent with (11.6) and (11.7), it must have associated marginal survival functions $\overline{F}_X(x) = P(X > x)$ and $\overline{F}_Y(y) = P(Y > y)$ and we must have

$$\overline{F}_Y(y)\exp[-\theta(y)x] = P(X > x, Y > y) = \overline{F}_X(x)\exp[-\tau(x)y], \tag{11.8}$$

where $\overline{F}_X(\cdot), \overline{F}_Y(\cdot), \theta(\cdot)$ and $\tau(\cdot)$ are unknown functions. This, of course, is just a special case of (11.5). If we take logarithms of both sides in (11.8) and define $\tilde{\phi}_2(y) = \log \overline{F}_Y(y)$ and $\tilde{\phi}_1(x) = \log \overline{F}_X(x)$ our functional equation becomes

$$\tilde{\phi}_2(y) - \theta(y)x = \tilde{\phi}_1(x) - \tau(x)y. \tag{11.9}$$

This is a special case of the Stephanos–Levi Civita–Suto functional equation (refer to Theorem 1.3). In order for (11.9) to hold we must have

$$\theta(y) = \alpha + \gamma y \tag{11.10}$$

and

$$\tau(x) = \beta + \gamma x \tag{11.11}$$

for some constants α, β, γ. Substituting this back in (11.8) we obtain the following expression for the joint survival function of (X, Y):

$$\bar{F}_{X,Y}(x, y) = P(X > x, Y > x) = \exp(\delta + \alpha x + \beta y + \gamma xy), \quad x > 0, \; y > 0. \tag{11.12}$$

Clearly δ must be 0. In order for (11.12) to represent a valid joint survival function we must have $\frac{\partial^2 \bar{F}(x,y)}{\partial x} \geq 0, \; \forall x, y > 0$. This means we must take $\alpha\beta \geq -\gamma$. In addition we need $\alpha, \beta < 0$, and $\gamma \leq 0$. Reparametrizing in terms of marginal scale parameters and an interaction parameter we have

$$\bar{F}_{X,Y}(x, y) = \exp\left[-\left(\frac{x}{\sigma_1} + \frac{y}{\sigma_2} + \theta \frac{xy}{\sigma_1 \sigma_2}\right)\right], \quad x > 0, \quad y > 0, \tag{11.13}$$

where $\sigma_1, \sigma_2 > 0$ and $0 \leq \theta \leq 1$. This is recognizable as Gumbel's type I bivariate exponential distribution (Gumbel (1960)). If we set $x = 0$ or $y = 0$ in (11.13) we find that the distribution has exponential marginals. As Gumbel noted the correlation is always nonpositive (analogous nonpositive correlation was encountered in the exponential conditionals distribution discussed in Section 4.4). In the present case we find

$$\rho(X, Y) = -1 + \int_0^\infty \frac{e^{-y}}{1 + \theta y} \, dy \tag{11.14}$$

and

$$-0.404 \leq \rho(X, Y) \leq 0. \tag{11.15}$$

Gumbel provided the following expressions for conditional densities, means and variances. For $y > 0$,

$$f_{X|Y}(x|y) = \frac{1}{\sigma_1}\left[\left(1 + \theta\frac{x}{\sigma_1}\right)\left(1 + \theta\frac{y}{\sigma_2}\right) - \theta\right]e^{-(1+\theta y/\sigma_2)x/\sigma_1}, \quad x > 0, \tag{11.16}$$

$$E(X|Y = y) = \sigma_1 \frac{1 + \theta + \frac{\theta y}{\sigma_2}}{\left(1 + \frac{\theta y}{\sigma_2}\right)^2}, \tag{11.17}$$

and

$$\text{var}(X|Y = y) = \sigma_1^2 \frac{(1 + \theta + \theta\frac{y}{\sigma_2})^2 - 2\theta^2}{\left(1 + \theta\frac{y}{\sigma_2}\right)^4}. \tag{11.18}$$

Nair and Nair (1988) characterized this Gumbel bivariate exponential distribution as the only one with the property that

$$E(X - x|X > x, Y > y) = E(X|Y > y), \quad \forall x, y > 0, \tag{11.19}$$

and

$$E(Y - y|X > x, Y > y) = E(Y|X > x), \quad \forall x, y > 0. \tag{11.20}$$

From our discussion, we may add to Nair and Nair's observation the fact that Gumbel's bivariate exponential distribution has exponential conditional survival functions (i.e. that (11.6) and (11.7) hold).

The above example was resolvable in part because the assumed conditional survival functions were available in closed form. Motivated by the discussion in the literature of generalized gamma functions we can introduce what we call generalized($\bar{\Phi}$) survival functions. For them we can often successively implement a conditional characterization program analogous to that which led to Gumbel's distribution.

Definition 11.1 (Generalized survival function). Let $\bar{\Phi}$ denote a specific survival function. The corresponding family of generalized ($\bar{\Phi}$) survival functions is of the form

$$\bar{F}(x; \mu, \sigma, \gamma, \delta) = \left[\bar{\Phi} \left(\left(\frac{x - \mu}{\sigma} \right)^\delta \right) \right]^\gamma, \tag{11.21}$$

where $\mu \in \mathbb{R}, \sigma > 0, \delta > 0, \gamma > 0$. □

Now let $\bar{\Phi}_1$ and $\bar{\Phi}_2$ be two survival functions. We seek to identify all bivariate distributions for (X, Y) such that for every y, the conditional survival function of X given $\{Y > y\}$ is a generalized ($\bar{\Phi}_1$) survival function and for every x, the conditional survival function of Y given $\{X > x\}$ is a generalized ($\bar{\Phi}_2$) survival function. Thus we ask that, for each $y \in S(Y)$,

$$P(X > x|Y > y) = \left[\bar{\Phi}_1 \left(\left(\frac{x - \mu_1(y)}{\sigma_1(y)} \right)^{\delta_1(y)} \right) \right]^{\gamma_1(y)} \tag{11.22}$$

and, for each $x \in S(X)$,

$$P(Y > y|X > x) = \left[\bar{\Phi}_2 \left(\left(\frac{y - \mu_2(x)}{\sigma_2(x)} \right)^{\delta_2(x)} \right) \right]^{\gamma_2(x)} \tag{11.23}$$

for some unknown functions $\mu_1(y), \sigma_1(y), \delta_1(y), \gamma_1(y), \mu_2(x), \sigma_2(x), \delta_2(x)$, and $\gamma_2(x)$.

The functional equation to be solved to determine models satisfying (11.22) and (11.23) is obtained by multiplying the right-hand sides of

(11.22) and (11.23) by the corresponding unknown marginal survival functions so that they may be equated (as was done in (11.5)). The tractability of the resulting functional equation depends on the nature of the survival functions $\bar{\Phi}_1$ and $\bar{\Phi}_2$, and depends on how many and which ones of the unknown functions $\mu_1(y), \ldots, \gamma_2(x)$ are assumed to be constants. There appears to be no general theory available but an interesting list of cases in which the program can be successfully carried out is provided in the next section.

Our exponential example corresponded to the choice $\bar{\Phi}_1(x) = \bar{\Phi}_2(x) = e^{-x}$, $x > 0$ and $\mu_1(y) = 0$, $\mu_2(x) = 0$, $\delta_1(y) = \delta_2(x) = \gamma_1(y) = \gamma_2(x) = 1$.

11.4   Examples of Distributions Characterized by Conditional Survival

11.4.1   Weibull Conditional Survival Functions

Suppose that (X, Y) has support $\mathbb{R}^+ \times \mathbb{R}^+$ and, for each $y > 0$,

$$P(X > x | Y > y) = \exp\{-[x/\sigma_1(y)]^{\gamma_1(y)}\}, \quad x > 0, \tag{11.24}$$

and, for each $x > 0$,

$$P(Y > y | X > x) = \exp\{-[y/\sigma_2(x)]^{\gamma_2(x)}\}, \quad y > 0. \tag{11.25}$$

If we multiply (11.24) by $\bar{F}_2(y)$ and equate it to (11.25) multiplied by $\bar{F}_1(x)$ and take logarithms we encounter the following functional equation:

$$\log \bar{F}_2(y) - \left[\frac{x}{\sigma_1(y)}\right]^{\gamma_1(y)} = \log \bar{F}_1(x) - \left[\frac{y}{\sigma_2(x)}\right]^{\gamma_2(x)}. \tag{11.26}$$

It is conjectured that no solution to (11.26) exists unless $\gamma_1(y)$ and $\gamma_2(x)$ are constant functions. If $\gamma_1(y) = \gamma_1$, $\forall y$ and $\gamma_2(x) = \gamma_2$ $\forall x$, then (11.26) is a special case of the Stephanos–Levi Civita–Suto equation (Theorem 1.3). It follows that

$$\sigma_1(y)^{\gamma_1} = (\alpha + \gamma \, y^{\gamma_2})^{-1},$$

$$\sigma_2(x)^{\gamma_2} = (\beta + \gamma \, x^{\gamma_1})^{-1},$$

and, eventually, that the joint survival function is of the form

$$\bar{F}(x, y) = \exp\left\{-\left[\left(\frac{x}{\sigma_1}\right)^{\gamma_1} + \left(\frac{y}{\sigma_2}\right)^{\gamma_2} + \theta\left(\frac{x}{\sigma_1}\right)^{\gamma_1}\left(\frac{y}{\sigma_2}\right)^{\gamma_2}\right]\right\}, \quad x > 0, \ y > 0. \tag{11.27}$$

where $\sigma_1, \sigma_2 > 0$ and $0 \leq \theta \leq 1$. The Gumbel bivariate exponential (11.23) corresponds to the choice $\gamma_1 = \gamma_2 = 1$ in (11.27).

It may be observed that these models exhibit negative dependence (and hence negative correlation). Negative correlation was also shown to be a feature of the exponential conditionals distribution discussed earlier and displayed in (4.3). In terms of survival modeling this can be viewed, in a sense, as the opposite of load sharing. After one component fails, the other appears to be invigorated rather than debilitated. Clearly this limitation on the achievable sign of the correlation coefficient must be taken into account in modeling efforts. Data sets with positive correlation will clearly not be appropriately fitted by Weibull conditionals models such as (11.27).

11.4.2   Logistic Conditional Survival Functions

Although survival models are usually associated with nonnegative random variables, there is nothing, in principle, in our development to stop us from considering random variables which can assume negative values. For example, we might postulate logistic conditional survival functions. Thus, for each $y \in \mathbb{R}$, we might assume

$$P(X > x|Y > y) = [1 + e^{(x-\mu_1(y))/\sigma_1(y)}]^{-1}, \quad x \in \mathbb{R}, \tag{11.28}$$

and, for each $x \in \mathbb{R}$,

$$P(Y > y|X > x) = [1 + e^{(y-\mu_2(x))/\sigma_2(x)}]^{-1}, \quad y \in \mathbb{R}. \tag{11.29}$$

If we seek a general solution we are led to an equation analogous to (11.26). Only the case $\sigma_1(y) = \sigma_1$ in (11.28) and $\sigma_2(x) = \sigma_2$ in (11.29) appears to be tractable. If we make this simplifying assumption we are led to the class of bivariate survival functions with logistic conditional survival functions given by

$$\bar{F}(x,y) = [1 + e^{(x-\mu_1)/\sigma_1} + e^{(y-\mu_2)/\sigma_2} + \theta e^{[(x-\mu_1)/\sigma_1 + (y-\mu_2)/\sigma_2]}]^{-1}, \tag{11.30}$$

where $\mu_1, \mu_2 \in \mathbb{R}^+$ and $\theta \in [0,2]$. The constraint $\theta \in [0,2]$ is needed to guarantee that $\dfrac{\partial^2 \bar{F}(x,y)}{\partial x \partial y} \geq 0$, $\forall x, y$.

11.4.3   Generalized Pareto Conditional Survival Functions

In this case we ask, that for each $y > 0$,

$$P(X > x|Y > y) = \left[1 + \left(\frac{x}{\sigma_1(y)}\right)^{c_1(y)}\right]^{-k_1(y)}, \quad x > 0, \tag{11.31}$$

and, for each $x > 0$,

$$P(Y > y|X > x) = \left[1 + \left(\frac{y}{\sigma_2(x)}\right)^{c_2(x)}\right]^{-k_2(x)}, \quad y > 0. \tag{11.32}$$

for positive functions $c_1(y), k_1(y), \sigma_1(y), c_2(x), k_2(x)$, and $\sigma_2(x)$. If (11.31) and (11.32) are both to hold, we must have

$$\bar{F}_2(y) \left[1 + \left(\frac{x}{\sigma_1(y)}\right)^{c_1(y)}\right]^{-k_1(y)} = \bar{F}_1(x) \left[1 + \left(\frac{y}{\sigma_2(x)}\right)^{c_2(x)}\right]^{-k_2(x)}.$$

$$(11.33)$$

If we introduce new functions

$$b_1(y) = \sigma_1(y)^{-c_1(y)}$$

and

$$b_2(x) = \sigma_2(x)^{-c_2(x)}$$

then (11.33) can be written as

$$\bar{F}_2(y)[1 + b_1(y)x^{c_1(y)}]^{-k_1(y)} = \bar{F}_1(x)[1 + b_2(x)y^{c_2(x)}]^{-k_2(x)}. \qquad (11.34)$$

The general class of solutions to (11.34) appears to be difficult to describe. There are, however, two special cases which are quite tractable:

(i) when $c_1(y) = c_1$ and $c_2(x) = c_2$ and ;

(ii) when $k_1(y) = k_1$ and $k_2(x) = k_2$.

In case (i), equation (11.34) reduces to one which is analogous to equation (4.1) of Arnold, Castillo, and Sarabia (1993d). It can be transformed to a form equivalent to an equation solved by Castillo and Galambos (1987b) (their equation (3.1)). Two families of solutions consequently exist. Substituting the solutions back into the expressions for the joint survival function (11.34) we obtain the following. In Family I,

$$\bar{F}(x,y) = \left[1 + \left(\frac{x}{\sigma_1}\right)^{c_1} + \left(\frac{y}{\sigma_2}\right)^{c_2} + \theta \left(\frac{x}{\sigma_1}\right)^{c_1} \left(\frac{y}{\sigma_2}\right)^{c_2}\right]^{-k}, \quad x,y > 0,$$

$$(11.35)$$

for positive constants $c_1, \sigma_1, c_2, \sigma_2, k$, and $\theta \in [0,2]$. Again the condition $\theta \in [0,2]$ is needed to ensure a positive density. This family of bivariate generalized Pareto distributions was first described in Durling (1970) (see also Arnold (1990) for a multivariate version).

The other family of solutions to (11.34) with $c_1(y) = c_1$ and $c_2(x) = c_2$ lead to joint survival functions of the form

$$\bar{F}(x,y) = \exp\left\{-\theta_1 \log\left[1 + \left(\frac{x}{\sigma_1}\right)^{c_1}\right] - \theta_2 \log\left[1 + \left(\frac{y}{\sigma_2}\right)^{c_2}\right] \right.$$
$$\left. -\theta_3 \log\left[1 + \left(\frac{x}{\sigma_1}\right)^{c_1}\right] \log\left[1 + \left(\frac{y}{\sigma_2}\right)^{c_2}\right]\right\}, x,y > 0, \quad (11.36)$$

for $\theta_1 > 0, \theta_2 > 0, \theta_3 \geq 0, \sigma_1 > 0, \sigma_2 > 0, c_1 > 0, c_2 > 0$.

Turning to case (ii), in which $k_1(y) = k_1$ and $k_2(x) = k_2$, it is not hard to verify that we must have $k_1 = k_2 = k$, and then the resulting functional equation to be solved may be transformed to one that is equivalent to (11.26). The only readily obtainable solutions will then have $c_1(y) = c_1$ and $c_2(x) = c_2$ and we will be led to solutions which are already included in the family (11.35). Thus the two parametric families (11.35) and (11.36) represent all the known bivariate distributions with generalized Pareto conditional survival functions.

11.4.4   *Extreme Conditional Survival Functions*

A smallest extreme value distribution has a survival function of the form

$$\bar{F}(x) = \exp[-e^{(x-\mu)/\sigma}], \quad -\infty < x < \infty, \tag{11.37}$$

where $\mu \in \mathbb{R}$ and $\sigma > 0$. In this context, we seek to identify bivariate distributions for (X, Y) for which all conditionals, of X given $Y \geq y$ and of Y given $X \geq x$, are members of the family (11.37). The reader is referred to Chapter 12 for a detailed discussion of distributions of this type.

11.5   Multivariate Extensions

A natural analog to Theorem 11.1 would involve the question of compatibility of the following k conditional survival functions:

$$P(X_i > x_i | \underline{X}_{(i)} > \underline{x}_{(i)}) = a_i(x_i, \underline{x}_{(i)}), \quad i = 1, 2, \ldots, k, \tag{11.38}$$

(recall $\underline{X}_{(i)}$ is $\underline{X}$ with X_i deleted, etc.). As usual when we write $\underline{a} > \underline{b}$ for two vectors it is to be interpreted as holding coordinatewise, i.e., $a_i > b_i$ for each coordinate i. The condition for compatibility is that for each $i \neq j$ the ratio $a_i(x_i; \underline{x}_{(i)})/a_j(x_j; \underline{x}_{(j)})$ should factor in the following manner:

$$\frac{a_i(x_i; \underline{x}_{(i)})}{a_j(x_j; \underline{x}_{(j)})} = \frac{u_j(\underline{x}_{(j)})}{u_i(\underline{x}_{(i)})}, \tag{11.39}$$

where the $u_j(\underline{x}_{(j)})$'s are $(k-1)$-dimensional survival functions.

As in Sections 11.3 and 11.4, the next step is to consider joint survival functions specified by the requirement that for each i, $P(X_i > x_i | \underline{X}_{(i)} > \underline{x}_{(i)})$ should belong to some particular parametric family of one-dimensional survival functions with parameters which might depend on $\underline{x}_{(i)}$.

For example, we might seek the most general class of k-dimensional survival functions with support $(0, \infty)^k$ such that for each i and for each $\underline{x}_{(i)} \in \mathbb{R}_+^{k-1}$,

$$P(X_i > x_i | \underline{X}_{(i)} > \underline{x}_{(i)}) = \exp[-\lambda_i(\underline{x}_{(i)})x_i], \quad x_i > 0, \tag{11.40}$$

for some positive functions $\lambda_i(\underline{x}_{(i)})$, $i = 1, 2, \ldots, k$. Multiply the expressions in (11.40) by $\bar{F}_{\underline{X}_{(i)}}(\underline{x}_{(i)})$, the $(k-1)$-dimensional survival functions which must exist for compatibility, and take logarithms, to obtain a system of Stephanos–Levi Civita–Suto functional equations whose only solutions lead to k-dimensional versions of (11.13) of the following form:

$$\bar{F}(x_1, \ldots, x_k) = \exp\left[-\sum_{\underline{s} \in \xi_k} \theta_{\underline{s}} \left(\prod_{j=1}^{k} x_j^{s_j} \right) \right], \quad \underline{x} > \underline{0}, \qquad (11.41)$$

where ξ_k is the set of vectors of 0's and 1's of dimension k with at least one coordinate being a 1. Constraints must be imposed in the $\theta_{\underline{s}}$'s which appear in (11.41) to guarantee that it represents a genuine survival function. Thus we must have $\theta_{\underline{s}} \geq 0$, $\forall \underline{s}$, and $\theta_{\underline{s}} > 0$, $\forall \underline{s}$, which include only one coordinate equal to 1. In addition, $\theta_{(1,1,00\ldots0)} \leq \theta_{(1,0\ldots0)} \theta_{(0,1,0\ldots0)}$, etc., since the bivariate marginals will necessarily be of the form (11.13) whose interaction parameter θ was constrained to be ≤ 1.

Gumbel (1960) discussed the model (11.41). He obtained it by marginal rather than conditional specification. It is known in the literature as a Gumbel type I multivariate exponential distribution.

Using analogous arguments we can identify the form of the k-dimensional analogs of the bivariate survival functions displayed in (11.27), (11.30), (11.35), and (11.36). For example, the analog to (11.35) is

$$\bar{F}(\underline{x}) = \left[1 + \sum_{\underline{s} \in \xi_k} \theta_{\underline{s}} \left(\prod_{j=1}^{k} x_j^{c_j s_j} \right) \right]^{-k}, \quad \underline{x} > \underline{0}, \qquad (11.42)$$

where the $\underline{\theta}_{\underline{s}}$'s are suitably constrained to guarantee that (11.42) is a valid k-dimensional survival function.

11.6   Conditional Distributions

The role played by survival functions in Sections 11.2–11.5 could instead be played by distribution functions. Thus we might ask whether two families of conditional distributions of the form

$$P(X \leq x | Y \leq y) = a(x, y) \qquad (11.43)$$

and

$$P(Y \leq y | X \leq x) = b(x, y) \qquad (11.44)$$

are compatible. As in Theorem 11.1 the answer is yes, provided that the ratio $a(x, y)/b(x, y)$ factors appropriately. Next we could ask about the nature of bivariate distributions that are constrained to have $P(X \leq x | Y \leq y)$ for each y and $P(Y \leq y | X \leq x)$ for each x belonging to specified parametric families of distributions.

Example 11.1 (Conditional power function distributions). Consider a random vector (X, Y) with $0 \leq X \leq 1, 0 \leq Y \leq 1$ and, for each $y \in (0, 1)$,

$$P(X \leq x | Y \leq y) = x^{\alpha_1(y)}, \quad 0 < x < 1, \tag{11.45}$$

while, for each $x \in (0, 1)$,

$$P(Y \leq y | X \leq x) = y^{\alpha_2(x)}, \quad 0 < y < 1. \tag{11.46}$$

Following a by now familiar program, we multiply (11.45) and (11.46) by the corresponding marginal distribution function (which must exist if (11.45) and (11.46) are to be compatible), equate them, and take logarithms to obtain

$$\log F_Y(y) + \alpha_1(y) \log x = \log F_X(x) + \alpha_2(x) \log y. \tag{11.47}$$

Solving this familiar functional equation (using Theorem 1.3) we find that

$$\alpha_1(y) = \alpha + \gamma \log y \tag{11.48}$$

and

$$\alpha_2(x) = \beta + \gamma \log x. \tag{11.49}$$

Substituting (11.48) and (11.49) in (11.47) we can identify $F_X(x)$ and $F_Y(y)$ and eventually obtain the following general expression for the joint distribution function of a random variable (X, Y), satisfying (11.45) and (11.46):

$$F_{X,Y}(x, y) = x^\alpha y^\beta e^{\gamma (\log x)(\log y)}, \quad 0 < x < 1, \quad 0 < y < 1. \tag{11.50}$$

In order for (11.50) to represent a genuine joint distribution function we must require that $\alpha > 0, \beta > 0$ and that γ be negative and satisfy

$$-\gamma \leq \alpha\beta. \tag{11.51}$$

$\square$

There is an alternative way to view the distribution (11.50) derived in Example 11.1. If (X, Y) has the distribution (11.50), then defining $U = -\log X$, $V = -\log Y$ we may recognize that (X, Y) has a Gumbel type I distribution with joint survival function displayed in (11.13). So one way to justify the constraint (11.51) is to recognize that it is equivalent to the constraint $\theta \leq 1$ which was needed in (11.13).

The close relation between (11.13) and (11.50), in which a distribution obtained via conditional survival specification (11.13) is intimately related to a distribution obtained via conditional distribution specification (11.50), suggests the possible existence of a "duality" between these conditional specification paradigms. And such is indeed the case. If a random vector (X, Y) satisfies (11.1) (conditional survival specification), then the random vector $(-X, -Y)$ satisfies (11.43) and (11.44) (conditional distribution specification). The advantage of considering (11.43) and (11.44) is that it allows us to readily focus on conditional distributions normally described via their distribution functions instead of their survival functions (such as the power function distribution as distinct from the exponential, Weibull, or Pareto distributions). In this case, of course, any distribution obtained using (11.43) and (11.44) could have been obtained using (11.1).

Multivariate extensions of the conditional distribution paradigm (11.43)–(11.44) can, of course, be formulated in a straightforward fashion or, alternatively, can be obtained via the transformation $\tilde{X} = -X$ from multivariate distributions obtained via conditional survival specification. Some of the bivariate distributions discussed in Chapter 12 are amenable to such extension to higher dimensions.

11.7   Conditional Proportional Hazards

In order to model the effects of covariates on survival, a popular model is the proportional hazards model. In it the survival function $\bar{F}_X(x)$ is assumed to have the form

$$\bar{F}_X(x) = \left[\bar{F}_0(x)\right]^\delta,$$
(11.52)

where δ depends on the values of the covariates and where $\bar{F}_0$ is the "baseline" survival function.

It seems reasonable to investigate the kinds of bivariate models obtainable by postulating that the conditional survival functions exhibit proportional hazards structure.

Thus we will consider a bivariate random variable (X, Y) and two specific survival functions $\bar{F}_1$ and $\bar{F}_2$. We seek to identify all possible joint survival functions for (X, Y) with the following properties:

(a) for each $y \in S(Y)$,

$$P(X > x | Y > y) = [\bar{F}_1(x)]^{\gamma_1(y)}, \quad \forall x \in S(X),$$
(11.53)

for some function $\gamma_1 : \mathbb{R} \to \mathbb{R}^+$; and

(b) for each $x \in S(X)$,

$$P(Y > y | X > x) = [\bar{F}_2(y)]^{\gamma_2(x)}, \quad \forall y \in S(Y),$$
(11.54)

for some function $\gamma_2 : \mathbb{R} \to \mathbb{R}^+$.

A model satisfying (11.53) and (11.54) will be called a conditional proportional hazards survival model. Note that, in the absolutely continuous case, we can introduce the corresponding conditional hazard functions:

$$h(x|Y > y) = \frac{d}{dx} \log P(X > x|Y > y) \tag{11.55}$$

and

$$h(y|X > x) = \frac{d}{dy} \log P(Y > y|X > x). \tag{11.56}$$

With this definition assuming the conditional formulation as in (11.53), (11.54), for two distinct values y_1 and y_2, the conditional hazard functions $h(x|Y > y_1)$ and $h(x|Y > y_2)$ differ only by a factor $\gamma(y_2)/\gamma(y_1)$, hence justifying the use of the term "proportional hazards." Note that

$$h(x|Y > y) = \gamma_1(y)h_1(x),$$

where $h_1(x)$ is the hazard function corresponding to the survival function $\bar{F}_1$. Analogously, we have

$$h(y|X > x) = \gamma_2(x)h_2(y),$$

where $h_2(y)$ is the hazard function corresponding to the survival function $\bar{F}_2$.

If (11.53) and (11.54) are to be compatible there must exist marginal survival functions $\bar{F}_X(x)$ and $\bar{F}_Y(y)$ and since we must have

$$P(X > x, Y > y) = P(Y > y)P(X > x|Y > y) = P(X > x)P(Y > y|X > x),$$

we are led to the following functional equation:

$$\bar{F}_Y(y)[\bar{F}_1(x)]^{\gamma_1(y)} = \bar{F}_X(x)[\bar{F}_2(y)]^{\gamma_2(x)}. \tag{11.57}$$

Here $\bar{F}_1$ and $\bar{F}_2$ are known and the other functions are unknown. Taking logarithms in (11.57) yields a Stephanos–Levi Civita–Suto functional equation. Eventually we conclude that, in order to satisfy (11.53) and (11.54), our joint survival function must be of the form

$$P(X > x, Y > x) = \exp\{\alpha \log \bar{F}_1(x) + \beta \log \bar{F}_2(y) + \gamma \log \bar{F}_1(x) \log \bar{F}_2(y)\}. \tag{11.58}$$

It is clear that (11.58) can be viewed as a marginal transformation of Gumbel's type I bivariate exponential distribution (Gumbel (1960)). Thus if (X, Y) satisfy (11.53) and (11.54), it must be the case that the transformed random vector

$$(U, V) = (-\log \bar{F}_1(X), -\log \bar{F}_2(Y)) \tag{11.59}$$

has a Gumbel bivariate survival function of the form (11.13).

Note that, since the Gumbel distribution has nonpositive correlation and since the marginal transformations in (11.59) are both monotone, the correlations in the model (11.58) are also nonpositive when they exist.

Remark 11.2 A parallel development is possible if we replace assumptions (11.53) and (11.54) by requirements of the form

$$P(X > x | Y = y) = [\bar{F}_1(x)]^{\gamma_1(y)}, \quad y \in S(Y), \tag{11.60}$$

and

$$P(Y > y | X = x) = [\bar{F}_2(y)]^{\gamma_2(y)}, \quad x \in S(X). \tag{11.61}$$

This will lead to a model which represents a marginal transformation of the negatively correlated exponential conditionals distribution (4.4) (see Arnold and Kim (1996) for details; see also Exercise 10.4 in Cox and Oates (1984)).

11.8   Conditional Accelerated Failure

A direct competitor of the proportional hazards paradigm as a model for survival mechanisms is the accelerated failure scheme. Covariates affect survival in the accelerated failure model via a time change or, equivalently, via a change of scale only, leaving the shape of the survival function unchanged. It is natural to try and develop conditional accelerated failure models parallel to the conditional proportional hazards models introduced in the previous section.

As usual let $\bar{F}_1$ and $\bar{F}_2$ be specific (baseline) survival functions. We now seek all joint distributions for (X, Y) where $X \geq 0, Y \geq 0$ such that:

(a) for each $y \in S(Y)$,

$$P(X > x | Y > y) = \bar{F}_1(\delta_1(y)x); \tag{11.62}$$

and

(b) for each $x \in S(X)$,

$$P(Y > y | X > x) = \bar{F}_2(\delta_2(x)y), \tag{11.63}$$

where $\delta_1 : S(Y) \to \mathbb{R}^+$ and $\delta_2 : S(X) \to \mathbb{R}^+$.

Assuming that corresponding marginal survival functions $\bar{F}_X(x), \bar{F}_Y(y)$ exist we are led to the following functional equation:

$$\bar{F}_Y(y)\bar{F}_1(\delta_1(y)x) = \bar{F}_X(x)\bar{F}_2(\delta_2(x)y), \tag{11.64}$$

where $\bar{F}_1, \bar{F}_2$ are known functions and the others are unknown. For certain very specific choices of $\bar{F}_1$ and $\bar{F}_2$ this can be solved easily. For example, it can be solved if $\bar{F}_1(x) = \exp[-(x/\sigma_1)^{\delta_1}]$ and $\bar{F}_2(y) = \exp[-(y/\sigma_2)^{\delta_2}]$ (the Weibull case). This doesn't provide us with any new models however since, as is well known, the Weibull model can be viewed as either an accelerated failure model or a proportional hazards model. Thus, the solution to (11.64) when $\bar{F}_1$ and $\bar{F}_2$ are Weibull is already subsumed in the family of models developed in Section 11.4.

Can we find other solutions to (11.64)?

To illustrate the difficulties inherent in such a quest, let us make the assumption that the functions appearing in (11.64) are suitably differentiable. Introduce new variables $u = \log x, v = \log y$ and new functions $\phi_1(u) = \log \bar{F}_X(u), \phi_2(v) = \log \bar{F}_Y(v), \tilde{\delta}_1(v) = \log \delta_1(y), \tilde{\delta}_2(u) = \log \delta_2(x), g_1(u) = \bar{F}_1(e^u)$, and $g_2(v) = \bar{F}_2(e^v)$.

Equation (11.64) now assumes the form

$$\tilde{\phi}_2(v) + g_1(u + \tilde{\delta}_1(v)) = \tilde{\phi}_1(u) + g_2(v + \tilde{\delta}_2(u)). \qquad (11.65)$$

Differentiating with respect to u and v yields

$$g_1''(u + \tilde{\delta}_1(v))\tilde{\delta}_1'(v) = g_2''(v + \tilde{\delta}_2(u))\tilde{\delta}_2'(u). \qquad (11.66)$$

This functional equation was studied by Narumi (1923) and discussed in Chapter 7 (see (7.14)). Only a very limited class of solutions can be found. It does include solutions that lead to a Weibull survival model (as we already know) but, apparently, no other solutions with simply described structure.

Remark 11.3 *A parallel development can be pursued beginning with a conditional accelerated failure model of the form:*

(a) *for each $y \in S(Y)$*

$$P(X > x|Y = y) = \bar{F}_1(\delta_1(y)x) \qquad (11.67)$$

for some function $\delta_1 : S(Y) \to \mathbb{R}^+$; and

(b) *for each $x \in S(X)$,*

$$P(Y > y|X = x) = \bar{F}_2(\delta_2(x)y) \qquad (11.68)$$

for some function $\delta_2 : S(X) \to \mathbb{R}^+$.

Unfortunately, in this formulation too, we are led to the same functional equation (11.64) which again we could solve in the Weibull case (rederiving material already discussed in Section 11.4) and for which we can only obtain complicated, difficult to interpret, non-Weibull solutions (arising from Narumi's general solution to (11.66)).

11.9   An Alternative Specification Paradigm

Specification of both families of conditional survival functions involves redundant information (that is why we had to check for consistency). Clearly,
instead of specifying $P(X > x|Y > y)$ for every x, y and $P(Y > y|X > x)$
for every x, y, it is enough to specify every function $P(X > x|Y > y)$ and
$P(Y > y|X > x)$ for just one value of x. Alternatively, some functional
of the family of survival functions $P(Y < y|X > x), x \in S(X)$, might be
adequate, in conjunction with knowledge of $P(X > x|Y > y)$ for every x, y,
to completely specify the joint distribution of (X, Y).

Based on our experience in Chapter 7, we may hope that one family
of conditional survival functions together with a conditional regression
specification might suffice.

Thus we seek all bivariate distributions such that for given functions
$a(x, y)$ and $\psi(x)$ we have

$$P(X > x|Y > y) = a(x, y), \quad (x, y) \in S(X) \times S(Y), \tag{11.69}$$

and

$$E(Y|X > x) = \psi(x), \quad x \in S(X). \tag{11.70}$$

We will say that $a(x, y)$ and $\psi(x)$ are compatible if there exists a joint
survival function $P(X > x, Y > y)$ satisfying (11.69) and (11.70).

Questions that arise in this context are:

(i) Under what conditions are functions $a(x, y)$ and $\psi(x)$ compatible?

(ii) If they are compatible, when do they determine a unique distribution?

(iii) For a given $a(x, y)$ can we identify the class of all compatible choices
for $\psi(x)$?

We will illustrate how these questions might be resolved in the case
in which (X, Y) is absolutely continuous and has as support the positive
quadrant (i.e., $x > 0, y > 0$). These are not unnatural restrictions for
survival models.

If $a(x, y)$ and $\psi(x)$ are to be compatible then there must exist a corresponding marginal survival function for Y which can be denoted by
$h(y)$ $[= P(Y > y)]$. For each $x > 0, \psi(x)$ may be obtained by integrat-
ing the conditional survival function $P(Y > y|X > x)$ with respect to y
over $[0, \infty)$. Thus for each $x > 0$,

$$
\begin{aligned}
\psi(x) &= \int_0^\infty P(Y > y|X > x) \, dy \\
&= \int_0^\infty P(X > x, Y > y)/P(X > x) \, dy \\
&= \int_0^\infty \frac{a(x, y)h(y)}{a(x, 0)h(0)} \, dy
\end{aligned}
$$

$$= \int_0^\infty \frac{a(x,y)}{a(x,0)} h(y) \, dy \tag{11.71}$$

since $h(0) = P(X > 0) = 1$. Thus $h(y)$ can be obtained by solving an integral equation with known kernel $a(x,y)$, i.e.,

$$\psi(x)a(x,0) = \int_0^\infty a(x,y)h(y) \, dy. \tag{11.72}$$

For certain choices of the kernel $a(x,y)$ (which is the specified conditional survival function) this equation can be solved.

The exponential case is perhaps the easiest to solve. For it we can answer all three questions (i), (ii), (iii) above.

Suppose that we assume an exponential conditional survival function described by

$$\begin{aligned} a(x,y) &= P(X > x | Y > y) \\ &= \exp[-(\alpha + \beta y)x], \quad x > 0, \quad y > 0, \end{aligned} \tag{11.73}$$

where $\alpha > 0$ and $\beta > 0$. Taking the limit as $y \to 0$, we find

$$a(x,0) = e^{-\alpha x}. \tag{11.74}$$

For any given conditional survival function $\psi(x)$ we need to determine the corresponding marginal survival function $h(y) = P(Y > y)$ by solving (11.72). Substituting (11.73) and (11.74), the equation to be solved assumes the form

$$\psi(x)e^{-\alpha x} = \int_0^\infty e^{-\alpha x - \beta x y} h(y) \, dy \tag{11.75}$$

or, equivalently,

$$\psi(x) = \int_0^\infty e^{-\beta x y} h(y) \, dy. \tag{11.76}$$

Immediately, we can see a role for Laplace transforms in solving our problem! Note that

$$\psi(0) = E(Y | X > 0) = E(Y) = \int_0^\infty h(y) \, dy.$$

Let us define a new density $\tilde{h}$ on $(0, \infty)$ by

$$\tilde{h}(y) = h(y)/\psi(0). \tag{11.77}$$

It follows from (11.75) and (11.76) that the Laplace transform of $\tilde{h}$, i.e.,

$$M_{\tilde{h}}(t) = \int_0^\infty e^{-ty} \tilde{h}(y) \, dy \tag{11.78}$$

satisfies

$$M_{\tilde{h}}(t) = \psi \left(\frac{t}{\beta} \right) / \psi(0). \tag{11.79}$$

Thus, provided that ψ is a completely monotone function (specifically a Laplace transform of a finite measure on $(0, \infty)$ with a decreasing density), then it uniquely determines $\tilde{h}$ by (11.79) which by normalization yields $h(y) \; [= P(Y > y)]$. In this case we are able to identify the form of every function ψ that is compatible with the family of conditional survival functions given by (11.73). We do need to check that the product $h(y)a(x, y)$ is a valid survival function.

Example 11.2 (Exponential conditional survival with $1/\psi(x)$ linear). Suppose that, in conjunction with exponential conditional survival (i.e., $a(x, y)$ given by (11.73)), we assume that

$$\psi(x) = (\gamma + \delta x)^{-1}. \tag{11.80}$$

Observe that this is a completely monotone function so we know that it will be compatible with (11.73). From (11.79) we will have, in this case,

$$\begin{aligned}
M_{\tilde{h}}(t) &= \frac{\left(\gamma + \delta \dfrac{t}{\beta} \right)^{-1}}{\gamma^{-1}} \\
&= \left(1 + \frac{\delta}{\beta\gamma} t \right)^{-1}.
\end{aligned} \tag{11.81}$$

We recognize this as the Laplace transform of an exponential density. So we can conclude that

$$\tilde{h}(y) = \frac{\beta\gamma}{\delta} \exp \left(-\frac{\beta\gamma}{\delta} y \right), \quad y > 0. \tag{11.82}$$

The survival function for Y, i.e., $h(y)$, is obtained by normalizing (11.82) to have the value 1 at $y = 0$. Thus

$$h(y) = P(Y > y) = \exp \left(-\frac{\beta\gamma}{\delta} y \right). \tag{11.83}$$

Consequently, the unique joint survival function with $P(X > x | Y > y)$ given by $a(x, y)$ in (11.73) and $E(Y | X > x)$ given by $\psi(x)$ in (11.80) is of the form

$$P(X > x, Y > y) = \exp[-\alpha x - \frac{\beta\gamma}{\delta} y - \beta x y], \quad x > 0, \; y > 0. \tag{11.84}$$

In order that (11.84) represents a valid joint survival function we need to impose the condition $0 < \delta/\gamma < \alpha$. We thus arrive at the Gumbel (I) bivariate exponential distribution (displayed earlier with a different parametrization in (11.13). $\qquad \square$

11.10   Bibliographic Notes

The key references for conditional survival models are Arnold (1995, 1996) and Arnold and Kim (1996).

Exercises

11.1 Consider the model with logistic conditional survival distributions given by (11.30). Obtain its correlation coefficient.

11.2 For any distribution F on $(0, \infty)$, define the odds-ratio function $\varphi(x) = F(x)/[1 - F(x)]$. Investigate the class of bivariate proportional conditional odds-ratio models. These are joint distributions $F(x, y)$ for which

$$\varphi(x|Y > y) = g_1(y)\varphi_1(x), \qquad x > 0, \ \forall y > 0,$$

and

$$\varphi(y|X > x) = g_2(x)\varphi_2(y), \qquad y > 0, \ \forall x > 0,$$

for some functions g_1 and g_2, where φ_1 and φ_2 are the marginal odds-ratio functions of $F(x, y)$.

11.3 Verify the assertions in Remark 11.2 assuming that the conditional survival functions satisfy (11.60) and (11.61).

11.4 Verify the assertions in Remark 11.3 assuming that the conditional survival functions satisfy (11.67) and (11.68).

11.5 Discuss the problem of identifying all bivariate distributions satisfying (11.69) and (11.70) in the case in which $S(X)$ and $S(Y)$ are finite sets.

12
Applications to Modeling Bivariate Extremes

12.1 Introduction

The Fisher–Tippet–Gnedenko models for univariate extremes are well understood, as are the corresponding multivariate extensions (see, for example, Galambos (1978, 1987) or Resnick (1987)). It should be remembered, however, that the multivariate extreme distributions discussed by these authors correspond to limiting distributions of normalized coordinatewise maxima of sequences of i.i.d. random vectors. Not many multivariate extreme data sets can be reasonably viewed as fitting into this paradigm. For example, if we observe maximum temperatures in the month at several different locations we have a multivariate extreme data set which does not seem to be necessarily explainable in terms of maxima of i.i.d. vectors. For such data sets, some role for univariate extreme distributions seems appropriate but it is not apparent whether it should be a marginal or a conditional role. And of course, even if the data were not generated by a process involving i.i.d. vectors, it might still be well fitted by a multivariate extreme distribution. In many cases, it seems appropriate to approach the problem of modeling multivariate extreme data sets using an augmented toolbox, not just the multivariate extreme models based on maxima of i.i.d. samples. In this chapter we will review some of the popular bivariate extreme models and compare them with certain conditionally specified bivariate extreme models (developed in the spirit of Chapters 4 and 11).

Attention is restricted to the bivariate case for ease of book-keeping. Multivariate extensions are of course appropriate and feasible.

A key difference between the classical models and the conditional specification models is to be found in the sign of the correlation between the variables in the models: nonnegative for classical models, non-positive for conditionally specified models.

In the Fisher–Tippet–Gnedenko classification, there are three types of univariate extreme distributions; the Gumbel, Weibull, and Frechet. Multivariate extensions of the three types could be considered. In this chapter, we limit discussion to bivariate versions of the Gumbel extreme value distribution ((12.1) below). Parallel developments could be pursued for the other two types.

12.2   Univariate and Bivariate Gumbel Distributions

Definition 12.1 (Univariate Gumbel distribution). The univariate Gumbel extreme value distribution (for maxima) has density of the form

$$f(x) = \frac{1}{\sigma} e^{-(x-\mu)/\sigma} \exp\left(-e^{-(x-\mu)/\sigma}\right), \qquad -\infty < x < \infty, \qquad (12.1)$$

where μ and σ are, respectively, location and scale parameters.   □

Certain bivariate extensions of (12.1) were introduced by Gumbel and Mustafi (1967). The models they introduced did have Gumbel marginals and indeed were legitimate bivariate extreme distributions in the sense of being possible limit laws for coordinatewise maxima of i.i.d. sequences of bivariate random variables. When these classic bivariate Gumbel distributions were introduced there was, however, no suggestion that they were selected in any optimal fashion from the vast array of possible bivariate extreme distributions with Gumbel marginals. A general form for such bivariate extreme distributions with Gumbel marginals is

$$F_{X,Y}(x,y) = \exp\left\{-\int_0^1 \min\left[f_1(s)e^{-x}, f_2(s)e^{-y}\right] ds\right\}, \qquad (12.2)$$

where $f_1(t)$ and $f_2(t)$ are non-negative Lebesgue integrable functions such that $\int_0^1 f_i(t)\, dt = 1$, $i = 1, 2$ (see, e.g., Resnick (1987), pp. 272).

The type I bivariate Gumbel distribution, as introduced in Gumbel and Mustafi (1967), has a joint distribution function of the form

$$F_{X,Y}(x,y) = \exp\left\{-e^{-(x-\mu_1)/\sigma_1} - e^{-(y-\mu_2)/\sigma_2} + \theta[e^{(x-\mu_1/\sigma_1} + e^{(y-\mu_2/\sigma_2}]^{-1}\right\},$$

$$(12.3)$$

where $\theta \geq 0$. The joint density is of the form

$$f_{X,Y}(x,y) = \frac{1}{\sigma_1 \sigma_2} g_1\left(\frac{x-\mu_1}{\sigma_1}, \frac{y-\mu_2}{\sigma_2}\right),$$

where

$$
\begin{aligned}
g_1(u, v) = & \left[\exp(-e^{-u} - e^{-v} - u - v + \theta(e^u + e^v)^{-1})\right] \\
& \times \left[1 - \theta(e^{2u} + e^{2v})(e^u + e^v)^{-2}\right. \\
& + 2\theta e^{2u+2v}(e^u + e^v)^{-3} \\
& \left. + \theta^2 e^{2u+2v}(e^u + e^v)^{-4}\right].
\end{aligned}
\tag{12.4}
$$

The Gumbel type II bivariate distribution (Gumbel and Mustafi, 1967) has the form

$$
F_{X,Y}(x, y) = \exp[-(e^{-(x-\mu_1)/(b\sigma_1)} + e^{-(y-\mu_2)/(b\sigma_2)})^b],
\tag{12.5}
$$

where $0 < b \leq 1$. In this case, the joint density is of the form

$$
f_{X,Y}(x, y) = \frac{1}{\sigma_1 \sigma_2} g_2 \left(\frac{x - \mu_1}{\sigma_1}, \frac{y - \mu_2}{\sigma_2}\right),
$$

where

$$
\begin{aligned}
g_2(u, v) = & \exp[-(e^{-u/b} + e^{-v/b})^b] \\
& \times [(e^{-u/b} + e^{-v/b})^{2b-2} e^{-u/b-v/b} \\
& + (b^{-1} - 1)(e^{-u/b} + e^{-v/b})^{b-2} e^{-u/b-v/b}].
\end{aligned}
\tag{12.6}
$$

Both of these bivariate Gumbel distributions qualify as legitimate bivariate extreme distributions (i.e., they can be written in the form (12.2)). For example, it is not difficult to verify the conditions of Galambos' (1987), Theorem 5.2.1).

For any bivariate extreme distibution (of the general form (12.2)), it can be shown that the coordinate random variables are associated. It then follows that all such distributions (including the two Gumbel–Mustafi models (12.3) and (12.5)) have non-negative correlations (see, e.g., Tiago de Oliveira (1962)).

Of course, in the real world, not all bivariate maxima data sets exhibit nonnegative correlation. In contrast to the classical modeling approach, the conditional specification route (to be illustrated in the next section) leads to models exhibiting correlations of the opposite sign.

12.3   Conditionally Specified Bivariate Gumbel Distributions

If we require that certain conditional distributions rather than marginals are of the Gumbel form, we encounter different distributions.

First suppose that we assume that for every y the conditional distribution of X given $Y = y$ is Gumbel($\mu_1(y), \sigma_1(y)$) and that for every x, the conditional distribution of Y given $X = x$ is Gumbel($\mu_2(x), \sigma_2(x)$).

To identify the most general class of distributions with such Gumbel conditionals we will need to solve the following functional equation:

$$
h_2(y)\frac{1}{\sigma_1(y)}\exp\left[-\frac{x-\mu_1(y)}{\sigma_1(y)} - e^{-(x-\mu_1(y))/\sigma_1(y)}\right]
$$
$$
= h_1(x)\frac{1}{\sigma_2(x)}\exp\left[-\frac{y-\mu_2(x)}{\sigma_2(x)} - e^{-(y-\mu_2(x))/\sigma_2(x)}\right], \tag{12.7}
$$

where h_1 and h_2 are the unknown marginal densities of X and Y, respectively. If we take logarithms in (12.7) and define new functions $g_2(y) = \log h_2(y) - \log \sigma_1(y)$ and $g_1(x) = \log h_1(x) - \log \sigma_2(x)$, our functional equation takes the simpler form

$$
g_2(y)-\frac{x-\mu_1(y)}{\sigma_1(y)} - e^{-(x-\mu_1(y))/\sigma_1(y)} = g_1(x)-\frac{y-\mu_2(x)}{\sigma_2(x)} - e^{-(y-\mu_2(x))/\sigma_2(x)}.
$$
$$\tag{12.8}$$

This functional equation appears to be difficult to solve. However, if we assume $\sigma_1(y) \equiv \sigma_1$ and $\sigma_2(x) \equiv \sigma_2$, it is readily solved. Other solutions would of course be of potential interest but, until they are found, the assumption of constant scale functions $(\sigma_1(y) \equiv \sigma_1, \sigma_2(x) \equiv \sigma_2)$ will still yield a flexible collection of models. With this assumption, (12.8) becomes

$$
g_2(y) - \tfrac{x}{\sigma_1} + \tfrac{\mu_1(y)}{\sigma_1} - \exp\left(-\tfrac{x}{\sigma_1} + \tfrac{\mu_1(y)}{\sigma_1}\right)
$$
$$
= g_1(x) - \tfrac{y}{\sigma_2} + \tfrac{\mu_2(x)}{\sigma_2} - \exp\left(-\tfrac{y}{\sigma_2} + \tfrac{\mu_2(x)}{\sigma_2}\right).
$$
$$\tag{12.9}$$

If we introduce the notation

$$
\tilde{g}_1(x) = g_1(x) + \frac{\mu_2(x)}{\sigma_2} + \frac{x}{\sigma_1},
$$
$$
\tilde{g}_2(y) = g_2(y) + \frac{\mu_1(y)}{\sigma_1} + \frac{y}{\sigma_2},
$$
$$
\psi_1(y) = e^{\mu_1(y)/\sigma_1},
$$
$$
\psi_2(x) = e^{\mu_2(x)/\sigma_2},
$$

then our equation can be written in the form

$$
\tilde{g}_2(y) - \psi_1(y)e^{-x/\sigma_1} = \tilde{g}_1(x) - \psi_2(x)e^{-y/\sigma_2}. \tag{12.10}
$$

This is a Stephanos–Levi Civita–Suto functional equation and can readily be solved. Using Theorem 1.3 we conclude that for some constants $\delta_1, \delta_2, \delta_3$ we must have

$$
\psi_1(y) = \delta_1 + \delta_3 e^{-y/\sigma_2} \tag{12.11}
$$

and
$$\psi_2(x) = \delta_2 + \delta_3 e^{-x/\sigma_1}. \tag{12.12}$$

Consequently,
$$\mu_1(y) = \sigma_1 \log(\delta_1 + \delta_3 e^{-y/\sigma_2}) \tag{12.13}$$

and
$$\mu_2(x) = \sigma_2 \log(\delta_2 + \delta_3 e^{-x/\sigma_1}). \tag{12.14}$$

Thus X given $Y = y$ belongs to the Gumbel$(\mu_1(y), \sigma_1)$ family, and Y given $X = x$ belongs to the Gumbel$(\mu_2(x), \sigma_2)$ family with $\mu_1(y)$ and $\mu_2(x)$ as given in (12.13) and (12.14). If we write the corresponding conditional distribution functions we see that

$$
\begin{aligned}
F_{X|Y}(x|y) &= \exp\left[-e^{-(x-\mu_1(y))/\sigma_1}\right] \\
&= \exp\left[-\delta_1 e^{-x/\sigma_1} - \delta_3 e^{-x/\sigma_1} e^{-y/\sigma_2}\right]. \tag{12.15}
\end{aligned}
$$

Analogously

$$F_{Y|X}(y|x) = \exp\left[-\delta_2 e^{-y/\sigma_2} - \delta_3 e^{-x/\sigma_1} e^{-y/\sigma_2}\right]. \tag{12.16}$$

From these expressions it is evident that the random variables

$$U = e^{-X/\sigma_1} \tag{12.17}$$

and

$$V = e^{-Y/\sigma_2} \tag{12.18}$$

have a joint distribution with exponential conditionals. From (4.14) the joint density of (U, V) will be of the form

$$f_{U,V}(u, v) = \exp(m_{00} - m_{10}u - m_{01}v - m_{11}uv)I(u > 0, \ v > 0). \tag{12.19}$$

From this, using (12.17) and (12.18), we can write the joint density of (X, Y) in the form

$$
\begin{aligned}
&f_{X,Y}(x, y) \\
&= \frac{\exp\left[-\frac{x}{\sigma_1} - \frac{y}{\sigma_2} + m_{00} - m_{10}e^{-x/\sigma_1} - m_{01}e^{-y/\sigma_2} - m_{11}e^{-x/\sigma_1 - y/\sigma_2}\right]}{\sigma_1\sigma_2},
\end{aligned}
$$
$$-\infty < x < \infty, \quad -\infty < y < \infty. \tag{12.20}$$

An alternative more easily interpretable parametrization is available. In (12.20) set $\mu_1 = \sigma_1 \log m_{10}$, $\mu_2 = \sigma_2 \log m_{01}$, and $\theta = m_{11}/(m_{10}m_{01})$. Also let $\tilde{m}_{00}$ denote a new normalizing constant. Our density then assumes the form

$$
\begin{aligned}
f_{X,Y}(x, y) = \frac{1}{\sigma_1\sigma_2} \exp\bigg[&-\frac{x - \mu_1}{\sigma_1} - \frac{y - \mu_2}{\sigma_2} + \tilde{m}_{00} \\
&-e^{-(x-\mu_1)/\sigma_1} - e^{-(y-\mu_2)/\sigma_2} - \theta e^{-(x-\mu_1)/\sigma_1 - (y-\mu_2)/\sigma_2}\bigg],
\end{aligned}
$$
$$\tag{12.21}$$

where $\mu_1, \mu_2 \in \mathbb{R}$ are location parameters, $\sigma_1, \sigma_2 > 0$ are scale parameters, and $\theta \geq 0$ is a dependency parameter.

For completeness we also write the corresponding conditional densities in this new parametrization:

$$f_{X|Y}(x|y) = \frac{1 + \theta e^{-(y-\mu_2)/\sigma_2}}{\sigma_1} e^{-(x-\mu_1)/\sigma_1} e^{-\left(1+\theta e^{-(y-\mu_2)/\sigma_2}\right) e^{-(x-\mu_1)/\sigma_1}},$$

(12.22)

$$f_{Y|X}(y|x) = \frac{1 + \theta e^{-(x-\mu_1)/\sigma_1}}{\sigma_2} e^{-(y-\mu_2)/\sigma_2} e^{-\left(1+\theta e^{-(x-\mu_1)/\sigma_1}\right) e^{-(xy-\mu_2)/\sigma_2}}.$$

(12.23)

It is not difficult to verify that model (12.21) is always totally negative of order 2 and consequently will exhibit nonpositive correlations. It is thus evident that model (12.21) is not a valid bivariate extreme distribution. Indeed, it does not even have Gumbel marginals.

The standarized form of (12.21) is

$$f(x,y) = k(\theta) \exp\left[-x - y - e^{-x} - e^{-y} - \theta e^{-x-y}\right],$$

(12.24)

where the normalizing constant is given by

$$k(\theta) = \theta e^{-1/\theta} / \left[-\mathrm{Ei}(1/\theta)\right],$$

(12.25)

in which

$$-\mathrm{Ei}(t) = \int_t^\infty \frac{e^{-u}}{u} du.$$

(12.26)

The marginal density of the standarized distribution is

$$f(x) = k(\theta) \frac{\exp[-x - e^{-x}]}{1 + \theta e^{-x}}.$$

(12.27)

A second conditional specification paradigm can be fruitfully employed in this Gumbel context. It will turn out to yield a bivariate distribution with nonpositive correlation also; but this time with Gumbel marginals. For it, we focus on conditional distributions rather than conditional densities (as in Section 11.6). Thus we seek to identify all joint distributions for (X, Y) such that for every real y, the conditional distribution of X given $(Y \leq y)$ is Gumbel$(\mu_1(y), \sigma_1(y))$ and for each x, the conditional distribution of Y given $(X \leq x)$ is Gumbel$(\mu_2(x), \sigma_2(x))$. Thus we will have

$$P(X \leq x | Y \leq y) = \exp[-e^{-(x-\mu_1(y))/\sigma_1(y)}]$$

(12.28)

and

$$P(Y \leq y | X \leq x) = \exp[-e^{-(y-\mu_2(x))/\sigma_2(x)}].$$

(12.29)

Since

$$P(X \leq x, Y \leq y) = P(X \leq x | Y \leq y) P(Y \leq y)$$
$$= P(Y \leq y | X \leq x) P(X \leq x),$$

the following functional equation holds:

$$\varphi_1(y) - e^{-(x-\mu_1(y))/\sigma_1(y)} = \varphi_2(x) - e^{-(y-\mu_2(x))/\sigma_2(x)}, \tag{12.30}$$

where $\phi_1(y) = \log P(Y \leq y)$ and $\phi_2(x) = \log P(X \leq x)$.

Even though (12.30) appears to be simpler than (12.8) it too has not been solved in general. As in the case of (12.8), considerable simplification occurs if we assume that $\sigma_1(y) \equiv \sigma_1$ and $\sigma_2(x) \equiv \sigma_2$.

Our equation simplifies to

$$\varphi_1(y) - e^{-x/\sigma_1} e^{\mu_1(y)/\sigma_1} = \varphi_2(x) - e^{-y/\sigma_2} e^{\mu_2(x)/\sigma_2}, \tag{12.31}$$

which is a functional equation of the SLCS form and is thus readily solvable (just as (12.9) was solvable). It follows that for some constants $\gamma_1, \gamma_2, \gamma_3$ we must have

$$e^{\mu_1(y)/\sigma_1} = \gamma_1 + \gamma_3 e^{-y/\sigma_2} \tag{12.32}$$

and

$$e^{\mu_2(x)/\sigma_2} = \gamma_2 + \gamma_3 e^{-x/\sigma_1}. \tag{12.33}$$

Substituting these back into (12.28) and (12.29) we find

$$P(X \leq x | Y \leq y) = \exp(-\gamma_1 e^{-x/\sigma_1} - \gamma_3 e^{-x/\sigma_1} e^{-y/\sigma_2}) \tag{12.34}$$

and

$$P(Y \leq y | X \leq x) = \exp(-\gamma_2 e^{-y/\sigma_2} - \gamma_3 e^{-x/\sigma_1} e^{-y/\sigma_2}). \tag{12.35}$$

Next let $y \to \infty$ in (12.34) to obtain the marginal distribution of X. This, combined with (12.35), yields the joint distribution of (X, Y):

$$F_{X,Y}(x, y) = \exp(-\gamma_1 e^{-x/\sigma_1} - \gamma_2 e^{-y/\sigma_2} - \gamma_3 e^{-x/\sigma_1} e^{-y/\sigma_2}). \tag{12.36}$$

If we make the transformation

$$U = \gamma_1 e^{-X/\sigma_1}, \tag{12.37}$$

$$V = \gamma_2 e^{-Y/\sigma_2} \tag{12.38}$$

in (12.36), we find that (U, V) has the bivariate exponential distribution of the first kind introduced by Gumbel (1960) (as discussed in Section 11.3).

Consequently, in order for (12.36) to be a proper bivariate distribution function, we must require that

$$\gamma_1 > 0; \quad \gamma_2 > 0,$$

and

$$0 \leq \gamma_3 \leq \gamma_1 \gamma_2.$$

If we define $\mu_1 = \sigma_1 \log \gamma_1$, $\mu_2 = \sigma_2 \log \gamma_2$, and $\theta = \gamma_3/(\gamma_1 \gamma_2)$ we may rewrite (12.36) in a more convenient form, viz.

$$F_{X,Y}(x,y) = \exp[-e^{-(x-\mu_1)/\sigma_1} - e^{-(y-\mu_2)/\sigma_2} - \theta e^{-(x-\mu_1)/\sigma_1 - (y-\mu_2)/\sigma_2}],$$

$$(12.39)$$

where $\mu_1, \mu_2 \in \mathbb{R}$, $\sigma_1, \sigma_2 > 0$, and $\theta \in [0,1]$. The density of (X,Y) is then given by

$$f_{X,Y}(x,y) = \frac{1}{\sigma_1 \sigma_2} g\left(\frac{x-\mu_1}{\sigma_1}, \frac{y-\mu_2}{\sigma_2}\right), \qquad (12.40)$$

where

$$\begin{aligned} g(u,v) &= \exp(-e^{-u} - e^{-v} - \theta e^{-u-v} - u - v) \\ &= \times[(1 + \theta e^{-u})(1 + \theta e^{-v}) - \theta]. \end{aligned}$$

It is not difficult to verify that model (12.39) exhibits negative quadrant dependence and, as a consequence, has nonpositive correlation. It does have Gumbel marginals but it is not a valid bivariate extreme distribution since it fails to satisfy the condition of Theorem 5.2.1 of Galambos (1987) (or we could just argue that it has correlation of the wrong sign).

12.4   Positive or Negative Correlation

The two Gumbel–Mustafi models (12.3) and (12.5) exhibit nonnegative correlation as does any bivariate extreme model (of the general form (12.2)). However, many bivariate data sets are not associated with maxima of sequences of i.i.d. random vectors even though marginally and/or conditionally a Gumbel model may fit quite well.

Quite often empirical extreme data are associated with dependent bivariate sequences. Unless the dependence is relatively weak, there is no reason to expect that classical bivariate extreme theory will apply in such settings and, consequently, no a priori argument in favor of nonnegative or nonpositive correlation.

The conditionally specified Gumbel models introduced in this chapter exhibit nonpositive correlation. Thus the Gumbel–Mustafi models and the conditionally specified models do not compete but, in fact, complement each other. Together they provide us with the ability to fit data sets exhibiting a broad spectrum of correlation structure, both negative and positive.

12.5   Density Contours

It is not easy to visualize densities from their algebraic formulations. Representative density contours and density plots are provided here (see Figures

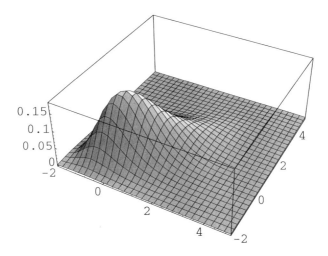

FIGURE 12.1. Density plot for the Gumbel type I model with $\theta = 0.9$.

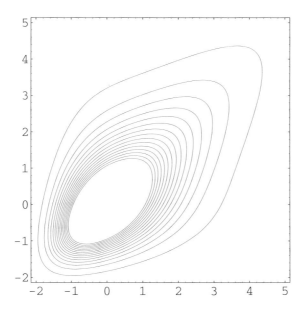

FIGURE 12.2. Density contour plot for the Gumbel type I model with $\theta = 0.9$.

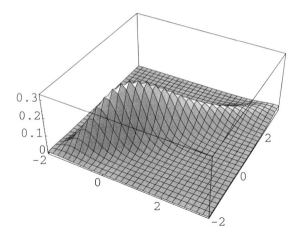

FIGURE 12.3. Density plot for the Gumbel type II model with $b = 0.3$.

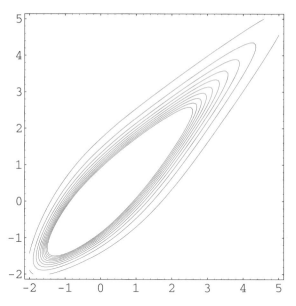

FIGURE 12.4. Density contour plot for the Gumbel type II model with $b = 0.3$.

12.1–12.8) for each of the four models introduced in this chapter, to help the reader visualize the nature of the dependency exhibited in each model. In all cases we have centered and standardized the density, i.e., we have set $\mu_1 = \mu_2 = 0$ and $\sigma_1 = \sigma_2 = 1$. These figures have been selected from a more extensive collection which may be found in Arnold, Castillo, and Sarabia (1998c).

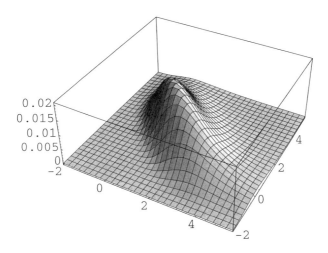

FIGURE 12.5. Density plot for model (12.21) with $\theta = 10$.

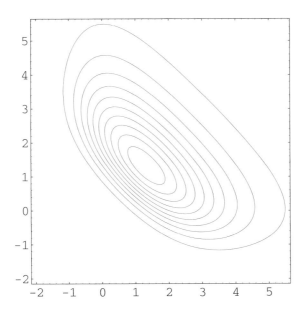

FIGURE 12.6. Density contour plot for model (12.21) with $\theta = 10$.

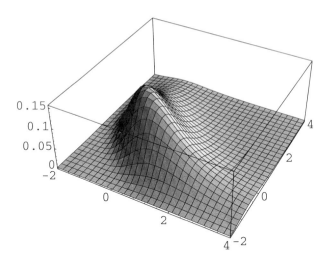

FIGURE 12.7. Density plot for model (12.39) with $\theta = 0.9$.

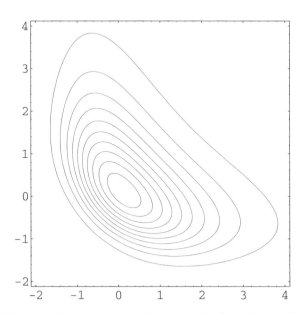

FIGURE 12.8. Density contour plot for model (12.39) with $\theta = 0.9$.

TABLE 12.1. Maximum wind speeds (mph) in Eastport and North Head for the period 1912–1948, from Simiu and Filliben (1975).

Year	Eastport	North Head	Year	Eastport	North Head
1912	53	69	1931	52	66
1913	41	65	1932	42	79
1914	54	70	1933	46	70
1915	49	63	1934	51	87
1916	60	73	1935	48	69
1917	54	65	1936	46	73
1918	52	68	1937	46	72
1919	56	65	1938	45	67
1920	48	57	1939	47	70
1921	39	95	1940	46	84
1922	57	60	1941	49	67
1923	46	68	1942	46	65
1924	46	64	1943	42	77
1925	51	70	1944	51	65
1926	38	73	1945	55	64
1927	50	65	1946	44	67
1928	45	66	1947	52	66
1929	50	63	1948	48	69
1930	48	66			

12.6 Maximal Wind Speeds

Simiu and Filliben (1975) presented data on annual maximal wind speeds at 21 locations in the United States of America. A convenient source for the data is Table 10.1 in Rice (1975). Approximately 40% of the 210 pairs of stations in this data set exhibit negative correlation so that the phenomenon is not an isolated one. As a representative example consider data from two stations, Eastport and North Head; see Table 12.1.

For each of the stations, Gumbel probability plots of the data are provided (Figures 12.9 and 12.10). Since the upper tails of theses plots exhibit little curvature we can reasonably assume, marginally, a Gumbel domain of attraction (see Castillo (1988) p. 173).

Since the data exhibit negative correlation it is reasonable to fit models (12.21) and (12.39) (rather than (12.3) and (12.5) which exhibit positive correlation). In addition a model with independent Gumbel marginals will be fitted. In the analysis, the data have been rescaled by dividing by 100 for computational convenience.

The parameters of model (12.39) are estimated using maximum likelihood while those of model (12.21) are estimated by maximizing the pseudo-

TABLE 12.2. Fitted Gumbel Models estimated (standard errors are shown in parentheses)

Parameters	Independent	Model (12.39)	Model (12.21)
μ_1	0.4604	0.4589	0.4333
	(0.0082)	(0.0083)	(0.0141)
σ_1	0.0470	0.0466	0.0415
	(0.0056)	(0.0055)	(0.0048)
μ_2	0.6622	0.6639	0.6372
	(0.0088)	(0.0091)	(0.0146)
σ_2	0.0513	0.0535	0.0501
	(0.0065)	(0.0067)	(0.0058)
θ	–	0.6864	1.8993
	–	(0.3509)	(1.4751)
$\log(L)$	108.103	110.992	114.025

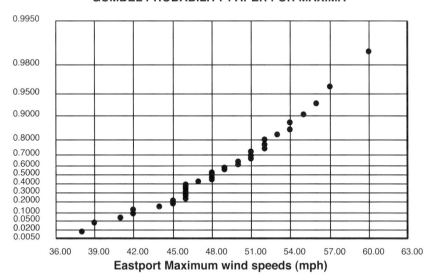

FIGURE 12.9. Eastport maximum wind speeds (mph) on Gumbel probability paper.

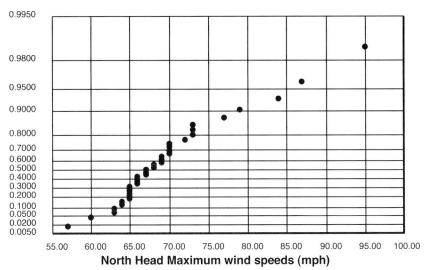

FIGURE 12.10. North Head maximum wind speeds (mph) on Gumbel probability paper.

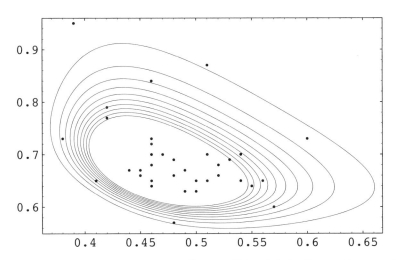

FIGURE 12.11. Density contour plot for model (12.21) for the maximum wind speed data.

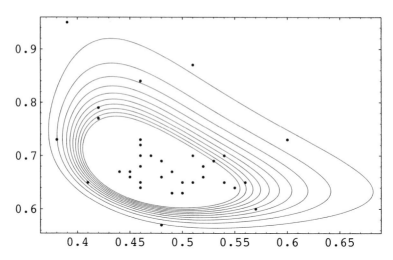

FIGURE 12.12. Density contour plot for model (12.39) for the maximum wind speed data.

likelihood:

$$\prod_{i=1}^{n} f_{X|Y}(x_i|y_i) f_{Y|X}(y_i|x_i). \tag{12.41}$$

Finally the simple independent marginal models are also estimated using maximum likelihood.

The corresponding parameter estimates together with estimated standard errors and corresponding values of the log-likelihoods are displayed in Table 12.2. For this data set it appears that model (12.21) provided a significantly improved fit when compared with model (12.39) or with the independence model. Density contour plots for the fitted versions of models (12.21) and (12.39) are provided in Figures 12.11 and 12.12.

12.7   More Flexibility Needed

Since both negative and positive dependence are routinely encountered in the real world, it seems desirable to have a flexible model capable of exhibiting correlations of both signs. It may be observed that all of the models discussed in this chapter involve a single dependency parameter. Models involving multiple dependency parameters might provide the flexibility desired. It must be remarked that it seems unlikely that we will be able to develop models exhibiting possible correlations of both signs via bivariate extreme or conditional specification arguments. The former are guaranteed to give nonnegative correlations while the latter seem to favor negative correlations in those cases where the corresponding functional equations are solvable. That is, unless we are satisfied with a cut-and-paste model, combining say (12.3) and (12.21), of the form

$$f(x,y;\theta) = \left[\exp(-e^{-u} - e^{-v} - u - v + \theta(e^u + e^v)^{-1})\right]$$
$$\times [1 - \theta(e^{2u} + e^{2v})(e^u + e^v)^{-2}$$
$$+ 2\theta e^{2u+2v}(e^u + e^v)^{-3} \hspace{2cm} (12.42)$$
$$+ \theta^2 e^{2u+2v}(e^u + e^v)^{-4}], \hspace{1cm} \text{if } \theta \geq 0,$$
$$= \exp[-x - y + \tilde{m}_{00} - e^{-x} - e^{-y} + \theta e^{-x-y}], \text{ if } \theta < 0.$$

12.8   Bibliographic Note

Bivariate extreme models are discussed in Arnold, Castillo, and Sarabia (1998c).

Exercises

12.1 Seek examples of bivariate extreme data, with negative correlation.

12.2 Confirm the assertions that models (12.21) and (12.39) are always negatively correlated?

12.3 Describe suitable k-variate extensions of the models (12.21) and (12.39). Discuss the corresponding correlation structures.

12.4 Derive models analogous to (12.21) and (12.39) for the Weibull and Frechet extreme value distributions (for definitions in the context of extreme value theory see, either Resnick (1987) or Arnold, Balakrishnan, and Nagaraja (1992)).

12.5 Are models (12.21) and (12.39) always unimodal?

12.6 We can construct families of bivariate distributions with Gumbel marginals with correlations of both positive and negative sign by applying marginal transformations to suitable families of bivariate uniform distributions (copulas). For example, we could begin with (U_1, U_2) having a Farlie–Gumbel–Morgenstern distribution

$$F_{U_1 U_2}(u_1, u_2) = u_1 u_2[1 + \theta(1 - u_1)(1 - u_2)], \quad 0 < u_1 < 1, \quad 0 < u_2 < 1,$$

where $\theta \in [-1, 1]$. Now define

$$X = -\log(-\log U_1)$$

and

$$Y = -\log(-\log U_2).$$

Confirm that (X, Y) has a bivariate Gumbel distribution with the sign of its correlation determined by the sign of θ.

13

Bayesian Analysis Using Conditionally Specified Models

13.1 Introduction

The standard Bayesian inference scenario involves data $\underline{X}$ whose distribution is governed by a family of densities $\{f(\underline{x}; \underline{\theta}) : \underline{\theta} \in \Theta\}$ where frequently Θ is of dimension, say k, greater than 2. A Bayesian analysis of such a problem will involve identification of a prior density for $\underline{\theta}$ which will be combined with the likelihood of the data to yield a posterior distribution suitable for inferences about $\underline{\theta}$. The use of informative priors (obtained from knowledgable expert(s)) is usually envisioned in such settings. Considerable modern Bayesian analysis has focussed on the frequently occurring case in which prior information is sparse or absent. In this context we encounter priors associated with adjectives such as: diffuse, noninformative, convenience, reference, etc. Much early Bayesian work was focussed on so-called conjugate priors. Such priors are convenient but lack flexibility for modeling informed prior belief. The classical scenario, in which most of the issues are already clearly visible, involves a sample from a normal distribution with unknown mean and variance.

We will use this normal example to illustrate the potential of conditionally specified models for providing more flexible but still manageable families of prior distributions for routine use. Subsequently, we discuss in more generality the role of conditionally specified distributions in Bayesian analysis. Special emphasis is placed on exponential family models. The normal example is examined in detail and a variety of specific applications are outlined.

13.2   Motivation from a Normal Example

Suppose that our data consists of n independent observations $X_1, X_2, \ldots, X_n$ with a common normal distribution. We will denote the common mean of the X_i's by μ and the common precision (the reciprocal of the variance) of the X_i's by τ. Throughout this chapter we will use "precisions" rather than "variances" to describe the scale parameters of normal random variables purely for book-keeping convenience (they allow us to talk of gamma distributions instead of "inverse" gamma distributions).

Both parameters are unknown and our parameter space is

$$\mathbb{R} \times \mathbb{R}^+ = \{(\mu, \tau) : -\infty < \mu < \infty, 0 < \tau < \infty\}.$$

The natural conjugate prior for (μ, τ) (see, e.g., deGroot (1970)) consists of densities of the form

$$f(\mu, \tau) \propto \exp\left(a \log \tau + b\tau + c\mu\tau + d\mu^2\tau\right), \tag{13.1}$$

where $a > 0, b < 0, c \in \mathbb{R}, d < 0$. Densities such us (13.1) have a gamma marginal density for τ and a conditional distribution for μ given τ that is normal with precision depending on τ. Several authors (e.g., Arnold and Press (1989a)) have remarked on the fact that there seems to be no compelling reason to expect prior beliefs to necessarily exhibit the specific dependence structure exhibited by (13.1).

In this setting the natural conjugate prior family for μ, assuming that τ is known, is the normal family. The natural conjugate prior family for τ, assuming μ is known, is the gamma family. Based on these observations, Arnold and Press (1989a) and others have advocated use of a joint prior for (μ, τ) with independent normal and gamma marginals, i.e., the family of densities of the form

$$f(\mu, \tau) \propto \exp\left(a \log \tau + b\tau + c\mu + d\mu^2\right), \tag{13.2}$$

where $a > 0, b < 0, c \in \mathbb{R}$, and $d < 0$. This is of course not a conjugate family and at first glance the advantage gained in ease of prior assessment can be expected to be offset by potential difficulties in analyzing the resulting posterior (but more on this later). Note also that (13.2) like (13.1) involves specific assumptions about the dependency (or lack thereof) between prior beliefs about τ and μ.

Moreover we are likely to be reluctant to forego the advantages of conjugacy. But which family of conjugate priors should be recommended? The tradeoff to be resolved here is between simplicity and ease of assessment versus complexity and an ability to match a diverse spectrum of potential prior beliefs. The classical "natural conjugate prior" is at the "simplicity" end of the spectrum. We argue that it generally lacks flexibility and frequently involves curious unjustifiable dependencies in the prior and posterior joint distributions of the model parameters. A more flexible prior family should generally be used. The models that form the central theme of this book, those specified in terms of conditional distributions, are candidates for use in this setting. As a small step in the direction of more flexible prior modeling, so-called conditionally conjugate priors can be advocated for use.

They, in a sense, represent a minimal desirable increase in flexibility over natural conjugate families. There is, as will be seen, a price to pay. And indeed it is a high price. The resulting conditionally conjugate families involve large number of hyperparameters. But of course, this is the very feature of conditionally conjugate priors that gives them the flexibility to model a broad spectrum of prior beliefs. They will not prove to be a panacea. Certain multimodal prior beliefs will be still poorly fitted even by conditionally conjugate families of priors. Here will be found scope for use of finite mixtures of priors to capture the multimodal features of prior beliefs.

Specific details of the proposed conditionally conjugate analysis of normal data will be provided in Section 13.6. In the immediately following sections we will review, to some degree, conjugate and convenience priors in general contexts and introduce conditionally conjugate families in some generality.

13.3   Priors with Convenient Posteriors

Suppose that we have data $\underline{X}$ whose distribution is governed by the family of densities $\{f(\underline{x}; \underline{\theta}) : \underline{\theta} \in \Theta\}$ where Θ is k-dimensional ($k > 1$).

Typically the informative prior used in this analysis is a member of a convenient family of priors (often chosen to be a conjugate family).

Definition 13.1 (Conjugate family). A family $\mathcal{F}$ of priors for $\underline{\theta}$ is said to be a conjugate family if any member of $\mathcal{F}$, when combined with the likelihood of the data, leads to a posterior density which is again a member of $\mathcal{F}$. □

But there are many conjugate families available in any situation. The most frequently used method of constructing a family of conjugate priors is to consider all possible posteriors corresponding to all possible hypothetical samples of all possible sizes from the given distribution beginning with a (possibly improper) uniform prior on Θ. For example, for samples from a normal (μ, τ) population (i.e., one with mean μ and precision τ) this approach leads to a conjugate prior family for (μ, τ) of the form (13.1).

In this example, the posterior distribution will again be of the "normal–gamma" form (13.1). As remarked earlier, this "natural" conjugate prior lacks flexibility for matching prior beliefs because of its paucity of hyperparameters and its inherent restrictions on the dependence between μ and τ. More flexible alternatives can and will be discussed in later sections.

Before developing such flexible alternatives, it is perhaps worthwhile to emphasize that the advantage of conjugate prior families resides not really in their conjugacy. Actually what is important is that the posterior be analytically tractable. It is not at all crucial that the prior and posterior should be in the same family. In the normal data example, perhaps other priors will yield "normal–gamma" posteriors. Instead of focussing on conjugacy, it is of evident interest to identify, in this setting, all possible priors which will lead to the chosen "convenient posteriors," namely the normal–gamma ones. More generally, for data corresponding to a given exponential family likelihood we might seek to identify all possible priors which could lead to posteriors in a given (quite possibly distinct) exponential family.

Consider data sets consisting of n observations (possibly vector valued) $\underline{X}^{(1)}, \ldots, \underline{X}^{(n)}$ from an m-parameter exponential family of the form

$$f(\underline{x}; \underline{\theta}) = \exp\left[\lambda(\underline{\theta}) + \sum_{j=0}^{m} \theta_j T_j(\underline{x})\right], \quad \underline{x} \in S(\underline{X}), \qquad (13.3)$$

where by convention $\theta_0 = 1$. The joint likelihood for the n observations is then given by

$$f(\underline{x}^{(1)}, \ldots, \underline{x}^{(n)}; \underline{\theta}) = \exp\left[n\lambda(\underline{\theta}) + \sum_{j=0}^{m} \theta_j \sum_{i=1}^{n} T_j(\underline{x}^{(i)})\right],$$
$$\underline{x}^{(i)} \in S(\underline{X}), \quad i = 1, 2, \ldots, n. \qquad (13.4)$$

It is assumed that the $T_j(\underline{x})$'s are linearly independent and nonconstant $(j = 1, 2, \ldots, m)$.

It may, if desired, be assumed that the $\underline{\theta}$'s are restricted to belong to the natural parameter space Θ, defined to include all $\underline{\theta}$'s for which (13.3) is integrable. Our goal is to determine the most general class of priors on Θ which will lead to posterior densities for $\underline{\theta}$ which belong to a prespecified ℓ-parameter exponential family of the form

$$f(\underline{\theta}; \underline{\eta}) = \exp\left[\nu(\underline{\eta}) + \sum_{k=0}^{\ell} \eta_k g_k(\underline{\theta})\right], \quad \underline{x}^{(i)} \in S(\underline{X}); \underline{\theta} \in \Theta_1, \qquad (13.5)$$

where by convention $\eta_0 = 1$ and where $\Theta_1 \subset \Theta$ (perhaps a proper subset). Thus we wish to identify the $\underline{\theta}$ marginal of a joint distribution for $(\underline{X}^{(1)}, \ldots, \underline{X}^{(n)}, \underline{\theta})$ which has conditional densities for $\underline{X}^{(1)}, \ldots, \underline{X}^{(n)}$ given $\underline{\theta}$ as in (13.4) and conditional densities of $\underline{\theta}$ given $\underline{X}^{(1)} = \underline{x}^{(1)}, \ldots, \underline{X}^{(n)} = \underline{X}^{(n)}$ of the form

$$f(\underline{\theta}|\underline{x}^{(1)}, \ldots, \underline{x}^{(n)}) = \exp\left[g_0(\underline{\theta}) + \sum_{k=1}^{\ell} \eta_k(\underline{x}^{(1)}, \ldots, \underline{x}^{(n)}) g_k(\underline{\theta})\right]$$
$$\times \exp\left[\nu(\eta(\underline{x}^{(1)}, \ldots, \underline{x}^{(n)}))\right], \quad \underline{\theta} \in \Theta_1, \qquad (13.6)$$

for some functions $\eta_k(\underline{x}^{(1)}, \ldots, \underline{x}^{(n)})$, $k = 1, 2, \ldots, \ell$.

But clearly we are dealing with a joint distribution with conditionals in exponential families. Using the material from Chapter 4 we can conclude that, if (13.4) and (13.6) are to hold for all $\underline{x}^{(1)}, \ldots, \underline{x}^{(n)}$ and $\underline{\theta}$, we must have a joint density of the form

$$f(\underline{x}^{(1)}, \ldots, \underline{x}^{(n)}; \underline{\theta}) = \exp\left[\sum_{j=0}^{m} \sum_{k=0}^{\ell} m_{jk} \left(\sum_{i=1}^{n} T_j(\underline{x}^{(i)}) g_k(\underline{\theta}) \right) \right],$$

$$\underline{\theta} \in \Theta_1, \quad \underline{x}^{(i)} \in S(\underline{X}), \quad i = 1, 2, \ldots, n,$$

(13.7)

where by our usual convention $T_0(\underline{x}^{(i)}) \equiv 1$ and $g_0(\underline{\theta}) \equiv 1$, for suitable choices of the real parameters $\{m_{ij}\}$. The constant m_{00} is determined so that the density integrates to 1. Densities of the form (13.7) indeed have all conditionals in the prescribed exponential families but, in fact, what we require is more constrained. We require that (13.4) hold exactly. This puts major constraints on the acceptable choices of the parameters in (13.7) and on the form of the functions $g_k(\underline{\theta})$.

Specifically, we must have

$$\theta_j = \sum_{k=1}^{\ell} m_{jk} g_k(\underline{\theta}), \quad j = 1, 2, \ldots, m.$$

(13.8)

This has serious implications regarding the attainability of desirable exponential families of posteriors. If we begin with the likelihood (13.4) and want to end up with posteriors in an exponential family of the form (13.5) whose $g_k(\underline{\theta})$'s do <u>not</u> satisfy (13.8), then we cannot attain this for any choice of prior. Generally speaking, the $g_k(\underline{\theta})$'s have to be linear combinations of the θ_j's, or we have no hope.

After imposing all the implied constraints, the joint density assumes the form (with new parameters)

$$f(\underline{x}^{(1)}, \ldots, \underline{x}^{(n)}; \underline{\theta})$$
$$= \exp\left[c + g_0(\underline{\theta}) + \sum_{k=1}^{\ell} b_k g_k(\underline{\theta}) + \sum_{i=1}^{n} T_0(\underline{x}^{(i)}) + \sum_{j=1}^{m} \theta_j \sum_{i=1}^{n} T_j(\underline{x}^{(i)}) \right],$$

$$\underline{\theta} \in \Theta_1, \quad \underline{x}^{(i)} \in S(X), \quad i = 1, 2, \ldots, n.$$

(13.9)

The corresponding prior is obtained by integrating out the $\underline{x}^{(i)}$'s in (13.9). Recalling that the likelihoods of the form (13.4) must integrate to 1, we find that the appropriate choice of the prior is of the form

$$f(\underline{\theta}) = \exp\left[c + g_0(\underline{\theta}) + \sum_{k=1}^{\ell} b_k g_k(\underline{\theta}) - n\lambda(\underline{\theta}) \right], \quad \underline{\theta} \in \Theta_1.$$

(13.10)

Note that the function $g_0(\underline{\theta})$, that appears in (13.10) and in the desired posterior density (13.5), can be quite arbitrary. In addition, there can be considerable flexibility in the choice of $g_k(\underline{\theta})$'s ($k \geq 1$), subject only to the contraint that (13.8) holds (i.e., that all the θ_j's can be expressed as linear combinations of the $g_k(\underline{\theta})$'s). If we wish to have a proper prior, we must impose a further restriction that $\exp[g_0(\underline{\theta}) + \sum_{k=1}^{\ell} b_k g_k(\underline{\theta}) - n\lambda(\underline{\theta})]$ be integrable over Θ_1.

Example 13.1 (Normal likelihood and normal posterior). For notational simplicity we assume a sample size of 1. So we have $X \sim N(\theta, 1)$ and we wish to determine the class of priors for which all posteriors for Θ given $X = x$ are members of the two-parameter (normal) exponential family:

$$f(\underline{\theta}; \underline{\eta}) = \exp\left[\nu(\underline{\eta}) + \eta_1\theta + \eta_2\theta^2\right]. \tag{13.11}$$

Here (using the notation of (13.5)) $g_0(\theta) = 0, g_1(\theta) = \theta, g_2(\theta) = \theta^2$. Since our likelihood assumes the form

$$f(x; \theta) = \exp\left[-\log\sqrt{2\pi} - \frac{\theta^2}{2} + \theta x - \frac{x^2}{2}\right], \tag{13.12}$$

we can, using the notation of (13.3), see that

$$\begin{aligned}
\lambda(\theta) &= \frac{-\theta^2}{2}, \\
T_0(x) &= -\log\sqrt{2\pi} - \frac{x^2}{2}, \\
T_1(x) &= x.
\end{aligned} \tag{13.13}$$

Immediately from (13.10), we see that the appropriate family of priors is of the form

$$f(\theta) = \exp\left[c + b_1\theta + b_2\theta^2 + \frac{\theta^2}{2}\right] = \exp[c + b_1\theta + \tilde{b}_2\theta^2]. \tag{13.14}$$

Thus, the two-parameter normal family of priors will be appropriate. $\square$

Example 13.2 (Normal likelihood and inverse Gaussian posterior). Again take one observation $X \sim N(\theta, 1)$. We wish to have posterior distributions which belong to the inverse Gaussian family.

$$f(\underline{\theta}; \underline{\eta}) = \exp[\nu(\underline{\eta}) - 3\log\theta/2 + \eta_1\theta + \eta_2\theta^{-1}], \quad \theta > 0. \tag{13.15}$$

Here $g_0(\theta) = -3\log\theta/2, g_1(\theta) = \theta, g_2(\theta) = \theta^{-1}$. Combining this with the likelihood (13.12) for which $\lambda(\theta) = -\theta^2/2$ we obtain using (13.10), the following family of priors (with inverse Gaussian posteriors):

$$f(\theta) = \exp[c - 3\log\theta/2 + b_1\theta + b_2\theta^{-1} + \frac{\theta^2}{2}]. \tag{13.16}$$

In this case, the family is not a conjugate prior family. $\square$

13.4   Conjugate Exponential Family Priors for Exponential Family Likelihoods

We envision data sets consisting of n observations (possibly vector valued) $\underline{x}^{(1)}, \ldots, \underline{x}^{(n)}$ from an m-parameter exponential family. After suitable reparametrization and introduction of appropriate sufficient statistics the corresponding likelihood of such a sample has the following representation:

$$f(\underline{x}; \underline{\theta}) = r_n(\underline{x}) \exp\left[\sum_{i=1}^{m} \theta_i s_i(\underline{x}) + n\lambda(\underline{\theta})\right], \qquad (13.17)$$

in which $\underline{x}$ denotes the full data set $\underline{x}^{(1)}, \ldots, \underline{x}^{(n)}$.

It is assumed, without loss of generality, that the $s_i(\underline{x})$'s are linearly independent and nonconstant and that the θ_i's are restricted to belong to the natural parameter space Θ, defined to include all $\underline{\theta}$'s for which (13.17) is integrable. All priors to be considered will be nonrestrictive, i.e., they will be positive on Θ. Consequently, since all functions of $\underline{\theta}$ will have the same domain, repeated mention of that domain will be unnecessary. When we say for all $\underline{\theta}$ we mean for all $\underline{\theta}$ in Θ.

The first problem to be addressed involves the identification of the most general t-parameter exponential family of priors for $\underline{\theta}$ which will be conjugate with respect to (13.17). There are mathematical advantages to allowing improper priors so we do not insist on integrability. A second question involves the description of all possible conjugate prior families.

We will use $\underline{b} = (b_1, \ldots, b_t)$ to denote the hyperparameters of the t-parameter conjugate exponential family of priors for the likelihood (13.17).

Our main result is:

Theorem 13.1 (Conjugacy in exponential families). *The most general t-parameter exponential family of prior distributions for $\underline{\theta} = (\theta_1, \ldots, \theta_m)$, that is conjugate with respect to likelihoods (13.17), is of the form*

$$f(\underline{\theta}|\underline{b}) = r_0(\underline{\theta}) \exp\left[\sum_{i=1}^{m} b_i\theta_i + b_{m+1}\lambda(\underline{\theta}) + \sum_{i=m+2}^{t} b_i s_i(\underline{\theta}) + \lambda_0(\underline{b})\right], \qquad t > m,$$

$$(13.18)$$

where $s_{m+2}(\underline{\theta}), \ldots, s_t(\underline{\theta})$ are arbitrary functions. If $t \leq m$, no such conjugate prior exists.   □

It is obvious that (13.18) does indeed form a conjugate prior family for likelihoods of the form (13.17). The posterior hyperparameter vector will be $(b_1 + s_1(\underline{x}), \ldots, b_m + s_m(\underline{x}), b_{m+1} + n, b_{m+2}, \ldots, b_t)$ (only the first $m+1$ hyperparameters are adjusted by the observations).

To verify that (13.18) gives the most general form of a conjugate exponential family of priors, we consider an arbitrary t-parameter exponential
family of the form

$$f(\underline{\theta}|\underline{b}) = r(\underline{\theta}) \exp\left[\sum_{i=1}^{t} b_i g_i(\underline{\theta}) + \lambda_0(\underline{b})\right]. \tag{13.19}$$

Consider a typical posterior kernel, obtained by combining (13.19) with
(13.17). Since (13.19) is required to be a conjugate family, we must have

$$r_n(\underline{x})r(\underline{\theta}) \exp\left[\sum_{i=1}^{t} b_i g_i(\underline{\theta}) + \lambda_0(\underline{b}) + \sum_{i=1}^{m} \theta_i s_i(\underline{x}) + n\lambda(\underline{\theta})\right]$$
$$\propto r(\underline{\theta}) \exp\left[\sum_{i=1}^{t} h_i(\underline{x}, \underline{b})g_i(\theta) + \lambda_0(\underline{h}(\underline{x}, \underline{b}))\right]. \tag{13.20}$$

But this is required to hold for all $\underline{x}$ for all $\underline{\theta}$ and for every $\underline{b}$. For any fixed
value of $\underline{b}$, (13.20) implies that a functional equation of the form

$$\sum_{k=1}^{m+t+1} \psi_k(\underline{\theta})\phi_k(\underline{x}) = 0$$

must be true. Such functional equations are readily solved (see Theorem
1.3) and we conclude that in the case of $t > m$,

$$\begin{aligned}
g_i(\underline{\theta}) &= \theta_i, \quad i = 1, 2, \ldots, m, \\
h_i(\underline{x}, \underline{b}) &= b_i + s_i(\underline{x}), \quad i = 1, 2, \ldots, m, \\
g_{m+1}(\underline{\theta}) &= \lambda(\underline{\theta}),
\end{aligned} \tag{13.21}$$

and

$$g_j(\underline{\theta}) \text{ are arbitrary functions}, \ j > m + 1.$$

Equation (13.18) then follows.

If we turn to the question of identifying more general (nonexponential
family) conjugate families we may argue as follows for likelihoods of the
form (13.17).

For any sample size n and data configuration $\underline{x}$ and any prior $f_0(\underline{\theta})$ the
posterior will be of the form

$$f(\underline{\theta}|x) \propto f_0(\underline{\theta}) \exp\left[\sum_{i=1}^{m} \theta_i s_i(\underline{x}) + n\lambda(\underline{\theta})\right]. \tag{13.22}$$

The class of all possible posteriors resulting from all possible hypothetical
samples of all possible sizes in conjunction with the prior $f_0(\underline{\theta})$ will necessarily be a conjugate family, since a further sample of size n_2 in conjunction
with a hypothetical sample of size n_1 must already have been included in

the list of hypothetical samples of size $n_1 + n_2$. Thus we are led to the conjugate prior family

$$f(\underline{\theta}|\underline{b}) = f_0(\underline{\theta}) \exp\left[\sum_{i=1}^{m} b_i\theta_i + b_{m+1}\lambda(\underline{\theta}) + \tilde{\lambda}_0(\underline{b})\right]. \qquad (13.23)$$

Of course, $f_0(\underline{\theta})$ could, in addition, be allowed to range over a $(t - m - 1)$-dimensional space yielding a richer still conjugate family. Since $f_0(\underline{\theta})$ was quite arbitrary, (13.23) does not necessarily represent an exponential family. Observe that the $t - m - 1$ hyperparameters associated with the family of possible $f_0(\underline{\theta})$'s are unaffected by the data. This is completely analogous to the hyperparameters $b_{m+2}, \ldots, b_t$ in the exponential conjugate prior (13.18) which are also unaffected (though eventually swamped) by the data.

13.5   Conditionally Specified Priors

Suppose that our data $\underline{X}$ has a likelihood

$$\{f(\underline{x}; \underline{\theta}) : \underline{\theta} \in \Theta \subset \mathbb{R}^k\}.$$

In order to specify our joint prior distribution of $\underline{\theta}$, we are faced with a problem of describing a k-dimensional density. Throughout this book we have argued that conditional specification is often a natural and convenient mode of visualization of such densities. In the bivariate case, this approach would involve characterizing the joint density of a random vector (θ_1, θ_2) by postulating the precise form, or perhaps the parametric form, of the two families of conditional densities associated with (θ_1, θ_2); i.e., the conditional densities of θ_1 given θ_2, for all θ_2's, and the conditional densities of θ_2 given θ_1, for all θ_1's. Taken into the prior assessment arena associated with our Bayesian inference problem, we would not question the investigator about his prior beliefs regarding the joint prior for $\underline{\theta}$, rather we would ask about prior beliefs about θ_1 given specific values of the other θ's, then about prior beliefs about θ_2 given specific values of the other θ's, etc. The advantage of this system is that we are only eliciting information about univariate distributions, a manifestly easier task than that of directly eliciting beliefs about multivariate distributions.

In k dimensions, suppose that for each coordinate θ_i of $\underline{\theta}$, if the other coordinates $\underline{\theta}_{(i)}$ ($\underline{\theta}$ with θ_i deleted) were known, a convenient conjugate prior family $f_i(\theta_i|\underline{\alpha}_{(i)})$, $\underline{\alpha}_{(i)} \in A_{(i)}$ is available. Here the $\underline{\alpha}_{(i)}$'s are hyperparameters. Under such circumstances it seems natural to use as a candidate family of prior distributions for $\underline{\theta}$, one which has the property that for each i, the conditional distribution of θ_i given $\underline{\theta}_{(i)}$ belongs to the family f_i. By construction such a flexible family will be a conjugate family. The simplest case involves exponential families. In it, each family of priors f_i (for θ_i

given $\underline{\theta}_{(i)}$), is an ℓ_i-parameter exponential family. The resulting conditionally conjugate family of densities is again an exponential family. However, it will have a large number of hyperparameters guaranteeing considerable flexibility for matching informed prior beliefs.

Specifically suppose that for each i, a natural conjugate prior for θ_i (assuming $\underline{\theta}_{(i)}$ were known) is available in the form of an ℓ_i-parameter exponential family:

$$f_i(\theta_i) \propto r_i(\theta_i) \exp\left[\sum_{j=1}^{\ell_i} \eta_{ij} T_{ij}(\theta_i)\right]. \tag{13.24}$$

A convenient family of joint priors for the full parameter vector $\underline{\theta}$ will consist of all k-dimensional densities with conditionals (of θ_i given $\underline{\theta}_{(i)}$, for every i) in the given exponential families (13.24). The resulting joint density for $\underline{\theta}$ from Theorem 4.1 is of the form

$$f(\underline{\theta}) = \left[\prod_{i=1}^{k} r_i(\theta_i)\right] \exp\left\{\sum_{j_1=0}^{\ell_1} \sum_{j_2=0}^{\ell_2} \cdots \sum_{j_k=0}^{\ell_k} m_{j_1 j_2 \cdots j_k} \left[\prod_{i=1}^{k} T_{i,j_i}(\theta_i)\right]\right\}, \tag{13.25}$$

where for notational convenience we have introduced the constant functions $T_{i0}(\theta_i) = 1$, $i = 1, 2, \ldots, k$. The parameter space of the family of densities (13.25) is of dimension $[\prod_{i=1}^{k}(m_i + 1)] - 1$ since $m_{00\ldots0}$ is determined as a function of the others to ensure that the density integrates to 1. It is readily verified that the family (13.25) will be a conjugate prior family for $\underline{\theta}$, and that the resulting posterior distributions will exhibit the same kind of conditional structure as did the prior distribution. Thus, using (13.25), both a priori and a posteriori we will have, for each i, the conditional density of θ_i given $\underline{\theta}_{(i)}$ being a member of the given ℓ_i-parameter exponential family (13.24).

Priors such as those displayed in (13.25) are called conditionally specified priors or conditionally conjugate priors.

An advantage of conditionally specified priors is that, by their construction, they are tailor-made for the Gibbs sampler. Simulation of pseudo-samples from (13.25) will involve only the need to devise appropriate simulation algorithms for the one-dimensional exponential families (13.24). Alternatively, densities such as (13.25) are frequently amenable to strategies involving rejection algorithms and/or importance sampling.

It should be remarked that the family of conditionally specified priors (13.25) includes as special cases two of its primary competitors. First, the natural conjugate prior family, obtained by beginning with a locally uniform prior ($\propto 1$) over the parameter space, and considering the resulting posteriors corresponding to all possible samples of all possible sizes from the given likelihood. This, taken as a family of priors, is by construction a conjugate family. It will be subsumed by (13.25). The second popular competitor would involve the assumption of independent priors for the co-ordinates θ_i of $\underline{\theta}$, using the families described in (13.24). Clearly such a distribution will have conditionals in the given exponential families (13.24) and so will be subsumed by (13.25). It can be obtained by setting many of the $m_{j_1,\ldots,j_k}$'s equal to zero to obtain the desired independence.

It also bears remarking that many of the hyperparameters in (13.25) will be unchanged when we use data to update from the prior to the posterior. As will be evident either from the preceding paragraph or from perusal of the examples in the next sections, a decision to only give nonzero values to those hyperparamters that <u>are</u> affected by the data will in fact bring us back to the natural conjugate family for $\underline{\theta}$. The additional hyperparameters, those unaffected by the data, have a role to play in providing flexibility for matching a broad spectrum of prior belief, not necessarily well described by the natural conjugate family.

Densities of any of the families of the form (13.25) are clearly nonnegative for any choice of the hyperparameters appearing in them. They are how-ever not guaranteed to be integrable unless the hyperparameters satisfy constraints which are sometimes quite complicated. It is usually deemed acceptable to have improper prior distributions but unacceptable to have improper posteriors. Typically if large amounts of data are available, the posterior distributions will be proper for most prior selections of the hy-perparameters. In practice a case by case determination of propriety will be necessary. Reference to Chapters 3 and 4 will be useful in some cases. It must be remarked that conditions sufficient for propriety of all posterior conditional distributions are sometimes readily checked. Here too, however, we must still check to determine whether the full joint posterior distribu-tion is proper, since proper conditionals do not, unfortunately, guarantee a proper joint distribution. This caveat is especially important if we plan to use the Gibbs sample for posterior simulation (the computer will only happily notice the propriety of the conditionals and will churn out superfi-cially acceptable results even when the joint distribution is improper). See Hobert and Casella (1996) for further discussion of this potential problem with the Gibbs sampler.

In order to pick a conditionally specified prior that will represent the informed expert's beliefs about the parameter $\underline{\theta}$, say, it will be necessary to request a considerable amount of information. Such information will, be-cause of human nature, undoubtedly be inconsistent. So we do not expect to find a conditionally specified prior that matches the provided information exactly; instead we seek a conditionally specified prior that is minimally discrepant from the given information. Typically our knowledge of the con-ditionally specified distribution (13.25) will be adequate to permit us to compute a variety of conditional moments and conditional percentiles as explicit functions of the hyperparameters. It is then possible to use a va-riety of optimization procedures to choose values of the hyperparameters that minimize the discrepancy between the elicited values of a spectrum of conditional moments and/or percentiles and their theoretical values as functions of the hyperparameters. Of course, the number of elicited conditional prior features must at least equal the number of hyperparameters in the prior family and, in practice, should be considerably larger. It will not be a quick process, but time invested in carefully choosing an approximation to the informed expert's true prior is surely well spent. Some examples of this kind of elicitation procedure are decribed in more detail in the following sections.

The conditionally specified priors described in this chapter do include as special cases the usual noninformative and the usual conjugate prior families. Consequently, any investigator already happy with such priors will be able to live comfortably with conditionally conjugate priors, he will just choose to not avail himself of their full flexibility.

It is important to emphasize that in using a family of conditionally specified priors we are not saying that the informed expert's true prior is a member of this family; we only say that the conditionally specified prior family will hopefully provide a flexible enough family to adequately approximate the expert's prior beliefs. As mentioned earlier, multimodal prior beliefs will perhaps be best modeled using finite mixtures of conditionally specified priors.

13.6   Normal Data

We will return to study in more detail the normal scenario introduced in Section 13.2.

The available data are n independent identically distributed random variables each normally distributed with mean μ and precision τ. The likelihood is of the form

$$f_{\underline{X}}(\underline{x}; \mu, \tau) = \frac{\tau^{n/2}}{(2\pi)^{n/2}} \exp\left[-\frac{\tau}{2} \sum_{i=1}^{n} (x_i - \mu)^2\right]. \tag{13.26}$$

If τ were known, a natural conjugate prior family for μ would be the normal family. If μ were known, a natural conjugate prior family for τ would be the gamma family. This suggests that an appropriate conjugate prior family for (μ, τ) (assuming both are unknown) would be one in which μ given τ is normally distributed for each τ, and τ given μ has a gamma distribution for each μ. The class of such gamma–normal distributions was discussed extensively in Section 4.8. They form an eight-parameter exponential family of distributions with densities of the form

$$f(\mu, \tau) \propto \exp\left[m_{10}\mu + m_{20}\mu^2 + m_{12}\mu \log \tau + m_{22}\mu^2 \log \tau\right]$$
$$\times \exp\left[m_{01}\tau + m_{02} \log \tau + m_{11}\mu\tau + m_{21}\mu^2\tau\right]. \tag{13.27}$$

For such a density we have:

(1) The conditional density of μ given τ is normal with mean

$$E(\mu|\tau) = \frac{-(m_{10} + m_{11}\tau + m_{12}\log\ \tau)}{2(m_{20} + m_{21}\tau + m_{22}\log\ \tau)} \tag{13.28}$$

and precision

$$1/\mathrm{var}(\mu|\tau) = -2(m_{20} + m_{21}\tau + m_{22}\log\ \tau). \tag{13.29}$$

(2) The conditional density of τ given μ is gamma with shape parameter $\alpha(\mu)$ and intensity parameter $\lambda(\mu)$, i.e.,

$$f(\tau|\mu) \propto \tau^{\alpha(\mu)-1}e^{-\lambda(\mu)\tau}, \tag{13.30}$$

with mean and variance

$$E(\tau|\mu) = \frac{1 + m_{02} + m_{12}\mu + m_{22}\mu^2}{-(m_{01} + m_{11}\mu + m_{21}\mu^2)}, \tag{13.31}$$

$$\mathrm{var}(\tau|\mu) = \frac{1 + m_{02} + m_{12}\mu + m_{22}\mu^2}{(m_{01} + m_{11}\mu + m_{21}\mu^2)^2}. \tag{13.32}$$

(3) Since the parameters in (13.29), (13.31) and (13.32) must be positive to yield proper conditional densities, natural constraints must be placed on the parameters in (13.27). Thus we must have

$$m_{21} < 0,\ \ m_{22} > 0,\ \ m_{01} < 0,\ \ m_{02} > -1,\ \ m_{12}^2 < 4m_{22}(m_{02}+1), \tag{13.33}$$

$$m_{20} + m_{22}\left[\log\left(-\frac{m_{22}}{m_{21}}\right) - 1\right] < 0,\ \ m_{11}^2 < 4m_{21}m_{01}. \tag{13.34}$$

If we propose to use densities like (13.27) as prior densities and if we are willing to accept improper priors and posteriors, then we need to impose no conditions on the parameters (or hyperparameters) in (13.27).

(4) The marginal densities for μ and τ associated with the joint density (13.27) are of the form

$$f(\mu) \propto \exp(m_{10}\mu + m_{20}\mu^2)\frac{\Gamma(m_{02} + m_{12}\mu + m_{22}\mu^2 + 1)}{[-(m_{01}+m_{11}\mu+m_{21}\mu^2)]^{(m_{02}+m_{12}\mu+m_{22}\mu^2+1)}}, \tag{13.35}$$

$$f(\tau) \propto \exp(m_{01}\tau + m_{02}\log\tau)$$

$$\times \exp\left(\frac{-(m_{10} + m_{11}\tau + m_{12}\log\tau)^2}{4(m_{20} + m_{21}\tau + m_{22}\log\tau)}\right)\sqrt{\frac{-\pi}{m_{20} + m_{21}\tau + m_{22}\log\tau}}I(\tau > 0). \tag{13.36}$$

TABLE 13.1. Adjustments in the parameters in the prior family (13.27), combined with likelihood (13.37).

Parameter	Prior value	Posterior value
m_{10}	m_{10}^*	m_{10}^*
m_{20}	m_{20}^*	m_{20}^*
m_{01}	m_{01}^*	$m_{01}^* - \frac{1}{2}\sum_{i=1}^n x_i^2$
m_{02}	m_{02}^*	$m_{02}^* + n/2$
m_{11}	m_{11}^*	$m_{11}^* + \sum_{i=1}^n x_i$
m_{12}	m_{12}^*	m_{12}^*
m_{21}	m_{21}^*	$m_{21}^* - n/2$
m_{22}	m_{22}^*	m_{22}^*

(5) The family (13.27) is indeed a conjugate prior family for normal likelihoods of the form (13.26). To verify this we rewrite the likelihood (13.26) in the more convenient form

$$f_{\underline{X}}(x; \mu, \tau) = (2\pi)^{-n/2} \exp\left[\frac{n}{2}\log\tau - \frac{\sum_{i=1}^n x_i^2}{2}\tau + \sum_{i=1}^n x_i\mu\tau - \frac{n}{2}\mu^2\tau\right].$$
(13.37)

A prior in the family (13.27) will yield, when combined with the likelihood (13.37), a posterior in the same family with prior and posterior parameters related as in Table 13.1.

Table 13.1 merits scrutiny to understand the nature of the proposed prior family (13.27) and its relation to families of priors used in more traditional analyses. First it is evident in Table 13.1 that four of the parameters, m_{10}, m_{20}, m_{12}, and m_{22}, those corresponding to the first factor in (13.27), once fixed in the prior, are unchanged by the data. They do not change, but their contribution to the posterior would eventually be swamped by large data sets. Their presence is needed to allow us the full flexibility of the gamma–normal prior family. Traditionally such flexibility has not been available.

Our model (13.27) subsumes two important cases:

1. *The classical prior distribution for* (μ, τ). This distribution has τ with a marginal gamma distribution and μ given τ distributed normally with its precision a scalar multiple of τ (see, e.g., DeGroot ((1970), p. 169). The deGroot priors correspond to the second factor in (13.27), that is, to

$$m_{10} = m_{20} = m_{12} = m_{22} = 0. \tag{13.38}$$

It is not at all evident that the dependence structure (between μ and τ), inherent in such a joint distribution, will necessarily adjust well with prior beliefs.

2. *The independence case.* A second approach, advocated by those who view marginal assessment of prior beliefs to be the most viable (see, e.g., Press (1982)), assumes independent gamma and normal marginals in (13.27). This corresponds to initially setting

$$m_{11} = m_{12} = m_{21} = m_{22} = 0. \tag{13.39}$$

It has been said that in such a case we do not have a conjugate prior since the resulting posterior will not have independent marginals. This is because the posterior values of m_{11} and m_{21} will no longer be zero (m_{12} and m_{22} remain zero since they are always unaffected by the data).

From our viewpoint both the classical priors with their unusual implied dependence structure and the independent marginal priors are within our conjugate prior family as are their corresponding posteriors. Consequently, any experimenter whose prior beliefs were adequately described by one or other of these restricted families will have no problem using the expanded family (13.27); his prior will be approximated by one of its members.

We must pay for the flexibility exhibited by our conditionally specified prior. In the normal case we have eight (hyper) parameters to assess. The earlier analyses rather arbitrarily set four of them equal to zero and just assessed the remaining four. It turns out that assessment of the eight hyperparameters is not as formidable a problem as we might fear, as we see in the next section.

13.6.1   Assessment of Appropriate Values for the Hyperparameters in the Normal Case

In this section we discuss the assessment of the prior hyperparameters. We consider the following methods:

1. *Matching conditional moments.* For a conditionally specified prior such as (13.27), it is natural to try to match conditional moments whose approximate values will be supplied by the knowledgable scientist who collected

the data. In our example, eight such conditional moments will suffice to determine all the hyperparameters. We propose, more generally, to ask the experimenter to provide prior values for more than eight conditional moments. We recognize that it is unlikely that such prior values will be consistent and what we propose is to select a prior of the form (13.27) that will have conditional moments that are minimally disparate from those provided a priori by the scientist.

Suppose that prior assessed values for the conditional means and variances are obtained for several different given choices of the precision τ and for several different given choices of the mean μ. Thus the experimenter provides his best guesses for the quantities (the subscript, A, denotes assessed value):

$$E_A(\mu|\tau_i) = \xi_i, \quad i = 1, 2, \ldots, m, \tag{13.40}$$

$$\text{var}_A(\mu|\tau_i) = \eta_i, \quad i = 1, 2, \ldots, m, \tag{13.41}$$

$$E_A(\tau|\mu_j) = \psi_j, \quad j = 1, 2, \ldots, \ell, \tag{13.42}$$

$$\text{var}_A(\tau|\mu_j) = \chi_j, \quad j = 1, 2, \ldots, \ell, \tag{13.43}$$

where $2m + 2\ell \geq 8$.

Note that the values $\{\tau_i\}_{i=1}^m$ and $\{\mu_j\}_{j=1}^\ell$ are known quantities. If indeed a density of the form (13.27) approximates the joint distribution of (μ, τ), then the values of the conditional moments in (13.40)–(13.43) will be well approximated by expressions derived from (13.28), (13.29), (13.31), and (13.32).

One possible approach, since exact equality is unlikely to be possible for any choice of the parameters $m_{10}, m_{20}, m_{01}, m_{02}, m_{11}, m_{12}, m_{21}$, and m_{22}, is to set up as an (admittedly somewhat arbitrary) objective function the sum of squared differences between the left- and right-hand sides of (13.28), (13.29), (13.31), and (13.32) [$2m + 2\ell$ terms in all] and, using a convenient optimization program, choose values of the parameters to minimize this objective function subject to constraints (13.33) and (13.34). The "assessed" prior would then be (13.27) with this choice of parameters.

An alternative simpler procedure is possible. If approximate equality is to hold in (13.28) and (13.29) then ξ_i/η_i will be approximately equal to the product of the right-hand sides of (13.28) and (13.29), a linear function of the parameters. Also η_i^{-1} will be approximately equal to the right-hand side of (13.29), again a linear function of the parameters. Turning to (13.31) and (13.32) we find that ψ_j/χ_j and ψ_j^2/χ_j will be well approximated by linear combinations of the parameters. Thus the following array of approximate linear relations should hold:

$$\xi_i/\eta_i \approx m_{10} + m_{11}\tau_i + m_{12}\log \tau_i, \tag{13.44}$$

$$-\eta_i^{-1}/2 \approx m_{20} + m_{21}\tau_i + m_{22}\log \tau_i, \tag{13.45}$$

$$-\psi_j/\chi_j \approx m_{01} + m_{11}\mu_j + m_{21}\mu_j^2, \tag{13.46}$$

$$(\psi_j^2/\chi_j) - 1 \approx m_{02} + m_{12}\mu_j + m_{22}\mu_j^2. \tag{13.47}$$

Least squares estimates of the eight parameters subject to constraints (13.33) and (13.34) are then obtainable by standard regression techniques. If the researcher is more certain about the prior conditional moments for some values of τ or μ, then weighted least squares could be used. It is these readily obtained values for the (hyper) parameters that we propose to use to determine the "assessed" prior. The density (13.27) with these assessed values of the parameters will have conditional moments not too disparate from those provided by the scientist.

An alternative approach would involve matching conditional percentiles. Since, for example, gamma percentiles are not describable in closed form, implementation of such an approach will be more challenging.

2. *Using diffuse or partially diffuse priors.* Utilization of conditionally specified priors such as those introduced in Section 13.5 involves assessment of many hyperparameters. It would not be uncommon to encounter an informed expert who honestly is unable to provide plausible values for the conditional means, variances, or percentiles utilized in the suggested prior assessment approach. In such situations it is quite reasonable to select and use values of the hyperparameters which reflect ignorance or diffuseness of prior information about the parameters.

If our informed expert expresses inability to provide any conditional moments of his prior, it would be appropriate to use a locally uniform joint prior for (μ, τ) which would correspond to the case in which all hyperparameters in (13.27) are set equal to 0 (with the possible exception of m_{02} which might be set equal to -1).

Accommodation of diffuse prior information in the family (13.27) does not thus appear to present any major problems.

3. *Using a fictitious sample.* The expert could begin with a diffuse prior. He then may "guess" a representative (fictitious) sample to be combined with the prior using Table 13.1. The resulting posterior hyperparameters then become the prior hyperparameters for subsequent analysis of the real data set.

13.6.2   Parameter Estimation in the Normal Case

Having assessed our prior values of the parameters, we may read off the corresponding posterior values of the parameters from Table 13.1. The posterior density will be of the gamma–normal form (i.e., (13.27)). If point estimates are desired they will be provided by $E(\mu|\underline{x})$, an estimate of the mean, by $E(\tau|\underline{x})$, an estimate of the precision τ, and by $E(\tau^{-1}|\underline{x})$, an estimate of the variance $\sigma^2 = 1/\tau$. The posterior distribution is a member of an exponential family, so numerical determination of these posterior expectations is not too difficult.

Two convenient alternatives are possible:

TABLE 13.2. Iris versicolor data: Sepal length in centimeters.

7.0	6.4	6.9	5.5	6.5	5.7	6.3	4.9	6.6	5.2	5.0	5.9	6.0
6.1	5.6	6.7	5.6	5.8	6.2	5.6	5.9	6.1	6.3	6.1	6.4	6.6
6.8	6.7	6.0	5.7	5.5	5.5	5.8	6.0	5.4	6.0	6.7	6.3	5.6
5.5	5.5	6.1	5.8	5.0	5.6	5.7	5.7	6.2	5.1	5.7		

- *Mode estimates.* We can use the mode of the posterior, i.e., solve the system of equations,

$$0 = m_{10} + 2(m_{20} + m_{21}\tau)\mu + m_{11}\tau + m_{12}\log\tau + 2m_{22}\mu\log\tau, \quad (13.48)$$

$$\tau = -\frac{m_{02} + \mu m_{12} + \mu^2 m_{22}}{(m_{01} + \mu m_{11} + \mu^2 m_{21})}. \quad (13.49)$$

Note that replacing τ from (13.49) in (13.48) we get an equation which depends only on μ.

- *Gibbs sampler estimates.* Since the density (13.27) has simple conditionals, a Gibbs sampler approach may be used to approximate the posterior moments. Thus to approximate $E(\tau^{-1}|\underline{x})$ we successively generate $\mu_1, \tau_1, \mu_2, \tau_2, \ldots, \mu_N, \tau_N$ using the posterior conditional distributions (with parameters given in (13.28)–(13.32)) and our approximation to $E(\tau^{-1}|\underline{x})$ will be $\sum_{k=1}^{N} 1/(N\tau_k)$ or perhaps $\sum_{k=N'+1}^{N'+N} 1/(N\tau_k)$ (if we allow time for the sampler to stabilize).

The Iris Data

To illustrate the above considerations we will reanalyze Fisher's (1936) famous Iris data.

The sepal lengths in the Iris versicolor data shown in Table 13.2 are plausibly approximately normally distributed.

Our model is normal with unknown mean μ and precision τ. The corresponding sufficient statistics assume the following values ($n = 50$):

$$\sum_{i=1}^{50} x_i = 296.8, \quad \sum_{i=1}^{50} x_i^2 = 1774.86.$$

For illustrative purposes we discuss three cases, that is, we assume that our knowledgable expert has supplied us with:

- *Case 1*: A fictitious (guessed) sample $\{6.5, 6.0, 5.8, 5.9, 6.1, 6.3, 6.2\}$ which will be combined with a diffuse prior using Table 13.1 to determine the prior hyperparameters. We obtain the posterior hyperparameter values by using Table 13.1 once more.

TABLE 13.3. Prior (Pr) and posterior (Pt) values for different expert assessments (Iris versicolor data).

	m_{10}	m_{20}	m_{01}	m_{02}	m_{11}	m_{12}	m_{21}	m_{22}
				Case 1				
Pr	0	-0.001	-131.02	3.5	42.8	0	-3.5	0.001
Pt	0	-0.001	-1018.5	28.5	339.6	0	-28.5	0.001
				Case 2				
Pr	-0.74	-8.93	-1020.3	28.13	332.7	-0.36	-29	8.76
Pt	-0.74	-8.93	-1907.7	53.13	629.5	-0.36	-54	8.76
				Case 3				
Pr	0	-0.001	-0.001	0	0	0	-0.001	0.001
Pt	0	-0.001	-887.4	25.	296.8	0	-25.001	0.001

TABLE 13.4. Bayesian estimates for different assessments (Iris versicolor data).

	Max. likelihood estimates		Mode estimates		Gibbs estimates		Numerical integration	
	$\hat{\mu}$	$\hat{\tau}$	$\hat{\mu}$	$\hat{\tau}$	$\hat{\mu}$	$\hat{\tau}$	$\hat{\mu}$	$\hat{\tau}$
Case 1	5.94	3.83	5.96	4.17	5.96	4.23	5.96	4.27
Case 2	5.94	3.83	5.94	4.84	5.94	4.86	5.94	4.87
Case 3	5.94	3.83	5.94	3.81	5.94	3.93	5.94	3.91

- *Case 2*: The following a priori conditional moments:

$$E[\mu|\tau = 3] = 5, \qquad E[\mu|\tau = 4] = 6, \qquad E[\mu|\tau = 5] = 6.1,$$
$$\text{var}[\mu|\tau = 3] = 0.006, \quad \text{var}[\mu|\tau = 4] = 0.004, \quad \text{var}[\mu|\tau = 5] = 0.003,$$
$$E[\tau|\mu = 5] = 3, \qquad E[\tau|\mu = 6] = 5, \qquad E[\tau|\mu = 7] = 4,$$
$$\text{var}[\tau|\mu = 5] = 0.02, \quad \text{var}[\tau|\mu = 6] = 0.03, \quad \text{var}[\tau|\mu = 7] = 0.04.$$

- *Case 3*: If we did not have an informed expert to aid us in the analysis we would undoubtedly begin with a diffuse prior. The simple use of Table 13.1 leads to the posterior.

In Table 13.3, the corresponding prior and posterior values of the hyperparameters are displayed (the fractional part 0.001 is used to avoid numerical problems with $m_{20} = 0$).

Estimates of μ and τ are then obtained by using maximum likelihood, the posterior mode, a Gibbs sampler simulation of the posterior density of the form (13.27), and by numerical integration of the posterior, with parameters as given in Table 13.3. The Gibbs sampler was iterated 10,300 times and the last 10,000 iterations were averaged to give the estimated marginal posterior means of μ and τ. The estimates obtained are given in Table 13.4.

13.7   Pareto Data

Suppose that the available data are n i.i.d. random variables each having a classical Pareto distribution with shape or inequality parameter α and precision parameter (the reciprocal of the scale parameter) τ. Thus the likelihood is of the form

$$
\begin{aligned}
f_{\underline{X}}(\underline{x}; \alpha, \tau) &= \prod_{i=1}^{n} \tau\alpha(\tau x_i)^{-(\alpha+1)} I(\tau x_i > 1) \\
&= \alpha^n \tau^{-n\alpha} \left(\prod_{i=1}^{n} x_i \right)^{-(\alpha+1)} I(\tau x_{1:n} > 1).
\end{aligned}
\tag{13.50}
$$

This can be conveniently rewritten in the form

$$
f_{\underline{X}}(\underline{x}; \alpha, \tau) = \exp\left[n \, \log \, \alpha - n\alpha \, \log \, \tau - \left(\sum_{i=1}^{n} \log x_i \right) (\alpha + 1) \right] I(\tau x_{1:n} > 1),
\tag{13.51}
$$

from which we obtain the maximum likelihood estimates

$$
\hat{\tau} = 1/\min(x_1, \ldots, x_n), \quad \hat{\alpha} = \frac{n}{n \, \log \, \hat{\tau} + \sum_{i=1}^{n} \log \, x_i},
\tag{13.52}
$$

which, for the sake of comparison, will be used later.

If τ were known, then a natural conjugate prior family of densities for α would be the gamma family. If α were known then a natural conjugate family of priors for τ would be the Pareto family. We are then led to consider as a conjugate prior family for (α, τ) (assuming both are unknown), one in which α given τ is gamma distributed for each τ and in which τ given α is Pareto distributed for each α. The corresponding six-parameter family of priors is then of the form

$$
\begin{aligned}
f(\alpha, \tau) &\propto \exp[m_{01} \log \tau + m_{21} \log \alpha \log \tau] \\
&\times \exp[m_{10}\alpha + m_{20} \log \alpha + m_{11}\alpha \log \tau] \, I(\tau c > 1),
\end{aligned}
\tag{13.53}
$$

where the two factors in the right-hand side refer to the hyperparameters which are unaffected by the data and those which are affected, respectively (see below). It is not difficult to verify that such densities do have Pareto and gamma conditionals.

For this density we have:

TABLE 13.5. Adjustments in the parameters in the prior (13.53) when combined with the likelihood (13.51).

Parameter	Prior value	Posterior value
m_{10}	m_{10}^*	$m_{10}^* - \sum_{i=1}^n \log x_i$
m_{20}	m_{20}^*	$m_{20}^* + n$
m_{01}	m_{01}^*	m_{01}^*
m_{11}	m_{11}^*	$m_{11}^* - n$
m_{21}	m_{21}^*	m_{21}^*
c	c^*	$\min(x_{1:n}, c^*)$

1. The conditional density of α given τ is gamma with shape parameter $\gamma(\tau)$ and intensity parameter $\lambda(\tau)$, i.e.,

$$f(\alpha|\tau) \propto \alpha^{\gamma(\tau)-1} e^{-\lambda(\tau)\alpha}, \qquad (13.54)$$

where the mean and variance are

$$E(\alpha|\tau) = -(1 + m_{20} + m_{21} \log \tau)/(m_{10} + m_{11} \log \tau), \qquad (13.55)$$

$$var(\alpha|\tau) = (1 + m_{20} + m_{21} \log \tau)/(m_{10} + m_{11} \log \tau)^2, \qquad (13.56)$$

2. The conditional density of τ given α is Pareto with shape or inequality parameter $\delta(\alpha)$ and precision parameter $\nu(\alpha)$, i.e.,

$$f(\tau|\alpha) \propto \nu(\alpha)\delta(\alpha) [\nu(\alpha)\tau]^{-(\delta(\alpha)+1)} I(\nu(\alpha)\tau > 1), \qquad (13.57)$$

where

$$\delta(\alpha) = -(1 + m_{01} + m_{11}\alpha + m_{21} \log \alpha), \qquad (13.58)$$

$$\nu(\alpha) = c. \qquad (13.59)$$

3. If we insist on proper prior densities, then there are constraints which must be imposed on the parameters in (13.53) to ensure that certain parameters appearing in the conditional densities, namely $\gamma(\tau)$, $\lambda(\tau)$,

$\nu(\tau)$, are always positive and $\delta(\tau) > -1$, to yield proper conditional densities. This implies

$$m_{21} > 0, \quad m_{11} < 0, \quad m_{10} < m_{11} \log c, \quad m_{20} > m_{21} \log c - 1,$$
$$c > 0, \quad -m_{01} + m_{21} [1 - \log(-m_{21}/m_{11})] > 0.$$
$$(13.60)$$

If we are willing to accept improper priors then no constraints are needed. It should be noted that (13.53) is not an exponential family of priors since the support of the density depends on one of the parameters (c).

4. The marginal densities for α and τ corresponding to the joint density (13.53) are of the following forms:

$$f(\alpha) \propto \exp(m_{10}\alpha + m_{20} \log \alpha) \frac{c^{-(m_{01} + m_{11}\alpha + m_{21} \log \alpha + 1)}}{-(m_{01} + m_{11}\alpha + m_{21} \log \alpha + 1)} I(\alpha > 0),$$
$$(13.61)$$

$$f(\tau) \propto \tau^{m_{01}} \frac{\Gamma(m_{20} + m_{21} \log \tau + 1)}{(-m_{10} - m_{11} \log \tau)^{m_{20} + m_{21} \log \tau + 1}} I(\tau c > 1). \quad (13.62)$$

5. The family (13.53) is readily verified to be a conjugate prior family for likelihoods of the form (13.50) (equivalently (13.51)). A prior from the family (13.53) will yield, when combined with the likelihood (13.51), a posterior again in the family (13.53) with prior and posterior (hyper) parameters related as in Table 13.5. It will be noted that two hyperparameters (m_{01} and m_{21}) are unaffected by the data. They appear in the first factor in (13.53).

Our model (13.53) includes:

1. *The "classical" conjugate prior family.* It was introduced by Lwin (1972). It corresponded to the case in which m_{01} and m_{21} were both arbitrarily set equal to 0.

2. *The independent gamma and Pareto priors.* These were suggested by Arnold and Press (1989a) and correspond to the choice $m_{11} = m_{21} = 0$ in (13.53).

Thus, the proposed flexible family includes the two most frequently proposed classes of priors.

13.7.1   Asessment of Appropriate Values for the Hyperparameters in the Pareto Case

The assessment of hyperparameters for the classical Pareto model will be achieved in a manner similar to that used in the normal case. We can use the following methods:

1. *Matching conditional moments and percentiles.* Conditional moments corresponding to the density (13.53) useful for this assessment are given in (13.55) and (13.56). When dealing with the conditional distribution of τ given α, a Pareto distribution, we are not guaranteed the existence of a mean for every choice of α. Instead we may try to find a prior by matching, as well as possible, conditional percentiles. We have, for densities of the form (13.53), the conditional p-percentiles given by

$$x_p[\tau|\alpha] = c^{-1}(1-p)^{1/(1+m_{01}+m_{11}\alpha+m_{21}\log\alpha)}. \tag{13.63}$$

The elicitation procedure would then involve asking the informed experimenter for his best guesses for quantitites of the form

$$\begin{align}
E(\alpha|\tau_i) &= \xi_i, \quad i = 1, 2, \ldots, m, \tag{13.64}\\
\mathrm{var}(\alpha|\tau_i) &= \eta_i, \quad i = 1, 2, \ldots, m, \tag{13.65}\\
x_{p_j}[\tau|\alpha_j] &= \chi_j(p_j), \quad j = 1, 2, \ldots, \ell, \tag{13.66}\\
x_{q_j}[\tau|\alpha_j] &= \chi_j(q_j), \quad j = 1, 2, \ldots, \ell. \tag{13.67}
\end{align}$$

Using arguments analogous to those used in Section 13.2 we will seek (hyper) parameters in (13.53) (i.e., $m_{10}, m_{20}, \ldots$) so that

$$\begin{align}
-\xi_i/\eta_i &\approx m_{10} + m_{11}\log\tau_i, \tag{13.68}\\
\xi_i^2/\eta_i &\approx 1 + m_{20} + m_{21}\log\tau_i, \tag{13.69}
\end{align}$$

$$\frac{\log\dfrac{1-p_j}{1-q_j}}{\log(\chi_j(p_j)/\chi_j(q_j))} \approx 1 + m_{01} + m_{11}\alpha_j + m_{21}\log\alpha_j. \tag{13.70}$$

Least-squares values of $m_{01}, m_{10}, m_{20}, m_{11}$, and m_{21} can be obtained using (13.68)–(13.70). Finally, we need to elicit the best guess for the minimum possible value of τ, this gives the elicited value of the reciprocal of c in (13.53). Note that a noninformed choice of c would correspond to a large value so that the posterior value of c would almost certainly be $x_{1:n}$.

2. *Using diffuse or partially diffuse priors.* The process is similar to the normal case.

3. *Using a fictitious sample.* The expert can also begin with a diffuse prior, guess a typical sample, and use expressions in Table 13.5 to calculate the prior hyperparameters (i.e., posterior hyperparameters corresponding to the fictitious sample).

13.7.2 Parameter Estimation in the Pareto Case

As in the classical normal case, we will be able to exploit the Gibbs sampler in studying the posterior distributions which, since they belong again to the family (13.53), have gamma and Pareto conditionals. Alternatively, posterior moments can be obtained by numerical integration.

The Annual Wage Data

To illustrate estimation procedures for Pareto data we will reanalyze the annual wage data discussed in Dyer (1981). The data set lists the annual wage in multiples of 100 U.S. dollars for 30 individuals.

Plausibly such a data set will be well described by a classical Pareto model (of the form (13.50)). The actual values of the data points are displayed in Table 13.6.

TABLE 13.6. Annual wage data (in multiples of 100 U.S. dollars).

112	154	119	108	112	156	123	103	115	107
125	119	128	132	107	151	103	104	116	140
108	105	158	104	119	111	101	157	112	115

The corresponding sufficient statistics are

$$x_{1:30} = 101, \quad \sum_{i=1}^{30} \log x_i = 143.523.$$

For illustrative purposes we discuss three cases. That is, we assume that our knowledgeable expert has supplied us with:

Case 1: A fictitious sample

$$\{110, 108, 112, 105, 122, 134, 117, 152, 131, 159, 121, 160, 143\},$$

which will be combined with a diffuse prior using Table 13.5 to yield the prior hyperparameter values in Table 13.7. Finally, Table 13.5 and the real data lead to the posterior values, which are also shown in Table 13.7.

Case 2: His best guess of the minimal possible value for τ, say 100 (this means that the prior choice of the hyperparameter c in (13.53) is 100), and the following a priori conditional moments and percentiles:

$$E[\alpha|\tau = 0.01] = 4.5, \qquad E[\alpha|\tau = 0.02] = 1,$$
$$\text{var}[\alpha|\tau = 0.01] = 0.6, \qquad \text{var}[\alpha|\tau = 0.02] = 0.03,$$
$$x_{0.1}[\tau|\alpha = 5] = 0.010003, \quad x_{0.2}[\tau|\alpha = 5] = 0.010006,$$
$$x_{0.9}[\tau|\alpha = 5] = 0.01006,$$

TABLE 13.7. Prior and posterior values for different expert assessments (wage data).

		m_{10}	m_{20}	m_{01}	m_{11}	m_{21}	c
Case 1	Prior	−63.02	12.1	0	−13.0	0.0	100
	Posterior	−206.55	42.1	0	−43.0	0.0	100
Case 2	Prior	−206.52	41.7	0	−43.12	1.52	100
	Posterior	−350.04	71.7	0	−73.12	1.52	100
Case 3	Prior	-0.005	-0.9	0	−0.0	0.0	100
	Posterior	−143.5	29.1	0	−30.0	0.0	100

TABLE 13.8. Bayesian estimates for different expert assessments (wage data).

	Max. likelihood estimates		Gibbs estimates		Numerical integration	
	$\hat{\alpha}$	$\hat{\tau}$	$\hat{\alpha}$	$\hat{\tau}$	$\hat{\alpha}$	$\hat{\tau}$
Case 1	5.918	0.00990	5.01	0.01	4.94	0.01
Case 2	5.918	0.00990	4.85	0.01	4.88	0.01
Case 3	5.918	0.00990	5.53	0.01	5.45	0.01

Case 3: Diffuse priors.

The assumed prior parameter values and the corresponding posterior values are shown in Table 13.7. The m_{ij} parameters in Case 2 were obtained by least squares.

Estimates of α and τ were then obtained by numerical integration and by using a Gibbs sampler simulation (using the last 10,000 of 10,300 iterations) from the posterior densities indicated in Table 13.7. The corresponding estimates are shown in Table 13.8.

13.8   Inverse Gaussian Data

A variety of parametrizations exist for the inverse Gaussian distribution. A convenient one for our purposes is provided by

$$f(x; \theta_1, \theta_2) = \sqrt{\frac{\theta_2}{\pi}} e^{2\sqrt{\theta_1 \theta_2}} x^{-3/2} e^{-\theta_1 x - \theta_2 x^{-1}} I(x > 0). \qquad (13.71)$$

If we have a sample of size n from the density (13.71) and if θ_2 were given, then a conjugate prior family for θ_1 is an exponential family of the form

$$f_1(\theta_1) \propto \exp(a\sqrt{\theta_1} + b\theta_1)I(\theta_1 > 0). \tag{13.72}$$

If θ_1 were given, a natural conjugate prior family for θ_2 is given by

$$f_2(\theta_2) \propto \exp(c\log\theta_2 + d\sqrt{\theta_2} + e\theta_2)I(\theta_2 > 0). \tag{13.73}$$

The natural requirement that our joint prior for (θ_1, θ_2) have conditionals in the families (13.72) and (13.73) leads to consideration of the following rich exponential family of joint priors with 11 hyper-parameters:

$$\begin{aligned} f(\theta_1, \theta_2) \propto\ & \exp(m_{10}\sqrt{\theta_1} + m_{20}\theta_1 + m_{01}\log\theta_2 \\ & + m_{02}\sqrt{\theta_2} + m_{03}\theta_2 + m_{11}\sqrt{\theta_1}\log\theta_2 \\ & + m_{12}\sqrt{\theta_1\theta_2} + m_{13}\sqrt{\theta_1}\theta_2 \\ & + m_{21}\theta_1\log\theta_2 + m_{22}\theta_1\sqrt{\theta_2} + m_{23}\theta_1\theta_2) \\ & \times I(\theta_1 > 0)I(\theta_2 > 0). \end{aligned} \tag{13.74}$$

13.9   Ratios of Gamma Scale Parameters

If we are comparing intensities of two independent Poisson processes, variances of independent normal samples, or comparing exponential distributions based on complete or censored samples, then we are interested in gamma random variables with known shape parameters and we wish to consider the ratio and/or difference of their precision parameters λ_1 and λ_2. Here, as in many other examples, it is not unreasonable to expect that prior beliefs about λ_1 and λ_2 are not necessarily independent.

Our data consist of two independent random variables (after reduction to sufficient statistics), X_1 and X_2 where $X_i \sim \Gamma(\alpha_i, \lambda_i)$, $i = 1, 2$. It is assumed that the α_i's are known. The likelihood of the data set (X_1, X_2) is thus

$$\begin{aligned} L(\lambda_1, \lambda_2) &= \lambda_1^{\alpha_1} x_1^{\alpha_1 - 1} e^{-\lambda_1 x_1} \lambda_2^{\alpha_2} x_2^{\alpha_2 - 1} e^{-\lambda_2 x_2} / \Gamma(\alpha_1)\Gamma(\alpha_2) \\ &\propto \exp\left(\alpha_1\ \log\ \lambda_1 - x_1\lambda_1 + \alpha_2\ \log\ \lambda_2 - x_2\lambda_2\right). \end{aligned} \tag{13.75}$$

We will call (13.75) the likelihood in terms of the original parametrization. We will also discuss a likelihood in terms of a transformed parametrization with parameters

$$\begin{aligned} \theta_1 &= \lambda_1/\lambda_2, \\ \theta_2 &= \lambda_2. \end{aligned} \tag{13.76}$$

Since our focus of interest is on the ratio λ_1/λ_2, the transformed parametrization (13.76) might actually be more natural. Using (13.76) our reparametrized likelihood becomes

$$L(\theta_1, \theta_2) \propto \exp\left[\alpha_1\ \log\ \theta_1 - x_1\theta_1\theta_2 + (\alpha_1 + \alpha_2)\log\ \theta_2 - x_2\theta_2\right]. \tag{13.77}$$

If our model is parametrized as in (13.75), we seek a flexible family of joint priors for (λ_1, λ_2) from which we will select a member to approximate the prior beliefs of the available informed expert(s). Note that if λ_2 were known then the conjugate family of priors for λ_1 (with likelihood (13.75)) would be a family of gamma distributions. Analogously, if λ_1 were known then the conjugate family of priors for λ_2 would again be the family of gamma distributions. It is then natural to consider joint prior distributions for (λ_1, λ_2) which have gamma conditionals. This conditionally conjugate prior family is an exponential family of the form

$$
\begin{aligned}
f(\lambda_1, \lambda_2) \propto\ & (\lambda_1 \lambda_2)^{-1} \exp\left(-m_{10}\lambda_1 - m_{01}\lambda_2 \right. \\
& + m_{20}\ \log \lambda_1 + m_{02}\ \log\ \lambda_2 \\
& + m_{11}\lambda_1\lambda_2 - m_{12}\lambda_1\ \log\ \lambda_2 \\
& \left. - m_{21}\lambda_2 \log\ \lambda_1 + m_{22}\ \log\ \lambda_1\ \log\ \lambda_2 \right),
\end{aligned} \tag{13.78}
$$

which has as its support the positive quadrant $\lambda_1 > 0, \lambda_2 > 0$. The class (13.78) is the most general class with all conditionals, of λ_1 given λ_2 and of λ_2 given λ_1, being gamma distributions. Specifically, we have

$$
\lambda_1 | \lambda_2 = \lambda_2^{(0)} \sim \Gamma(m_{20} - m_{21}\lambda_2^{(0)} + m_{22}\ \log\ \lambda_2^{(0)}, m_{10} - m_{11}\lambda_2^{(0)} + m_{12}\ \log\ \lambda_2^{(0)}), \tag{13.79}
$$

and

$$
\lambda_2 | \lambda_1 = \lambda_1^{(0)} \sim \Gamma(m_{02} - m_{12}\lambda_1^{(0)} + m_{22}\ \log\ \lambda_1^{(0)}, m_{01} - m_{11}\lambda_1^{(0)} + m_{21}\ \log\ \lambda_1^{(0)}). \tag{13.80}
$$

In order to guarantee integrability of the density (13.78), certain constraints must be placed on the hyperparameters (the m_{ij}'s) in (13.78) as discussed in Chapter 4. We are generally willing to accept improper priors but we typically insist on proper posteriors. To assure this we need to assure that the data are of sufficient richness to make the posterior values of m_{10}, m_{01}, m_{20}, and m_{02} large enough to satisfy the constraints listed in Chapter 4. Of course, if the prior itself is proper (i.e., if it initially has sufficiently large values for m_{10}, m_{01}, m_{20}, and m_{02}) the posterior will necessarily be proper for any realizations of $\underline{X}$.

Since one of the posterior simulation strategies to be used involves use of the Gibbs sampler, propriety of the posterior distribution is essential (Hobert and Casella (1996)).

It is clear that when a prior of the form (13.78) is combined with the likelihood (13.75), the resulting posterior distribution is a member of the same family of priors. This reconfirms our assertion that such conditionally specified priors are indeed conjugate priors. In fact only four of the hyperparameters in (13.78) are changed by the data, namely m_{10}, m_{01}, m_{20}, and m_{02}. The usual prior, involving independent gamma distributions for λ_1, λ_2, is included in the family (13.78). It corresponds to the case in which only the data-affected hyperparameters (m_{10}, m_{01}, m_{20}, and m_{02}) are given nonzero values. The standard noninformative prior (see, e.g., Berger (1985), p. 85) is also included in (13.78). It corresponds to the case in which _all_ m_{ij}'s are zero except m_{20} and m_{02} which are set equal to -1. Thus (13.78) is a sufficiently rich family to include the usual choices but posesses additional flexibility to match a broader spectrum of prior beliefs than the usual prior families.

A key feature of the family (13.78), as with all conditionally conjugate prior families, is that both its prior and posterior distributions are conditionally specified. That is (in both prior and posterior) the conditional distribution of λ_1, given any value of λ_2, is a gamma distribution and the conditional distribution of λ_2, given any value of λ_1, is again a gamma distribution.

Simulation of realizations from (13.78) is thus readily achievable using a Gibbs sampler technique, in a by now familiar manner.

If $(\lambda_1^{(1)}, \lambda_2^{(1)}), (\lambda_1^{(2)}, \lambda_2^{(2)}), \ldots, (\lambda_1^{(N)}, \lambda_2^{(N)})$ is a large simulated realization of variables with density (13.78) then, for any function $g : \mathbb{R}^{+2} \to \mathbb{R}$,

$$g(\lambda_1^{(1)}, \lambda_2^{(1)}), g(\lambda_1^{(2)}, \lambda_2^{(2)}), \ldots, g(\lambda_1^{(N)}, \lambda_2^{(N)})$$

provides a simulated sample from the density of the variable $g(\lambda_1, \lambda_2)$. In particular, interest might be focussed on $g(\lambda_1, \lambda_2) = \lambda_1/\lambda_2$. Instead we might be interested in $g(\lambda_1, \lambda_2) = \lambda_1 - \lambda_2$. We can analyze both parametric functions without additional difficulties.

This is in sharp contrast to the situation encountered in a classical non-Bayesian framework. In that setting, attention is focussed on λ_1/λ_2 since $\lambda_1 - \lambda_2$ is markedly more difficult for them to deal with.

Use of the Gibbs sampler to simulate realizations from (13.78) is not obligatory. A readily available alternative for computing posterior moments involves the use of importance sampling (see Appendix A.3). A reasonable approach would draw samples from independent gamma densities for λ_1 and λ_2 and weight them by the corresponding ratio of densities.

If attention is definitely to be focussed on λ_1/λ_2 it is worth considering the possiblity of transforming parameters using (13.76) before analysis (recall that (13.76) introduced new parameters $\theta_1 = \lambda_1/\lambda_2$ and $\theta_2 = \lambda_2$). A conditionally conjugate prior for (θ_1, θ_2) may be readily determined. For the likelihood (13.77), given θ_2, the gamma family is a conjugate family for θ_1 while, given θ_1, the gamma family is conjugate for θ_2. Thus the conditionally conjugate joint prior for (θ_1, θ_2), with the likelihood (13.77), is also a gamma conditionals distribution, i.e.,

$$\begin{aligned} f(\theta_1, \theta_2) \ \propto \ & (\theta_1 \theta_2)^{-1} \exp(-m_{10}\theta_1 - m_{01}\theta_2 \\ & + m_{20} \ \log \ \theta_1 + m_{02} \ \log \ \theta_2 \\ & + m_{11}\theta_1\theta_2 - m_{12}\theta_1 \ \log \ \theta_2 \\ & - m_{21}\theta_2 \ \log \ \theta_1 + m_{22} \ \log \ \theta_1 \ \log \ \theta_2). \quad (13.81) \end{aligned}$$

The corresponding gamma conditionals for θ_1 and θ_2 have parameters as indicated in (13.79) and (13.80) (with θ's replacing the λ's). The family (13.81) provides a flexible family of priors for (θ_1, θ_2) paralleling the flexible family of priors for (λ_1, λ_2) provided by (13.78). They are not equivalent. If (λ_1, λ_2) has a gamma conditionals distribution then (θ_1, θ_2) (defined by (13.76)) has a distribution that is readily evaluated via Jacobians and will not have a gamma conditionals distribution. Similarly if (θ_1, θ_2) has a gamma conditionals distribution, then (λ_1, λ_2) will not have such a distribution. The choice between the flexible families of priors (13.78) and (13.81) will probably depend on which parametrization is most easily visualized by the informed expert, i.e., the one for which hyperparameter elicitation will be most straightforward. Neither parametrization will possess computational advantages, so the choice will be either based on prior eliciation advantages or on some feeling that one parametrization is "more natural" than the other. Observe that if we use the gamma conditionals prior for (θ_1, θ_2) and combine it with the likelihood (13.77), only 4 hyperparameters will be affected by the data. In this case the data-affected hyperparameters are m_{01}, m_{20}, m_{02} and m_{11} (a slightly different list from that associated with a gamma conditionals prior for (λ_1, λ_2)). Independent marginal priors for θ_1 and θ_2 and vague priors for them can be accommodated by suitable choices of the m_{ij}'s.

It must be emphasized, once more, that in using either of the families (13.78) or (13.81) we are not implying that the informed expert's true prior is a member of either of the families. We are only agreeing to approximate his true prior distribution with a member of one or the other flexible families of priors.

To accomplish this, we elicit values for the conditional means and variances of the parameters λ_1 and λ_2. If the elicited values are

$$
\begin{aligned}
\xi_i^{(1)} &= E(\lambda_1 | \lambda_2 = \lambda_2^{(i)}), \quad i = 1, 2, \dots, \ell_1, \\
\xi_i^{(2)} &= \operatorname{var}(\lambda_1 | \lambda_2 = \lambda_2^{(i)}), \quad i = 1, 2, \dots, \ell_1, \\
\eta_i^{(1)} &= E(\lambda_2 | \lambda_1 = \lambda_1^{(i)}), \quad i = 1, 2, \dots, \ell_2,
\end{aligned}
$$

and

$$
\eta_i^{(2)} = \operatorname{var}(\lambda_2 | \lambda_1 = \lambda_1^{(i)}), \quad i = 1, 2, \dots, \ell_2,
$$

then, paralleling the technique used in Section 13.6.1 for normal data, we will set up the following system of linear equations in the m_{ij}'s:

$$
\begin{aligned}
\xi_i^{(1)} / \xi_i^{(2)} &= c_{10} - c_{11} \lambda_2^{(i)} + c_{12} \log \lambda_2^{(i)}, \quad i = 1, 2, \dots, \ell_1, \\
(\xi_i^{(1)})^2 / \xi_i^{(2)} &= c_{20} - c_{21} \lambda_2^{(i)} + c_{22} \log \lambda_2^{(i)}, \quad i = 1, 2, \dots, \ell_1, \\
\eta_i^{(1)} / \eta_i^{(2)} &= c_{01} - c_{11} \lambda_1^{(i)} + c_{21} \log \lambda_1^{(i)}, \quad i = 1, 2, \dots, \ell_2,
\end{aligned}
$$

and

$$(\eta_i^{(1)})^2/\eta_i^{(2)} = c_{02} - c_{12}\lambda_1^{(i)} + c_{22}\log\lambda_1^{(i)}, \quad i = 1, 2, \ldots, \ell_2.$$

Appropriate values of the m_{ij}'s will then be found using a standard least-squares or regression program. These are viewed as the elicited values of the hyperparameters and determine the conditionally conjugate prior to be used in subsequent analysis.

13.10   Comparison of Normal Means

It is not uncommon to be faced with a problem of comparing means from independent normal samples. Indeed, the problem is almost the canonical introductory problem in statistical methods textbooks. Under the name analysis of variance, we ask whether or not all the means are equal. We perform multiple comparisons, estimate contrasts, etc. And we routinely assume variance homogeneity, to avoid Behrens–Fisher-type "problems." Fiducial probabilists were less concerned about variance homogeneity but their viewpoint (despite the weight and influence of R. A. Fisher) never really was accepted by mainstream applied statisticians. Bayesian analysts were undaunted by variance heterogeneity. It just meant more parameters in the model, more complicated priors and posteriors, and a larger computer account in order to process, at least approximately, the data. The current analysis accepts this Bayesian thesis.

Suppose we have independent samples from k normal populations, i.e.,

$$X_{ij} \sim N(\mu_i, \tau_i), \quad i = 1, 2, \ldots, k, \quad j = 1, 2, \ldots, n_i, \tag{13.82}$$

(here $\tau_i = 1/\sigma_i^2$ denotes the precision of the ith distribution). In this setting we focus interest, as is often done, on the μ_i's, regarding the unknown τ_i's as nuisance parameters. The likelihood of our data set (13.82) will involve $2k$ parameters. If all parameters but, say, μ_j were known, then a natural conjugate prior for μ_j would be normal. If all parameters but, say, τ_ℓ were known, then a natural conjugate prior for τ_ℓ would be a gamma distribution. A fully flexible conditionally specified joint prior for $(\mu_1, \ldots, \mu_k, \tau_1, \ldots, \tau_k)$ would be one in which the conditional distributions of each μ_i, given all the remaining $2k-1$ parameters, is normal and the conditional distribution of each τ_j, given all the remaining $2k-1$ parameters, is gamma. The resulting family of joint priors is (cf. Section 8.6):

$$
f(\underline{\mu}, \underline{\tau}) = (\tau_1\tau_2\ldots\tau_k)^{-1}\exp\left\{\sum_{j_1=0}^{2}\sum_{j_2=0}^{2}\cdots\sum_{j_k=0}^{2}\sum_{j_1'=0}^{2}\sum_{j_2'=0}^{2}\cdots\right.
$$
$$
\left.\cdots\sum_{j_k'=0}^{2}\left[m_{\underline{j},\underline{j'}}\prod_{i=1}^{k}q_{ij_i}(\mu_i)\prod_{i'=1}^{k}q'_{i'j'_{i'}}(\tau_{i'})\right]\right\}, \tag{13.83}
$$

where
$$q_{i0}(\mu_i) = 1,$$
$$q_{i1}(\mu_i) = \mu_i,$$
$$q_{i2}(\mu_i) = \mu_i^2,$$
$$q'_{i'0}(\tau_{i'}) = 1,$$
$$q'_{i'1}(\tau_{i'}) = -\tau_{i'},$$
$$q'_{i'2}(\tau_{i'}) = \log \tau_{i'}.$$

There are thus $3^{2k} - 1$ hyperparameters (the $m_{j,j'}$'s) in this prior (the constant $m_{0,0}$ is determined by the other m's to ensure that the density integrates to 1). The traditional informative prior for this problem has most of these $3^{2k} - 1$ hyperparameters set equal to zero. The only hyperparameters given nonzero values are those $4k$ hyperparameters which are affected by the data. The traditional prior is thus conjugate but severely restricted in its ability to match prior beliefs. To elicit appropriate values for the array of $3^{2k} - 1$ hyperparameters, we propose to request the informed expert to provide values for prior conditional means and precisions of each μ_i, given a spectrum of specific values of $\underline{\mu}_{(i)}$ ($\underline{\mu}$ with μ_i deleted), and $\underline{\tau}$ and of each $\tau_{i'}$ given a spectrum of specific values of $\underline{\tau}_{(i')}$ and $\underline{\mu}$. These, in a manner parallel to that described in Section 13.6 for the case $k = 1$, yield a collection of linear relations that should hold among the hyperparameters. Typically no solution exists, since our expert is not infallible and will usually give inconsistent a priori values for conditional moments. We choose hyperparameters to be minimally discrepant from the given information in the sense of being a least-squares solution. As mentioned earlier, only $4k$ of these parameters will have different values in the posterior distribution from those values held in the prior distribution.

Assuming that appropriate prior hyperparameters can be obtained and that $4k$ of them can be updated using the data to obtain posterior hyperparameters, we would then use the Gibbs sampler to generate realizations $(\underline{\mu}^{(k)}, \underline{\tau}^{(k)})$, $k = 1, 2, \ldots, N$, from the posterior distribution, after discarding the initial iterations. We can then study the approximate posterior distribution of $\sum_{i=1}^{k}(\mu_i - \bar{\mu})^2$ in order to decide whether there is evidence for differences among the μ_i's, etc.

We can illustrate this kind of analysis using an example in which $k = 2$. Note that, since we are <u>not</u>, assuming $\tau_1 = \tau_2$ ($=$ variance homogeneity), we are dealing with a Behrens–Fisher problem, well known to be troublesome from a classical view point.

Example 13.3 (Basal metabolism data). Our data set is a much analyzed one described in Snedecor and Cochran (1967, p. 118), based on a 1940 Ph.D. Thesis of Charlotte Young, and reproduced in Table 13.9. The goal is to compare basal metabolism of college women under two different sleep regimes.

We wish to specify a conditionally conjugate joint prior for $(\mu_1, \mu_2, \tau_1, \tau_2)$, utilize the data in Table 13.9 to obtain the corresponding (still conditionally

TABLE 13.9. Basal Metabolism of 26 College Women (Calories per square meter per hour)

7 or more hours of sleep		6 or less hours of sleep	
1. 35.3	9. 33.3	1. 32.5	7. 34.6
2. 35.9	10. 33.6	2. 34.0	8. 33.5
3. 37.2	11. 37.9	3. 34.4	9. 33.6
4. 33.0	12. 35.6	4. 31.8	10. 31.5
5. 31.9	13. 29.0	5. 35.0	11. 33.8
6. 33.7	14. 33.7	6. 34.6	
7. 36.0	15. 35.7	$\Sigma X_{2j} = 369.3$	
8. 35.0	$\Sigma X_{1j} = 516.8$		
$n_1 = 15, \bar{X}_1 = 34.45$ cal./sq. m./hr.		$n_2 = 11, \bar{X}_2 = 33.57$ cal./sq. m./hr.	

conjugate) posterior for $(\mu_1, \mu_2, \tau_1, \tau_2)$, and then we wish to consider the approximate posterior distribution of the difference between means $\nu \overset{\Delta}{=} \mu_1 - \mu_2$. In addition, we will look at the approximate posterior distribution of $\xi \overset{\Delta}{=} \tau_1/\tau_2$ to verify whether we are indeed in a Behrens–Fisher setting, i.e., a setting in which $\xi \neq 1$. Our conditionally conjugate prior family of joint densities for $(\mu_1, \mu_2, \tau_1, \tau_2)$ is of the following form (cf. (13.83)):

$$f(\mu_1, \mu_2, \tau_1, \tau_2) \propto (\tau_1 \tau_2)^{-1} \exp[m_{1000}\mu_1 + m_{0100}\mu_2 - m_{0010}\tau_1$$

$$- m_{0001}\,\tau_2 + \ldots + m_{2222}\mu_1^2\mu_2^2 \log\tau_1 \log\tau_2], \tag{13.84}$$

involving $3^4 - 1 = 80$ hyperparameters. Only the eight hyperparameters

$$m_{0010},\quad m_{0001},\quad m_{0020},\quad m_{0002},\quad m_{1010},\quad m_{0101},\quad m_{2010},\text{ and } m_{0201}$$

will be changed from prior to posterior by the likelihood of the data set in Table 13.9. The classical Bayesian analysis of this data set would give nonzero values to some or all of these eight hyperparameters and set the remaining 72 equal to 0. We have the (awesome!) additional flexibility provided by the 80 hyperparameter family.

We will illustrate with an application to the metabolism data assuming diffuse prior information, i.e., all m's set equal to zero in (13.84). For comparison, reference can be made to Arnold, Castillo, and Sarabia (1997) where two alternatives are considered (using slightly different notation), namely:

(i) Independent conjugate priors for each parameter (the only nonzero m's in (13.84) are m_{1000}, m_{0100}, m_{0010}, m_{0001}, m_{2000}, m_{0200}, m_{0020}, and m_{0002}); and

(ii) A classical analysis that assumes that only the hyperparameters that will be affected by the data are nonzero (i.e., m_{0010}, m_{0001}, m_{0020}, m_{0002}, m_{1010}, m_{0101}, m_{2010}, and m_{0201}).

In the diffuse prior case, our prior is of the form

$$f(\underline{\mu}, \underline{\tau}) \propto (\tau_1 \tau_2)^{-1} \text{ if } \tau_1, \tau_2 > 0 \text{ and } -\infty < \mu_1, \mu_2 < \infty. \quad (13.85)$$

The posterior distribution becomes

$$
\begin{aligned}
f(\underline{\mu}, \underline{\tau}|\text{Data}) \propto (\tau_1 \tau_2)^{-1} \exp \Bigg(& \frac{n_1}{2} \log \tau_1 + \frac{n_2}{2} \log \tau_2 \\
& -\tau_1 \frac{1}{2} \sum_{j=1}^{n_1} x_{1j}^2 - \tau_2 \frac{1}{2} \sum_{j=1}^{n_2} x_{2j}^2 + \mu_1 \tau_1 \sum_{j=1}^{n_1} x_{1j} \\
& + \mu_2 \tau_2 \sum_{j=1}^{n_2} x_{2j} - \frac{n_1}{2} \mu_1^2 \tau_1 - \frac{n_2}{2} \mu_2^2 \tau_2 \Bigg),
\end{aligned}
\quad (13.86)
$$

and the posterior conditional distributions to be used in the Gibbs sampler are:

$$\mu_1|\tau_1 \sim N\left(\mu = \frac{1}{n_1} \sum_{j=1}^{n_1} x_{1j}; \sigma^2 = \frac{1}{n_1 \tau_1}\right),$$

$$\mu_2|\tau_2 \sim N\left(\mu = \frac{1}{n_2} \sum_{j=1}^{n_2} x_{2j}; \sigma^2 = \frac{1}{n_2 \tau_2}\right),$$

$$\tau_1|\mu_1 \sim \Gamma\left(\frac{n_1}{2}; \frac{1}{2} \sum_{j=1}^{n_1} x_{1j}^2 - \mu_1 \sum_{j=1}^{n_1} x_{1j} + \mu_1^2 \frac{n_1}{2}\right),$$

$$\tau_2|\mu_2 \sim \Gamma\left(\frac{n_2}{2}; \frac{1}{2} \sum_{j=1}^{n_2} x_{2j}^2 - \mu_2 \sum_{j=1}^{n_2} x_{2j} + \mu_2^2 \frac{n_2}{2}\right).$$

Using the data from Table 13.9, the nonzero posterior hyperparameters are

$$
\begin{aligned}
m_{0020} &= 15/2, \\
m_{0002} &= 11/2, \\
m_{0010} &= -8937.4, \\
m_{0001} &= -6206, \\
m_{1010} &= 516.8, \\
m_{0101} &= 369.3, \\
m_{2010} &= -15/2, \\
m_{0201} &= -11/2.
\end{aligned}
$$

Using these posterior hyperparameters, simulated approximate posterior distributions of $\nu = \mu_1 - \mu_2$ and of $\xi = \tau_1/\tau_2$ were obtained using the

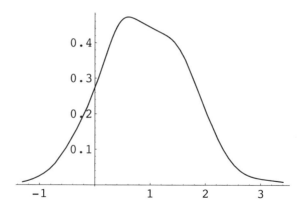

FIGURE 13.1. Diffuse priors: Simulated density of $\mu_1 - \mu_2$ using the Gibbs sampler with 500 replications and 300 starting runs.

Gibbs sampler with 800 iterations discarding the first 300. To display the results of this simulation we have used kernel density estimates to obtain smooth curves.

We have used the kernel estimation expression

$$\hat{f}(x) = \frac{1}{hn} \sum_{i=1}^{n} \frac{1}{\sqrt{2\pi}} \exp(-(x - x_i)^2/(2h^2)),$$

where

$$h = 1.06 \frac{\sigma}{n^{1/5}}$$

in the symmetric case and

$$h = 0.9 \frac{\sigma}{n^{1/5}}$$

in the nonsymmetric case, as suggested in Silverman (1986). The resulting approximate posterior densities for $\nu = \mu_1 - \mu_2$ and $\xi = \tau_1/\tau_2$ are shown in Figures 13.1 and 13.2.

The corresponding approximate posterior means and variances are

$$
\begin{aligned}
E(\nu) &= 0.910, \\
var(\nu) &= 0.591, \\
E(\xi) &= 0.372, \\
var(\xi) &= 0.096.
\end{aligned}
$$

It is clear from Figure 13.1 that, for this data set, $\mu_1 - \mu_2$ is slightly positive (more sleep associated with higher metabolism) although the treatment difference might well be considered not to be significant (a 95% interval for ν would include $\nu = 0$). It is also clear, since τ_1/τ_2 appears to be clearly less than 1, that indeed we were right in not assuming equal variances. We were indeed confronted by a Behrens–Fisher situation.   □

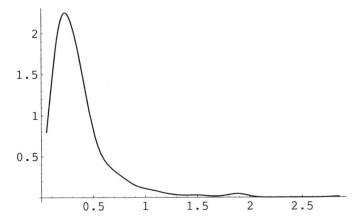

FIGURE 13.2. Diffuse priors: Simulated density of δ_1/δ_2 using the Gibbs sampler with 500 replications and 300 starting runs.

13.11   Regression

In principle there is no reason not to utilize the conditional specification approach in more complicated, yet still classical, modeling situations such as simple linear regression.

Suppose that n independent observations $X_1, X_2, \ldots, X_n$ are available and that their distribution is well described by a simple linear regression model, i.e., for each i

$$X_i \sim N(\alpha + \beta t_i, \sigma^2), \tag{13.87}$$

where the t_i's are known quantities (values of the "independent" variable) and the parameters α, β, and σ^2 are unknown. The natural parameter space is $\mathbb{R} \times \mathbb{R} \times \mathbb{R}^+$; corresponding to $\alpha \in \mathbb{R}$, $\beta \in \mathbb{R}$, and $\sigma^2 \in \mathbb{R}^+$. As usual we will reparametrize in terms of the precision $\tau \ (= 1/\sigma^2) > 0$. If α and β were known, a natural conjugate prior for τ would be a gamma distribution. If β and τ were known, a natural conjugate prior for α would be normal, and if α and τ were known, a normal prior for β would be natural. Thus we are led to consider the family of joint distributions for (α, β, τ) with normal–normal–gamma conditionals; in the sense that α given β and τ is normal, β given α and τ is normal, and τ given α and β has a gamma distribution. It is not difficult to write the general form of such densities as follows:

$$
\begin{aligned}
f(\alpha, \beta, \tau) \propto \exp\big\{ & m_{100}\alpha + m_{200}\alpha^2 + m_{010}\beta + m_{020}\beta^2 + m_{110}\alpha\beta \\
& + m_{120}\alpha\beta^2 + m_{210}\alpha^2\beta + m_{102}\alpha\log\tau + m_{012}\beta\log\tau \\
& + m_{220}\alpha^2\beta^2 + m_{202}\alpha^2\log\tau + m_{022}\beta^2\log\tau + m_{112}\alpha\beta\log\tau \\
& - m_{121}\alpha\beta^2\tau + m_{122}\alpha\beta^2\log\tau - m_{211}\alpha^2\beta\tau \\
& + m_{212}\alpha^2\beta\log\tau + m_{222}\alpha^2\beta^2\log\tau \big\} \\
& \times \exp\big\{ -m_{001}\tau + m_{002}\log\tau - m_{101}\alpha\tau - m_{011}\beta\tau - m_{201}\alpha^2\tau \\
& - m_{021}\beta^2\tau - m_{111}\alpha\beta\tau \big\},
\end{aligned}
$$

$$(13.88)$$

where the two factors in the right-hand side refer to the hyperparameters which are unaffected by the data and those which are affected, respectively. For this density we have:

1. The conditional distributions of α given β, τ is $N(\mu_1(\beta, \tau), \sigma_1^2(\beta, \tau))$, where

$$
\begin{aligned}
-2\mu_1(\beta, \tau)r(\beta, \tau) &= (m_{100} + m_{110}\beta + m_{120}\beta^2) \\
&\quad - (m_{101} + m_{111}\beta + m_{121}\beta^2)\tau \qquad (13.89) \\
&\quad + (m_{102} + m_{112}\beta + m_{122}\beta^2)\log\tau, \\
r(\beta, \tau) &= m_{200} + m_{210}\beta + m_{220}\beta^2 \\
&\quad - (m_{201} + m_{211}\beta + m_{221}\beta^2)\tau \qquad (13.90) \\
&\quad + (m_{202} + m_{212}\beta + m_{222}\beta^2)\log\tau, \\
\frac{-1}{2\sigma_1^2(\beta, \tau)} &= r(\beta, \tau). \qquad (13.91)
\end{aligned}
$$

2. The conditional distributions of β given α, τ is $N(\mu_2(\alpha, \tau), \sigma_2^2(\alpha, \tau))$, where

$$
\begin{aligned}
-2\mu_2(\alpha, \tau)s(\beta, \tau) &= (m_{010} + m_{110}\alpha + m_{210}\alpha^2 - m_{111}\alpha^2\tau) \\
&\quad + (m_{112}\alpha + m_{012} + m_{212}\alpha^2)\log\tau, \quad (13.92) \\
s(\beta, \tau) &= m_{020} + m_{120}\alpha + m_{220}\alpha^2 \\
&\quad - (m_{021} + m_{121}\alpha + m_{221}\alpha^2)\tau \qquad (13.93) \\
&\quad + (m_{122}\alpha + m_{022} + m_{222}\alpha^2)\log\tau, \\
\frac{-1}{2\sigma_2^2(\alpha, \tau)} &= s(\beta, \tau). \qquad (13.94)
\end{aligned}
$$

3. The conditional distributions of τ given α, β is $\Gamma(a(\alpha, \beta), \lambda(\alpha, \beta))$, where

$$
\begin{aligned}
a(\alpha, \beta) &= m_{002} + (m_{102} + m_{112}\beta + m_{122}\beta^2)\alpha + m_{012}\beta \\
&\quad + (m_{202} + m_{212}\beta + m_{222}\beta^2)\alpha^2 + m_{022}\beta^2 + 1, (13.95)
\end{aligned}
$$

TABLE 13.10. Adjustments in the parameters in the prior family (13.88), combined with likelihood (13.97).

Parameter	Prior value	Posterior value
m_{002}	m_{001}^*	$m_{001}^* + n/2$
m_{001}	m_{002}^*	$m_{002}^* + \dfrac{\sum_{i=1}^{n} x_i^2}{2}$
m_{011}	m_{011}^*	$m_{011}^* - \sum_{i=1}^{n} x_i t_i$
m_{021}	m_{021}^*	$m_{021}^* + \dfrac{\sum_{i=1}^{n} t_i^2}{2}$
m_{101}	m_{101}^*	$m_{101}^* - \sum_{i=1}^{n} x_i$
m_{111}	m_{111}^*	$m_{111}^* + \sum_{i=1}^{n} t_i$
m_{201}	m_{201}^*	$m_{201}^* + n/2$

$$
\begin{aligned}
\lambda(\alpha,\beta) \;=\; & -\left(m_{001} + (m_{101} + m_{111}\beta + m_{121}\beta^2)\alpha - m_{011}\beta \right.\\
& \left. - (m_{201} + m_{211}\beta + m_{221}\beta^2)\alpha^2 - m_{021}\beta^2\right). \qquad (13.96)
\end{aligned}
$$

The likelihood function can be written as

$$
\begin{aligned}
L(\alpha,\beta,\tau) \propto \exp\Big[& n\log\tau - \tau/2\sum x_i^2/2 - \alpha^2 n\tau/2 - \beta^2\tau\sum t_i^2/2 \\
& + \alpha\tau\sum x_i + \beta\tau\left(\sum x_i t_i - \alpha\sum t_i\right)\Big].
\end{aligned} \qquad (13.97)
$$

This combined with the family of priors (13.88), leads to a posterior of the same form, whose hyperparameters are given in Table 13.10.

Our model includes:

1. *The classical approach.* In this approach all hyperparameters in (13.88) that are not affected by the data are set (rather arbitrarily) equal to zero. The resulting seven-parameter conjugate prior family can be identified with the second factor in (13.88) and might be judged to be adequate in some circumstances.

2. *The independent approach.* A second approach would be to insist that the usual conjugate prior family and the independent marginals prior

family be included. This results in an eleven-parameter family. The hyperparameters included in this situation are: m_{100}, m_{200}, m_{010}, m_{020}, m_{001}, m_{002}, m_{101}, m_{011}, m_{201}, m_{021}, m_{111}.

3. *A simple flexible family.* One approach would assume that (α, β) given τ has a classical bivariate normal conditional density and that τ given (α, β) has a gamma distribution. This can be accomplished by setting the following list of nine m_{ijk}'s equal to zero in (13.88): m_{120}, m_{210}, m_{220}, m_{121}, m_{211}, m_{221}, m_{122}, m_{212}, m_{222}. The resulting 17 parameters conjugate prior family may be judged to be adequate in some circumstances.

13.11.1   Assessment of Appropriate Values for the Hyperparameters

The somewhat daunting formula (13.88) involves 26 hyperparameters! Only seven of these are affected by the data (they are m_{001}, m_{002}, m_{101}, m_{011}, m_{201}, m_{021}, and m_{111}). In a manner parallel to that described in Section 13.6.1, only simple least-squares programs are needed to determine values of the hyperparameters essentially concordant with the values of the conditional first and second moments supplied by the informed expert.

It is of course possible to restrict certain of the hyperparameters in (13.88) to be zero to yield a simpler but still conjugate prior family.

More general linear models involving even more hyperparameters will almost inevitably result in the informed expert's judgments being diffuse, to be reflected by setting all or many of the prior hyperparameters equal to zero.

13.12   The 2 × 2 Contingency Table

When comparing two drugs, a common scenario involves n_i subjects receiving treatment i of whom x_i experience relief from symptoms, where $i = 1, 2$. The basic data, realizations of two independent binomial (n_i, p_i) random variables, are often displayed in a 2 × 2 contingency table. Interest frequently is directed to the ratio of the corresponding odds ratios, i.e., the cross-product ratio

$$\psi(\underline{p}) = \frac{p_1(1 - p_2)}{p_2(1 - p_1)}. \tag{13.98}$$

A natural conjugate prior for p_1, assuming p_2 is known, is a beta prior. The same is, of course, true for p_2 assuming p_1 is known. The corresponding conditionally conjugate joint prior for (p_1, p_2) will have beta conditionals and will be of the form

$$
\begin{aligned}
f(p_1, p_2) \;=\; & [p_1(1-p_1)p_2(1-p_2)]^{-1} \\
& \times \exp\left[m_{11} \log p_1 \log p_2 + m_{12} \log p_1 \log(1-p_2) \right. \\
& + m_{21} \log(1-p_1) \log p_2 + m_{22} \log(1-p_1) \log(1-p_2) \\
& + m_{10} \log p_1 + m_{20} \log(1-p_1) \\
& + m_{01} \log p_2 + m_{02} \log(1-p_2) + m_{00} \big] \\
& \times I(0 < p_1 < 1) I(0 < p_2 < 1).
\end{aligned}
\tag{13.99}
$$

Since our likelihood is

$$
\ell(p_1, p_2) \propto p_1^{x_1}(1-p_1)^{n_1-x_1} p_2^{x_2}(1-p_2)^{n_2-x_2}
\tag{13.100}
$$

it is evident that the posterior density (combining (13.99) and (13.100)) will again be in the family (13.99) with only four of the hyperparameters (namely $m_{10}, m_{20}, m_{01}, m_{02}$) being affected by the data. Note that the natural conjugate joint prior would have independent marginals. It may be argued, however, that when we are comparing drugs in an experiment such as this, our prior beliefs about the efficacies of the drugs are unlikely to be independent. The conditionally conjugate prior allows us to accommodate dependent as well as independent prior beliefs. Using the prior (13.99), the resulting posterior will also have beta conditionals and, consequently, simulated realizations from the posterior distribution of the cross-product ratio (13.98) are readily obtained using the Gibbs sampler. For details, see Arnold and Thoni (1997).

13.13 Multinomial Data

Suppose that our data consists of the results of n independent trials each with $k+1$ possible outcomes $1, 2, \ldots, k+1$. For $i = 1, 2, \ldots, k$, let X_i denote the number of outcomes of type i observed in the n trials. Then $\underline{X}$ has a multinomial distribution with parameters n and $\underline{p} = (p_1, p_2, \ldots, p_k)$. Based on $\underline{X}$, we wish to make inferences about $\underline{p}$ (note that $\sum_{i=1}^{k} p_i < 1$ and for convenience we define $p_{k+1} = 1 - \sum_{i=1}^{k} p_i$). If $\underline{p}_{(1)}$ (i.e., $(p_2, \ldots, p_k)$) were known, the natural conjugate prior for p_1 would be a scaled beta density.

Considerations such as this will lead us to a joint prior for $\underline{p}$ which has scaled Beta conditionals, i.e. such that for $i = 1, 2, \ldots, k$,

$$
p_i | \underline{p}_{(i)} \sim \left(1 - \sum_{j=1; j \neq i}^{k} p_j \right) \times beta\left(a_i(p_i), b_i(p_{(i)}) \right), \quad i = 1, 2, \ldots, k,
$$

where $p_i > 0$; $i = 1, 2, \ldots, k$, and $p_1 + \ldots + p_k < 1$.

Such distributions were discussed by James (1975) and were described in two dimensions in Section 5.9.

A general form for such densities may be written as

$$f(p_1, \ldots, p_k) \propto \prod_{i=1}^{k} p_i^{\alpha_i - 1} \left[1 - (p_1 + \ldots + p_k) \right]^{\alpha_{k+1} - 1} e^{\phi(p_1, \ldots, p_k)}, \quad (13.101)$$

where

$$\phi(p_1, \ldots, p_k) = \sum_{i<j} a_{ij} \log p_i \log p_j + \sum_{i<j<k} a_{ijk} \log p_i \log p_j \log p_k \\ + \ldots + a_{12\ldots k} \log p_1 \ldots \log p_k. \quad (13.102)$$

If p has a scaled beta conditionals distribution, i.e., has (13.101) as its joint density, we write

$$\underline{p} \sim \text{SBC}(\underline{\alpha}, A) \quad (13.103)$$

(here $\underline{\alpha}$ is of dimension $k + 1$).

Note that if $A \equiv 0$ this reduces to the standard Dirichlet density, often used as a prior in multinomial settings. Since our likelihood is of the form

$$L(\underline{p}) \propto \prod_{i=1}^{k} p_i^{x_i} \left(1 - \sum_{i=1}^{k} p_i \right)^{n - \sum_{i=1}^{k} x_i} I\left(p_i > 0, \forall i, \sum_{i=1}^{k} p_i < 1 \right),$$

it follows immediately that the family (13.103) is a conjugate family and that the posterior distribution of p given $\underline{X} = \underline{x}$ will be in the same family. Specifically we will have, introducing the notation $x_{k+1} = n - \sum_{i=1}^{k} x_i$ and $\underline{\tilde{x}} = (\underline{x}, x_{k+1})$,

$$\underline{p} | \underline{X} = \underline{x} \sim \text{SBC}(\underline{\alpha} + \underline{\tilde{x}}, A). \quad (13.104)$$

Gibbs sampler simulations using the posterior density will be readily acomplished since simulation of univariate scaled beta variables is a straightforward exercise.

13.14   Change Point Problems

In a variety of situations, abrupt change can occur in stochastic mechanisms generating data. In such settings, we are often interested in determining whether a change has occurred in a series of observations and, if we decide that there has been a change, we would like to determine when it occurred.

Such change point problems are routinely encountered in quality control, economic analysis, etc. Indeed it is difficult to envision situations in which such problems will not occur and be of interest. Any effort to generate a sequence of i.i.d. observations would surely be a potential candidate for a change point analysis!

Putting on our Bayesian hats and picking up our conditionally specified tool box, we can approach such problems quite confidently. We will illustrate with a very simple example: More realistic examples will require more book-keeping, more prior elicitation, and more complicated algorithms but, in general, no new insights.

Suppose that we have a sequence of n independent observations for which

$$X_i \sim \text{Poisson}(\lambda t_i), \quad i = 1, 2, \ldots, k,$$

and

$$X_i \sim \text{Poisson}(\alpha \lambda t_i), \quad i = k+1, \ k+2, \ldots, n,$$

where $\lambda > 0, \alpha > 0$, and $k \in \{1, \ldots, n\}$ are unknown parameters. Note that if $n = k$, then by convention $\alpha = 1$; and, of course, in such a situation, no "change" occurred. The likelihood function is of the form

$$L(\alpha, \lambda, k) \propto \lambda^{\sum_{i=1}^{n} x_i} \alpha^{\sum_{i=k+1}^{n} x_i} e^{-\lambda \sum_{i=1}^{k} t_i - \alpha \lambda \sum_{i=k+1}^{n} t_i}. \tag{13.105}$$

If λ and k were known, a conjugate prior for α would be a gamma distribution. Similarly, if α and k were known, a conjugate prior for λ would be a gamma distribution. If α and λ were both known then a conjugate prior family of densities for k is of the form

$$f(k; \underline{c}) \propto \prod_{j=k+1}^{n} c_j, \quad k = 1, 2, \ldots, n, \tag{13.106}$$

where $\underline{c} = (c_2, \ldots, c_n)$ is a vector of nonnegative hyperparameters. Just to have a name for it, we will call (13.106) a change point distribution. We then will use as our general prior for (α, λ, k) one which has gamma and "change point" distributions as conditionals. The posterior will be in the same family and simulations from the posterior will only require ability to simulate gamma, and "change point" variables. Note that it is not uncommon, in face of the lack of prior information, to choose k to be a priori a uniform random variable. This can be accommodated in our model since the "change point" distribution reduces to the uniform distribution when $\underline{c} \equiv \underline{1}$.

13.15   Bivariate Normal

A final example of a potentially useful conditionally conjugate prior will take us right back to Bhattacharyya's normal conditionals density. Suppose $(X_1, Y_1), \ldots, (X_n, Y_n)$ are i.i.d. bivariate normal random variables with mean vector (μ_X, μ_Y) and known covariance Σ. An appropriate conditionally conjugate prior for (μ_X, μ_Y) would of course be the normal conditionals density (4.31). It will give us more flexibility for matching prior beliefs about (μ_X, μ_Y) than does the usual classical bivariate normal prior.

13.16   No Free Lunch

The enormous number of parameters present in high-dimensional conditionally specified priors is the source of their flexibility but, in practice, will pose insurmountable elicitation problems unless some simplifying structure is imposed. Some acceptable hierarchy of nested submodels must be developed to facilitate elicitation of appropriate prior values of parameters to match the informed experts' beliefs. Almost inevitably, many of the available parameters will be set to zero in applications (without going all the way back to the natural conjugate priors). Even without such a hierarchical structuring, the conditionally specified approach retains its utility since it does include the "usual" vague and conjugate priors and the "independent marginals" priors as special cases and it provides simple algorithms for dealing with them.

13.17   Bibliographic Notes

Section 13.3 is based on Arnold, Castillo, and Sarabia (1996a). Section 13.4 covers material from Arnold, Castillo, and Sarabia (1993c).

Conditionally specified priors are discussed in a series of papers by Arnold, Castillo, and Sarabia (1997, 1998a, 1998b).

The material on multinomial data and change point problems has not appeared elsewhere.

Exercises

13.1 Suppose a series $\underline{z} = \{z_1, z_2, \ldots, z_n\}$ is generated by a stationary autoregressive model AR(1),

$$z_t = \phi z_{t-1} + a_t, \quad t = 2, 3, \ldots,$$

where $|\phi| < 1$ and the error terms a_t are i.i.d. $N(0, \sigma^2)$ observations. Our aim is to use Bayesian techniques for inference regarding the parameters (ϕ, τ), where $\tau = \sigma^{-2}$ is the precision.

(a) Prove that the likelihood function of the process is given by

$$f(\underline{z}|\phi, \tau) = (2\pi)^{-n/2} \tau^{n/2} (1 - \phi^2)^{1/2}$$
$$\times \exp\left\{-\frac{\tau}{2}\left[(1 - \phi^2)z_1^2 + \sum_{t=2}^{n}(z_t - \phi z_{t-1})^2\right]\right\}.$$

(b) If ϕ is a known parameter, prove that the gamma distribution is conjugate for τ. Prove that if τ is known then

$$f(\phi; \alpha, \beta, \gamma) = k(\alpha, \beta, \gamma)(1 - \phi^2)^\alpha \exp(-\beta\phi - \gamma\phi^2)I(-1 < \phi < 1)$$

is a conjugate density for ϕ.

(c) Find the most general density for (ϕ, τ) such that $\tau|\phi$ is gamma, and $\phi|\tau$ is $f(\phi; \alpha, \beta, \gamma)$. Prove that the obtained distribution is conjugate for the AR(1) process.

(d) Show how the density $f(\phi; \alpha, \beta, \gamma)$, given in (d), can be simulated in order to implement the Gibbs sampling method.

13.2 Assume a random sample $X_1, \ldots, X_n$ from a location and shift exponential distribution with pdf,

$$f(x; \lambda, \mu) = \lambda e^{-\lambda(x-\mu)}, \quad \text{if} \quad x > \mu.$$

We are interested in Bayesian inference for the parameters (λ, μ) using conjugate conditionally specified priors.

(a) If μ is a known parameter, show that the gamma distribution is a conjugate distribution for λ. If now λ is known, show that the truncated exponential distribution with pdf,

$$f(\mu; a, b, c) = k(a, b, c)e^{c\mu}, \quad \text{if} \quad a < \mu < b,$$

where a, b, c are parameters and $k(a, b, c)$ is the normalizing constant, is a conjugate prior distribution for μ.

(b) Obtain the most general bivariate distribution with gamma and truncated exponential conditionals. Calculate the marginals, and the conditional means and variances.

(c) Propose a method for elicitation of hyperparameters in the gamma-truncated exponential conditionals distribution.

13.3 In the problem of Bayesian inference of ratios of gamma scale parameters, study the elicitation of hyperparameters, using conditional moments.
(Arnold, Castillo, and Sarabia (1998a).)

13.4 Consider a normal distribution where the standard deviation is proportional to the mean,

$$X \sim N(\mu, \sigma^2 = \mu^2/\lambda), \quad \lambda > 0.$$

This multiplicative model appears in processes where the measurement error increases with the mean value. The likelihood function is

$$f(x; \mu, \lambda) = \frac{\lambda^{n/2}}{|\mu|^n (2\pi)^{n/2}} \exp\left[-\frac{\lambda}{2\mu^2} \sum_{i=1}^{n} (x_i - \mu)^2\right].$$

(a) If μ is known, prove that the gamma distribution is a conjugate distribution for λ.

(b) If λ is known, prove that the generalized inverse normal distribution (Robert (1991)) with pdf,

$$f(x; \alpha, \eta, \tau) = \frac{k(\alpha, \eta, \tau)}{|x|^{\alpha}} \exp\left[-\frac{1}{2\tau^2}\left(\frac{1}{x} - \eta\right)^2\right],$$

where $\alpha > 1$, $\tau > 0$, and $k(\alpha, \eta, \tau)$ norming constant, is a conjugate prior distribution for μ.

(c) Obtain the most general distribution with gamma and generalized inverse normal conditionals. Prove that this distribution is conjugate for the likelihood $f(x; \mu, \lambda)$. Discuss the problem of assessment of hyperparameters.

13.5 Consider the conjugate prior distribution (13.98) for a 2×2 contingency table. Suggest a method for the assessment of the hyperparameters.

13.6 Let (X, Y) a bivariate normal distribution with $E(X) = \mu_1$, $E(Y) = \mu_2$ known, $V(X) = V(Y) = \sigma^2$, and $\rho(X, Y) = \rho$. Obtain a bivariate conjugate prior distribution for the parameters (τ, ρ), where $\tau = 1/\sigma^2$, based on conditional specification.

13.7 Suppose that data are available from a $\Gamma(\alpha, \lambda)$ distribution. Identify the family of conditionally conjugate prior densities for this problem. Which hyperparameters are affected by the data?

13.8 Consider the improper joint density

$$f(x, y) = e^{-xy} I(x > 0, y > 0).$$

Since this density has proper conditional densities it may be used to generate a sequence $X, Y, X, Y, \ldots$ using a Gibbs sampler algorithm. Try this to see what happens. Discuss the behavior of the related Markov chain(s). (Cf. Hobert and Casella (1996).)

13.9 Consider a "random effects" one-way classification. Here our data are of the form

$$Y_{ij} = \mu + A_i + \epsilon_{ij}, \quad i = 1, 2, \ldots, k, \quad j = 1, 2, \ldots, n_i,$$

where the A_i's are i.i.d. $N(0, \sigma_\tau^2)$ variables and the ϵ_{ij}'s are i.i.d. $N(0, \sigma^2)$ variables independent of the A_i's. Assume, for simplicity, that $\mu = 0$. Discuss an appropriate conditionally conjugate analysis of this problem.

13.10 Suppose that $\underline{X}_1, \underline{X}_2, \ldots, \underline{X}_n$ are i.i.d. $N^{(k)}(\underline{\theta}, \Sigma_0)$ random vectors where, for simplicity, Σ_0 is known. What kinds of prior densities for $\underline{\theta}$ will always yield normal posterior densities for $\underline{\theta}$?

(Bischoff (1993).)

14
Conditional Specification Versus Simultaneous Equation Models

14.1 Introduction

Conditional specification (CS) has been the focus of discussion throughout this book. In two dimensions, we model the joint distribution of (X_1, X_2) by discussing the distributions of X_1 associated with different values of X_2 and the distribution of X_2 associated with different values of X_1.

In the economics literature, it is more common to address such modeling issues using what is known as the simultaneous equation (SE) formulation. An enormous corpus of literature (beginning with Haavelmo (1943)) is available on this topic. In the SE formulation, X_1 is viewed as a function of X_2 with some associated error and X_2 is viewed as a function of X_1 with some associated error. From the beginning, Haavelmo recognized that the SE formulation was not amenable to direct interpretation in conditional terms. Much of the literature in fact involves models derived from the SE formulation that do have some parameters that are interpretable in conditional terms. Conditional specification models may be considered to be viable alternatives to SE models. Their advantage is that they admit ready interpretation of the parameters in the model. The issues involved in any CS versus SE comparison are well illustrated in the linear normal case, which undoubtedly is the most commonly studied case in the economics literature. Our discussion will, for simplicity, be concerned with the bivariate case. In applications, higher-dimensional examples are most commonly encountered.

14.2   Two Superficially Similar Models

Suppose that we wish to describe the joint distribution of a random vector (X_1, X_2).

Definition 14.1 (CS linear normal model in two dimensions). A conditionally specified model for (X_1, X_2) with normal linear dependence is a model such that for some constants μ_1, μ_2, β_{12}, β_{21}, σ_1^2, and σ_2^2, the conditional distributions of (X_1, X_2) are given by

$$X_1 | X_2 = x_2 \sim N(\mu_1 + \beta_{12}x_2, \sigma_1^2), \tag{14.1}$$

$$X_2 | X_1 = x_1 \sim N(\mu_2 + \beta_{21}x_1, \sigma_2^2). \tag{14.2}$$

$\square$

A closely related linear SE model can be defined as follows:

Definition 14.2 (SE linear normal model in two dimensions). A linear SE model in two dimensions with normal errors is a model such that, for certain constants μ_1, μ_2, β_{12}, β_{21}, σ_1^2, and σ_2^2 the random vector (X_1, X_2) is related to a normal random vector $(\varepsilon_1, \varepsilon_2)$ by the relations

$$X_1 = \mu_1 + \beta_{12}X_2 + \varepsilon_1, \tag{14.3}$$

$$X_2 = \mu_2 + \beta_{21}X_1 + \varepsilon_2, \tag{14.4}$$

in which $\varepsilon_1 \sim N(0, \sigma_1^2)$ and $\varepsilon_2 \sim N(0, \sigma_2^2)$. $\square$

Note that in these models exogenous variables have not been considered.

We have deliberately chosen our notation to make the models as superficially similar as possible.

From (14.1) we know that if we are given $X_2 = x_2$, then conditionally X_1 is normal with mean $\mu_1 + \beta_{12}x_2$ and variance σ_1^2. From (14.3) it appears that if we know X_2 to be x_2, say, then X_1 would be normal with mean $\mu_1 + \beta_{12}x_2$ and variance σ_1^2. Analogous, precise, or vague statements about the distribution of X_2 given (or "knowing") $X_2 = x_2$ can be made. But there is clearly something wrong with our interpretation of the CS model here. We have not even prescribed the joint distribution of the ε_i's in (14.3)–(14.4). We have only specified its marginals. Undoubtedly, the joint distribution of (X_1, X_2) will depend on the full joint distribution of $(\varepsilon_1, \varepsilon_2)$, not just its marginals.

And, of course, we need to know the joint distribution of (X_1, X_2) in order to determine the nature of the corresponding conditional distributions (of X_1 given $X_2 = x_2$ and of X_2 given $X_1 = x_1$). For many choices of the joint distribution of $(\varepsilon_1, \varepsilon_2)$, even though the marginals of ε_1 and ε_2 are normal, the derived conditional distributions of X_1 given X_2 and X_2 given X_1 will not be normal. So the CS and SE models are generally different breeds of cats.

But potential confusion definitely does exist if we make the assumption that is usually made about the joint distribution of $(\varepsilon_1, \varepsilon_2)$. Typically, it is assumed that $(\varepsilon_1, \varepsilon_2)$ has a classical bivariate normal distribution. Such an assumption will indeed imply that (X_1, X_2) itself has a classical bivariate normal distribution.

But, even if $(\varepsilon_1, \varepsilon_2)$ are chosen to be independent (a not uncommon choice), the derived conditional distributions do not coincide with those in the conditional specification model (14.1)–(14.2) (except in the trivial case in which $\beta_{12} = \beta_{21} = 0$). Thus, although models (14.1)–(14.2) and (14.3)–(14.4) are superficially similar, they are very different and must consequently be visualized differently when introspecting about their appropriateness to model any real-world configuration of variables.

Thus it is not appropriate to try to justify the SE model (14.3)–(14.4) using simple arguments about conditional distributions of each variable given the other. Such justification is appropriate for the CS model (14.1)–(14.2). Any argument for the SE model must be based on some statements about an existing functional relationship between observable random variables (X_1, X_2) and unobservable random variables $(\varepsilon_1, \varepsilon_2)$ with classical bivariate normal structure. Haavelmo (1943) knew this and so, of course, the mistaken conditional interpretation is avoided in the literature. The price that inevitably must be paid is that it becomes exceedingly difficult to interpret the parameters $\mu_1, \mu_2, \beta_{12}, \beta_{21}, \sigma_1^2$, and σ_2^2 that appear in the SE formulation. It is much easier to just come up front and assume from the beginning that (X_1, X_2) has a bivariate normal distribution. If we insist on choosing our model based on our perception of likely forms for cross sections of the joint density of (X_1, X_2) (i.e., conditional densities), then only the conditional specification route seems justified.

And, of course, with an assumption of normal conditionals, without insisting on linear regressions and constant conditional variances, we would be led to Bhattacharyya's normal conditionals distribution (3.26) or its multivariate extension (8.16) as suitable models for our data.

14.3   General CS and SE Models

Definition 14.3 (General CS model). A general CS model for a bivariate random variable (X_1, X_2) is a model of the form

$$X_1 | X_2 = x_2 \sim F_{x_2}(x_1; \underline{\theta}) \tag{14.5}$$

and

$$X_2 | X_1 = x_1 \sim F_{x_1}(x_2; \underline{\theta}), \tag{14.6}$$

where $\{F_{x_1}\}$ and $\{F_{x_2}\}$ are known indexed families of distributions depending on some or all of the parameters $\underline{\theta}$. $\square$

As we have seen in earlier chapters, the indexed families of distributions referred to in (14.5)–(14.6) must be carefully selected to guarantee the existence of a well-defined model.

Definition 14.4 (General SE model).   A general SE model for a bivariate random variable (X_1, X_2) is a model that assumes the existence of functions g_1 and g_2 such that

$$g_1(X_1, X_2, \underline{\theta}) = \varepsilon_1 \tag{14.7}$$

and

$$g_2(X_1, X_2, \underline{\theta}) = \varepsilon_2, \tag{14.8}$$

where $(\varepsilon_1, \varepsilon_2)$ have a known joint distribution. An invertibility assumption is desirable in (14.7)–(14.8). We usually assume that it is possible from (14.7)–(14.8) to solve for (X_1, X_2) as functions of $\varepsilon_1, \varepsilon_2$, and $\underline{\theta}$. Except for this requirement g_1 and g_2 can be relatively arbitrary. In practice, g_1 and g_2 are chosen to be of a relatively simple form (often linear), and it is not unusual to assume that ε_1 and ε_2 are independent.   □

Several examples of such CS and SE specifications will be sketched in the following sections.

14.4   Linear Normal Models

We will now carefully analyze the two models (14.1)–(14.2) (the CS model) and (14.3)–(14.4) (the SE model). They are reasonably both called linear normal models since both are built using linear functional relationships and normal distributions. As has been remarked, the models are superficially very similar. It turns out that both formulations lead to classical bivariate normal distributions for (X_1, X_2) with parameters that involve $\mu_1, \mu_2, \beta_{12}, \beta_{21}, \sigma_1^2,$ and σ_2^2.

The potential for confusion in the two models lies in the fact that the same parameters $\mu_1, \mu_2, \beta_{12}, \beta_{21}, \sigma_1^2,$ and σ_2^2 appear in both models. But the roles played by the symbols in the two distributions are different. For example, σ_1^2 in CS is a conditional variance (the conditional variance of X_1 given $X_2 = x_2$). In SE, σ_1^2 is the variance of ε_1, the conditional variance of X_1 given $X_2 = x_2$ in SE is not σ_1^2.

First consider the CS model (14.1)–(14.2). According to Theorem 1.2, for compatibility we need to check whether we can factor the ratio of conditional densities in the form

$$\frac{f_{X_1|X_2}(x_1|x_2)}{f_{X_2|X_1}(x_2|x_1)} = u(x_1)v(x_2),$$

where $u(x_1)$ is integrable. For this to be true we must have

$$\frac{\beta_{12}}{\sigma_1^2} = \frac{\beta_{21}}{\sigma_2^2} \text{ (to be able to factor)} \tag{14.9}$$

and

$$|\beta_{12}\beta_{21}| < 1 \text{ (for integrability) .} \tag{14.10}$$

The function $u(x_1)$, suitably normalized yields the marginal $f_{X_1}(x_1)$ density which in this case is normal. Thus the joint distribution of (X_1, X_2) determined by (14.1)–(14.2) will be the classical bivariate normal. Specifically,

$$(X_1, X_2) \sim N\left((\nu_1, \nu_2), \begin{pmatrix} \tau_1^2 & \delta\tau_1\tau_2 \\ \delta\tau_1\tau_2 & \tau_2^2 \end{pmatrix}\right), \tag{14.11}$$

where

$$\nu_1 = \frac{\mu_1 + \beta_{12}\mu_2}{1 - \beta_{12}\beta_{21}}, \tag{14.12}$$

$$\nu_2 = \frac{\mu_2 + \beta_{21}\mu_1}{1 - \beta_{12}\beta_{21}}, \tag{14.13}$$

$$\tau_1^2 = \sigma_1^2/(1 - \beta_{12}\beta_{21}), \tag{14.14}$$

$$\tau_2^2 = \sigma_2^2/(1 - \beta_{12}\beta_{21}), \tag{14.15}$$

and

$$\delta = \text{sgn}(\beta_{12})\sqrt{\beta_{12}\beta_{21}}. \tag{14.16}$$

Perhaps the easiest way to get (14.11)–(14.16) is to use the well-known formula for conditional means and variances and covariances of a bivariate normal (14.11), and equate them to the conditional means and variances given in (14.1)–(14.2).

The reader will recall, from Chapter 3, Bhattacharyya's assertion that normal conditionals and linear regressions will inevitably lead to such a classical bivariate normal model.

However, it must be recalled that we must insist that (14.9) and (14.10) hold in our specification of the joint density of (X_1, X_2) (i.e., in (14.12)–(14.16)). There are consequently really only five parameters, say, $\mu_1, \mu_2, \sigma_1^2, \sigma_2^2$, and β_{12} (since β_{21} is a function of the others). Consequently, model (14.1)–(14.2) would be better written as

$$X_1|X_2 = x_2 \sim N(\mu_1 + \beta_{12}x_2, \sigma_1^2), \tag{14.17}$$

$$X_2|X_1 = x_1 \sim N(\mu_2 + \beta_{12}\sigma_2^2 x_1/\sigma_1^2, \sigma_2^2), \tag{14.18}$$

where

$$|\beta_{12}| < \frac{\sigma_1}{\sigma_2}. \tag{14.19}$$

It is of course well known that the parameter space of the classical bivariate normal distribution is of dimension 5. Thus our conditional specification model coincides with the usual bivariate normal model. The interrelationship is made clear by the fact that we can solve for $\mu_1, \mu_2, \sigma_1^2, \sigma_2^2$, and β_{12} in terms of the means, variances, and covariances of (X_1, X_2) as follows:

$$\mu_1 = \frac{E(X_2)\mathrm{Var}(X_1) - E(X_2)\mathrm{Cov}(X_1, X_2)}{\mathrm{Var}(X_2)}, \tag{14.20}$$

$$\mu_2 = \frac{E(X_1)\mathrm{Var}(X_2) - E(X_1)\mathrm{Cov}(X_1, X_2)}{\mathrm{Var}(X_1)}, \tag{14.21}$$

$$\sigma_1^2 = \mathrm{Var}(X_1) - \frac{\mathrm{Cov}(X_1, X_2)^2}{\mathrm{Var}(X_2)}, \tag{14.22}$$

$$\sigma_2^2 = \mathrm{Var}(X_2) - \frac{\mathrm{Cov}(X_1, X_2)^2}{\mathrm{Var}(X_1)}, \tag{14.23}$$

$$\beta_{12} = \frac{\mathrm{Cov}(X_1, X_2)}{\mathrm{Var}(X_1)}. \tag{14.24}$$

Using these relationships, method of moments, or equivalently maximum likelihood estimates of the parameters are thus readily available.

In summary, assumptions (14.1) and (14.2) lead to a well-defined classical bivariate normal model (14.11)–(14.16) provided that: (i) $\beta_{12} = \beta_{21}\sigma_1^2/\sigma_2^2$ and (ii) $|\beta_{12}\beta_{21}| < 1$; or provided we rewrite the model in the form (14.17)–(14.18) with constraint (14.19).

Now let us consider the SE model (14.3)–(14.4).

As remarked in Section 14.2, we must completely specify the joint distribution of $(\varepsilon_1, \varepsilon_2)$ in order to have a well-defined model; specification of only the marginal distributions of ε_1 and ε_2 is not adequate. We will assume that $(\varepsilon_1, \varepsilon_2)$ has a classical bivariate normal distribution. Thus $(\varepsilon_1, \varepsilon_2) \sim N(\underline{0}, \Sigma)$, where

$$\Sigma = \begin{pmatrix} \sigma_1^2 & \rho\sigma_1\sigma_2 \\ \rho\sigma_1\sigma_2 & \sigma_2^2 \end{pmatrix}, \tag{14.25}$$

and we assume that

$$X_1 = \mu_1 + \beta_{12}X_2 + \varepsilon_1 \tag{14.26}$$

and

$$X_2 = \mu_2 + \beta_{21}X_1 + \varepsilon_2. \tag{14.27}$$

In matrix notation, we have

$$A\begin{pmatrix} X_1 \\ X_2 \end{pmatrix} = \begin{pmatrix} \mu_1 \\ \mu_2 \end{pmatrix} + \begin{pmatrix} \varepsilon_1 \\ \varepsilon_2 \end{pmatrix}, \tag{14.28}$$

where

$$A = \begin{pmatrix} 1 & -\beta_{12} \\ -\beta_{21} & 1 \end{pmatrix}. \tag{14.29}$$

To be able to solve (14.28) to get $\underline{X}$ as a function of $\underline{\varepsilon}$, we must assume that A is nonsingular, i.e., that

$$\beta_{12}\beta_{21} \neq 1. \tag{14.30}$$

No other assumptions are necessary. It then follows that

$$\begin{pmatrix} X_1 \\ X_2 \end{pmatrix} = A^{-1}\begin{pmatrix} \mu_1 \\ \mu_2 \end{pmatrix} + A^{-1}\begin{pmatrix} \varepsilon_1 \\ \varepsilon_2 \end{pmatrix} \tag{14.31}$$

and, consequently, that $\underline{X}$ is classical bivariate normal, i.e., that

$$\underline{X} \sim N\left(A^{-1}\begin{pmatrix} \mu_1 \\ \mu_2 \end{pmatrix}, \ A^{-1}\Sigma(A^{-1})'\right). \tag{14.32}$$

If we denote the means, variances, and correlation of $\underline{X}$ by $\tilde{\nu}_1, \tilde{\nu}_2, \tilde{\tau}_1^2, \tilde{\tau}_2^2$, and $\tilde{\delta}$, respectively, to parallel the notation used in (14.11) for the CS model, we find from (14.32) that

$$\tilde{\nu}_1 = \frac{\mu_1 + \beta_{12}\mu_2}{1 - \beta_{12}\beta_{21}}, \tag{14.33}$$

$$\tilde{\nu}_2 = \frac{\mu_2 + \beta_{21}\mu_1}{1 - \beta_{12}\beta_{21}}, \tag{14.34}$$

$$\tilde{\tau}_1^2 = \frac{2\rho\sigma_1\sigma_2\beta_{12} + \sigma_1^2 + \beta_{12}^2\sigma_2^2}{(1 - \beta_{12}\beta_{21})^2}, \tag{14.35}$$

$$\tilde{\tau}_2^2 = \frac{2\rho\sigma_1\sigma_2\beta_{21} + \sigma_2^2 + \beta_{21}^2\sigma_1^2}{(1 - \beta_{12}\beta_{21})^2}, \tag{14.36}$$

and

$$\tilde{\delta} = \frac{(1 + \beta_{12}\beta_{21})\rho\sigma_1\sigma_2 + \beta_{21}\sigma_1^2 + \beta_{12}\sigma_2^2}{\sqrt{[2\rho\sigma_1\sigma_2\beta_{12} + \sigma_1^2 + \beta_{12}^2\sigma_2^2][2\rho\sigma_1\sigma_2\beta_{21} + \sigma_2^2 + \beta_{21}^2\sigma_1^2]}}. \tag{14.37}$$

Again we remark that, distinct from the CS case, β_{12} and β_{21} can differ in sign and are only constrained by the requirement that their product should not equal 1.

If we compare (14.11)–(14.16) with (14.33)–(14.37), it is clear that the models differ. The means coincide but the variances and covariances are different and the constraints on the β's are different. Indeed, as is well known, in the SE model the parameters $\mu_1, \mu_2, \beta_{12}, \beta_{21}, \sigma_1^2, \sigma_2^2$, and ρ are not identifiable. The natural bivariate normal parameter space is of dimension 5 not 7. If we give arbitrary fixed values to β_{12} and β_{21} in the SE model, it is possible to set up a 1 to 1 correspondence between the parameters in (14.11)–(14.16) and the remaining five parameters in (14.33)–(14.34).

Some simplification is encountered if we assume that ε_1 and ε_2 are independent, i.e., that $\rho = 0$ in (14.25). The simplified means, variances, and correlation are

$$\overset{\approx}{\nu}_1 = \frac{\mu_1 + \beta_{12}\mu_2}{(1 - \beta_{12}\beta_{21})}, \tag{14.38}$$

$$\overset{\approx}{\nu}_2 = \frac{\mu_2 + \beta_{21}\mu_1}{(1 - \beta_{12}\beta_{21})}, \tag{14.39}$$

$$\overset{\approx}{\tau}_1^2 = \frac{\sigma_1^2 + \beta_{12}^2\sigma_2^2}{(1 - \beta_{12}\beta_{21})^2}, \tag{14.40}$$

$$\overset{\approx}{\tau}_2^2 = \frac{\sigma_2^2 + \beta_{21}^2\sigma_1^2}{(1 - \beta_{12}\beta_{21})^2}, \tag{14.41}$$

and

$$\overset{\approx}{\delta} = \frac{\beta_{21}\sigma_1^2 + \beta_{12}\sigma_2^2}{\sqrt{(\sigma_1^2 + \beta_{12}^2\sigma_2^2)(\sigma_2^2 + \beta_{21}^2\sigma_1^2)}}. \tag{14.42}$$

Indeed since the moments given in (14.38)–(14.42) span the entire five-dimensional parameter space of the classical bivariate normal distribution there is no mathematical reason to introduce the parameter ρ in the model, it does not enrich the model. There, of course, might be a valid theoretical reason for having a structural model involving dependent ε's. Even the simplified SE model (with $\rho = 0$) is however not identifiable since it involves six, not five, parameters.

In summary, though they seem to involve the same parameters the CS and SE linear normal models are clearly distinct. The key point is that $\mu_1, \mu_2, \beta_{12}, \beta_{21}, \sigma_1^2$, and σ_2^2 play different roles in the two models. In CS, $\mu_1 + \beta_{12}x_2$ is the conditional mean of X_1 given $X_2 = x_2$. In SE it is not. In CS, σ_1^2 is the conditional variance of X_1 given $X_2 = x_2$. In SE, it is not. In addition the β's have different constraints in the two set-ups.

14.5   Nonlinear Normal Models

In this section we discuss what happens if we replace the linear regression and constant conditional variance assumptions of (14.1)–(14.2) and the parallel linear assumptions in (14.3)–(14.4) by nonlinear conditions. In this arena the CS and SE specification lead us in completely different directions, as we shall see.

In a general CS configuration, we would seek models exhibiting conditional nonlinear normal structure.

Definition 14.5 (Nonlinear normal CS models). A CS nonlinear normal model is a model such that there exist possibly quite general functions $\mu_1(x_2), \sigma_1(x_2), \mu_2(x_1)$, and $\sigma_2(x_1)$ such that for each x_2,

$$X_1|X_2 = x_2 \sim N(\mu_1(x_2), \sigma_1^2(x_2)) \tag{14.43}$$

and for each x_1

$$X_2|X_1 = x_1 \sim N(\mu_2(x_1), \sigma_2^2(x_1)). \qquad (14.44)$$

□

But from our discussion in Chapter 3 we know how to characterize distributions satisfying (14.43)–(14.44). The classical bivariate normal model is included but so are other interesting models as detailed in Chapter 3.

In order for the CS model to be valid the functions $\mu_1(\cdot), \mu_2(\cdot), \sigma_1(\cdot)$ and $\sigma_2(\cdot)$ must have very specific forms as displayed in (3.28)–(3.31). No other forms for the functions $\mu_1(\cdot), \mu_2(\cdot), \sigma_1(\cdot)$ and $\sigma_2(\cdot)$ are acceptable. If we turn to the parallel SE model we find that a wider variety of forms for $\mu_1(\cdot), \mu_2(\cdot), \sigma_1(\cdot)$, and $\sigma_2(\cdot)$ are acceptable.

The SE model that most closely parallels the CS model (14.43)–(14.44) (and is most likely to be confused with it) is the nonlinear normal SE model.

Definition 14.6 (Nonlinear normal SE models). A nonlinear normal SE model is a model involving functions $\mu_1(\cdot), \mu_2(\cdot), \sigma_1(\cdot)$ and $\sigma_2(\cdot)$ such that

$$\frac{X_1 - \mu_1(X_2)}{\sigma_1(X_2)} \overset{d}{=} \varepsilon_1 \qquad (14.45)$$

and

$$\frac{X_2 - \mu_2(X_1)}{\sigma_2(X_1)} \overset{d}{=} \varepsilon_2, \qquad (14.46)$$

where $(\varepsilon_1, \varepsilon_2)$ has a bivariate normal distribution (perhaps assuming that $\varepsilon_1, \varepsilon_2$ are independent standard normal variables). This is the general normal SE model.

□

In order for (14.45) and (14.46) to represent a valid model, all that is required is that the transformation

$$\varepsilon_1 = \frac{x_1 - \mu_1(x_2)}{\sigma_1(x_2)},$$

$$\varepsilon_2 = \frac{x_2 - \mu_2(x_1)}{\sigma_2(x_1)}, \qquad (14.47)$$

be invertible. This will be true (although it may not be easy to check) for a broad spectrum of choices for the mean and standard deviation functions $(\mu_1(\cdot), \mu_2(\cdot), \sigma_1(\cdot)$ and $\sigma_2(\cdot))$. If we assume differentiability and define

$$J(x_1, x_2) = \begin{vmatrix} \dfrac{\partial}{\partial x_1} \dfrac{x_1 - \mu_1(x_2)}{\sigma_1(x_2)} & \dfrac{\partial}{\partial x_2} \dfrac{x_1 - \mu_1(x_2)}{\sigma_1(x_2)} \\ \dfrac{\partial}{\partial x_1} \dfrac{x_2 - \mu_2(x_1)}{\sigma_2(x_1)} & \dfrac{\partial}{\partial x_2} \dfrac{x_2 - \mu_2(x_1)}{\sigma_2(x_1)} \end{vmatrix}, \qquad (14.48)$$

then the resulting joint density for (Y_1, Y_2) is of the form

$$f_{X_1, X_2}(x_1, x_2) = |J(x_1, x_2)| f_{\underline{\varepsilon}}\left(\frac{x_1 - \mu_1(x_2)}{\sigma_1(x_2)}, \; \frac{x_2 - \mu_2(x_1)}{\sigma_2(x_1)}\right). \qquad (14.49)$$

Even though $\underline{\varepsilon}$ was chosen to have a classical bivariate normal distribution, the density of (X_1, X_2) in (14.49) does not generally have normal marginals or conditionals. (X_1, X_2) are only structurally related to the normal variables $(\varepsilon_1, \varepsilon_2)$.

However the general structural equation model (14.45), (14.46) can be made extremely general by permiting nonnormal choices for the joint distribution of $(\varepsilon_1, \varepsilon_2)$. In fact, since the transformation (14.47) is assumed to be invertible, we can choose the joint distribution of $\underline{\varepsilon}$ to guarantee that $\underline{X}$ has any absolutely continuous joint density!

We can pick the joint distribution of $\underline{\varepsilon}$ to guarantee that $\underline{Y}$ will be classical bivariate normal or indeed, if we wish, of the normal conditionals form with parameters as in (3.26). In this somewhat awkward sense the SE model subsumes and extends the CS model, provided we allow $(\varepsilon_1, \varepsilon_2)$ to have "contrived" distributions. Usually, as remarked earlier, $(\varepsilon_1, \varepsilon_2)$ are assumed to have a classical bivariate normal distribution and the mean and standard deviation functions in (14.45), (14.46) are assumed to have relatively simple forms (e.g., linear, bilinear, quadratic, biquadratic, etc.). Such models, unless they reduce to (14.1)–(14.2), will typically fail to have normal marginals or conditionals.

In summary, the CS and SE models in the nonlinear, just as in the linear case, are markedly different. The functions $\mu_1(\cdot), \mu_2(\cdot), \sigma_1(\cdot)$, and $\sigma_2(\cdot)$ play different roles in the two models and are subject to different constraints.

14.6   Pareto CS and SE Models

Recall that we say that X has a Pareto distribution with inequality parameter α and scale parameter σ, and write $X \sim P(\alpha, \sigma)$ if

$$f_X(x) = \frac{\alpha}{\sigma}\left(1 + \frac{x}{\sigma}\right)^{-(\alpha+1)} I(x > 0). \qquad (14.50)$$

Definition 14.7 (Conditionally specified Pareto model). A conditionally specified Pareto(α) model for (X_1, X_2) is a model such that for each $x_2 > 0$,

$$X_1 | X_2 = x_2 \sim P(\alpha, \sigma_1(x_2)) \qquad (14.51)$$

and for each $x_1 > 0$

$$X_2 | X_1 = x_1 \sim P(\alpha, \sigma_2(x_1)). \qquad (14.52)$$

$\square$

From the discussion of Chapter 5, we know that such distributions must have corresponding joint density of the form

$$f_{X_1,X_2}(x_1, x_2) = (m_{00} + m_{10}x_1 + m_{01}x_2 + m_{11}x_1x_2)^{-(\alpha+1)}I(x_1 > 0, x_2 > 0). \tag{14.53}$$

Thus the class of conditionally specified Pareto(α) distributions is severely restricted in scope. It does include some densities with Pareto marginals and conditionals.

The class of SE Pareto(α) models is much richer.

Definition 14.8 (SE Pareto model). A SE Pareto model is defined by equations of the form

$$X_1/\sigma_1(X_2) = \varepsilon_1 \tag{14.54}$$

and

$$X_2/\sigma_2(X_1) = \varepsilon_2, \tag{14.55}$$

where $(\varepsilon_1, \varepsilon_2)$ have a joint distribution with Pareto($\alpha, 1$) marginals. The ε_i's might be assumed i.i.d. To ensure that (14.54)–(14.55) will lead to a well-defined model we only need to insist that the transformation defined by (14.54)–(14.55) be invertible.   □

A broad spectrum of choices for the functions $\sigma_1(\cdot)$ and $\sigma_2(\cdot)$ is thus available. The associated joint distributions for (X_1, X_2) will generally not be of the CS form (14.53), though it is possible to contrive a tailor-made dependent joint distribution for $(\varepsilon_1, \varepsilon_2)$ which will lead to the CS model.

For the CS Pareto(α) model (14.53) it is true that

$$X_1 \left(\frac{m_{10} + m_{11}X_2}{m_{00} + m_{01}X_2} \right) \sim P(\alpha, 1) \tag{14.56}$$

and

$$X_2 \left(\frac{m_{01} + m_{11}X_1}{m_{00} + m_{10}X_1} \right) \sim P(\alpha, 1). \tag{14.57}$$

Of course the $P(\alpha, 1)$ random variables appearing in (14.56) and (14.57) are not independent. It may be of interest to see what kind of distribution we will encounter for (X_1, X_2) when (14.56)–(14.57) are regarded as a SE specification of (X_1, X_2) involving independent Pareto($\alpha, 1$) random variables, say ε_1 and ε_2.

Since the transformation from $(\varepsilon_1, \varepsilon_2)$ to (X_1, X_2) associated with (14.56) and (14.57) is invertible (recall $\varepsilon_1 > 0$, $\varepsilon_2 > 0$, $X_1 > 0$, $X_2 > 0$) we can determine its Jacobian, i.e.,

$$
J(x_1, x_2) = \begin{vmatrix} \dfrac{m_{10} + m_{11}x_2}{m_{00} + m_{01}x_2} & \dfrac{(m_{00}m_{11} - m_{01}m_{10})x_1}{(m_{00} + m_{01}x_2)^2} \\[2ex] \dfrac{(m_{00}m_{11} - m_{01}m_{10})x_2}{(m_{00} + m_{10}x_1)^2} & \dfrac{m_{01} + m_{11}x_1}{m_{00} + m_{10}x_1} \end{vmatrix}
$$

$$
= \frac{(m_{00} + m_{10}x_1 + m_{01}x_2 + m_{11}x_1x_2)(\tilde{m}_{00} + \tilde{m}_{10}x_1 + \tilde{m}_{01}x_2 + \tilde{m}_{11}x_1x_2)}{(m_{00} + m_{10}x_1)^2(m_{00} + m_{01}x_2)^2},
$$

$$(14.58)$$

where

$$
\begin{aligned}
\tilde{m}_{00} &= m_{00}m_{01}m_{10}, \\
\tilde{m}_{10} &= m_{00}m_{10}m_{11}, \\
\tilde{m}_{01} &= m_{00}m_{01}m_{11}, \\
\tilde{m}_{11} &= m_{01}m_{10}m_{11}.
\end{aligned}
$$

$$(14.59)$$

Thus the joint density of (X_1, X_2) is given by

$$
f_{X_1,X_2}(x_1, x_2) = \alpha^2 J(x_1, x_2) \left[1 + \frac{x_1(m_{10} + m_{11}x_2)}{m_{00} + m_{01}x_2} \right]^{-(\alpha+1)}
$$

$$(14.60)$$

$$
\times \left[1 + \frac{x_2(m_{01} + m_{11}x_1)}{m_{00} + m_{10}x_1} \right]^{-(\alpha+1)}.
$$

It may be observed that this density does not have Pareto marginals nor does it have Pareto conditionals.

As in the normal case, for Pareto models CS and SE lead in general to different distributions: highly restricted in nature, in the case of CS, quite general in the case of SE.

We finish this section by illustrating an SE Pareto(α) model with scale functions that are not bilinear. Suppose that

$$
\begin{pmatrix} X_2^2 X_1 \\ X_1^2 X_2 \end{pmatrix} \stackrel{d}{=} \begin{pmatrix} \varepsilon_1 \\ \varepsilon_2 \end{pmatrix},
$$

$$(14.61)$$

where $\varepsilon_1, \varepsilon_2$ are i.i.d. Pareto($\alpha, 1$) random variables. In this case, we can explicitly solve for X_1, X_2:

$$
X_1 = \sqrt[3]{\frac{\varepsilon_1^2}{\varepsilon_2}},
$$

$$
X_2 = \sqrt[3]{\frac{\varepsilon_2^2}{\varepsilon_1}}.
$$

The Jacobian of the transformation in (14.61) is

$$
J(x_1, x_2) = \begin{vmatrix} x_2^2 & 2x_1x_2 \\ 2x_1x_2 & x_1^2 \end{vmatrix} = -3x_1^2x_2^2
$$

and thus the joint density of (X_1, X_2) is

$$f_{X_1,X_2}(x_1, x_2) = 3\alpha^2 x_1^2 x_2^2 \left[(1 + x_1 x_2^2)(1 + x_1^2 x_2)\right]^{-(\alpha+1)} I(x_1 > 0, x_2 > 0). \tag{14.62}$$

Of course, in this case, there is no analogous CS model to compare with (since we cannot have $X_1|X_2 = x_2 \sim P(\alpha, x_2^{-2})$ and $X_2|X_1 = x_1 \sim P(\alpha, x_1^{-2})$).

14.7 Discrete Models

There is, in principle, no reason to restrict discussion to absolutely continuous distributions. We have seen in earlier chapters, a variety of conditionally specified discrete distributions. Analogous SE models can be formulated but as we will show with an example, they are often of limited utility since the resulting set of possible values for (X_1, X_2) is often unusual in structure. Our example involves geometric distributions. From the discussion in Section 4.12 we know that the general form of a geometric conditionals density is

$$f_{X_1,X_2}(x_1, x_2) \propto q_1^{x_1} q_2^{x_2} q_3^{x_1 x_2}, \quad x_1, x_2 = 0, 1, 2, \ldots, \tag{14.63}$$

where $q_1, q_2 \in (0, 1)$ and $q_3 \in (0, 1]$.

A geometric SE model could be of the form

$$g_1(X_1, X_2) = \varepsilon_1 \tag{14.64}$$

and

$$g_2(X_1, X_2) = \varepsilon_2, \tag{14.65}$$

where the ε_i's are independent geometric random variables and the transformation defining the ε's in terms of the X's is invertible. As a specific example we may consider

$$X_2^2 X_1 - 1 = \varepsilon_1, \tag{14.66}$$

$$X_1^2 X_2 - 1 = \varepsilon_2, \tag{14.67}$$

where $\varepsilon_1 \sim \mathcal{G}(p_1)$ and $\varepsilon_2 \sim \mathcal{G}(p_2)$ and the ε's are independent.

The transformation (14.66)–(14.67) is indeed invertible. Using the notation $q_i = 1 - p_i$, it is not difficult to verify that the joint density (X_1, X_2) defined by (14.66)–(14.67), is given by

$$f_{X_1,X_2}(x_1, x_2) = \left(\frac{p_1 p_2}{q_1 q_2}\right) q_1^{x_2^2 x_1} q_2^{x_1^2 x_2} I(x_1^2 x_2 \in \mathbb{N}, x_2^2 x_1 \in \mathbb{N}), \tag{14.68}$$

where $\mathbb{N}$ denotes the natural numbers.

Equation (14.68) describes a well-defined joint discrete density function. But it differs markedly from the CS geometric model (14.63). A major difference is to be found in the support sets of the models, i.e., the sets of possible values for (X_1, X_2) prescribed by the models. In the CS model the possible values of (X_1, X_2) are all pairs of nonnegative integers. In the SE model (14.68), possible values of (X_1, X_2) are of the form

$$\left(\sqrt[3]{\frac{m^2}{n}}, \sqrt[3]{\frac{n^2}{m}} \right),$$

where m and n are natural numbers. It is hard to imagine situations in which a model with such an unusual support set would be plausible. Similar anomalous support sets will be encountered for many discrete SE models and, consequently, the usefulness of discrete SE models is likely to be severely curtailed.

14.8   Higher-Dimensional Models

It is necessary and natural to consider extensions to higher dimensions. The extension of CS models to higher dimensions has been documented extensively in Chapters 8 and 10.

There is no difficulty in extending our SE models to k dimensions.

Definition 14.9 (Multidimensional SE models).   The model $\underline{X} = (X_1, \ldots, X_k)$ is a k-dimensional SE model if it is related to $\underline{\varepsilon} = (\varepsilon_1, \ldots, \varepsilon_k)$ by the set of equations

$$g_i(X_1, X_2, \ldots, X_k, \underline{\theta}) = \varepsilon_i; \quad i = 1, 2 \ldots, k, \qquad (14.69)$$

where $\underline{\varepsilon}$ is assumed to have a known distribution (not infrequently with the ε_i's i.i.d.). In order for (14.69) to lead to a well-defined model it is only necessary that the transformation (14.69), which defines $\underline{\varepsilon}$ in terms of $\underline{X}$, be invertible.   □

The case in which the g_i's are linear functions and the ε's are normal has received considerable attention and has proved to be a useful flexible model in many economic applications. However, the caveats raised in earlier sections, about the difficulty in interpreting the parameters of such structural models, remain cause for concern in k dimensions just as in two-dimensions.

14.9   Bibliographic Note

This chapter is based on Arnold, Castillo, and Sarabia (1998b).

Exercises

14.1 Verify the expressions given for the parameters in the joint density (14.11).

14.2 Verify the expressions given for the parameters in the SE model (14.32).

14.3 Compare exponential CS and SE models. For the CS model $X_1|X_2 = x_2 \sim \exp(\mu(x_2))$ and $X_2|X_1 = x_1 \sim \exp(\nu(x_1))$. For the SE model $\mu(X_2)X_1 = \epsilon_1$ and $\nu(X_1)X_2 = \epsilon_2$, where (ϵ_1, ϵ_2) has a joint distribution with standard exponential marginals.

14.4 Determine the form of joint density for (ϵ_1, ϵ_2) to ensure that the SE Pareto model (14.54)–(14.55) is of the CS form (14.53).

14.5 Consider the following nonlinear normal SE model

$$\sqrt{1 + aX_2^2}\,X_1 = \epsilon_1,$$

$$\sqrt{1 + bX_1^2}\,X_2 = \epsilon_2,$$

where ϵ_1, ϵ_2 are i.i.d. $N(0, 1)$. How is this model related to the centered normal conditionals model (3.51)?

15
Paella

15.1 Introduction

In this chapter we will gather together a selection of topics related to conditional specification (CS). Either because of their tangential relation to our main theme or because of their sometimes preliminary state of development, they have been collected in this chapter. Some of them promise considerable future development. They are presented in no particular order; tasty ingredients, thoroughly mixed, as a paella should be.

15.2 Diatomic Conditionals and Stop-Loss Transforms

Diatomic distributions sound more exotic than Bernoulli distributions but of course they really aren't. Instead of being random variables with possible values $0, 1$, they have two possible values. Naturally a location and scale transform will reduce any diatomic random variable to a Bernoulli variable. So, for example, questions about the correlation between diatomic variables will reduce to questions about correlated Bernoulli variables.

Our concern is with diatomic random variables X, Y with possible values $\{x_1, x_2\}$ and $\{y_1, y_2\}$, respectively, with the convention that $x_1 < x_2$ and $y_1 < y_2$. As in Chapter 1 we denote the corresponding conditional probability distribution matrices, now 2×2, by A and B, where $a_{ij} = P(X =$

$x_i|Y = y_j)$ and $b_{ij} = P(Y = y_j|X = x_i)$. In the present case we can write, using notation suggested by Hurlimann (1993),

$$A = \begin{pmatrix} a & 1-\alpha \\ 1-a & \alpha \end{pmatrix} \qquad (15.1)$$

and

$$B = \begin{pmatrix} b & 1-b \\ 1-\beta & \beta \end{pmatrix}, \qquad (15.2)$$

where $a, b, \alpha, \beta \in [0, 1]$ (usually in $(0, 1)$). We must have at least one non-zero entry in each row and column of A and B.

Four parameters apparently are involved. There are really only three since we know that, for compatibility, we must have equal cross-product ratios for A and B. This follows from Theorem 2.1, if $a, b, \alpha, \beta \in (0, 1)$. If any of the numbers a, b, α, β is zero we have to interpret the cross-product ratio as taking values in $[0, \infty) \cup \{\infty\}$. Since we never have two zeros in the same row or column, we do not ever encounter an undefined cross-product ratio (i.e., we never see $0/0$). Thus we always will have, for compatibility,

$$\frac{a\alpha}{b\beta} = \frac{(1-a)(1-\alpha)}{(1-b)(1-\beta)}. \qquad (15.3)$$

For future reference we note the values of the determinants of A and B:

$$|A| = a + \alpha - 1, \qquad (15.4)$$

$$|B| = b + \beta - 1. \qquad (15.5)$$

When A, B are compatible, we will denote the corresponding marginal densities by

$$\underline{\pi} = (\pi, 1 - \pi) \qquad (15.6)$$

and

$$\underline{\eta} = (\eta, 1 - \eta), \qquad (15.7)$$

where $\pi, \eta \in (0, 1)$. If A and B are compatible, they must have the same incidence sets and then the corresponding marginal density for X, i.e., $\underline{\pi}$, is readily determined. We have

$$\pi = \frac{\dfrac{a}{1-a}(1-\beta)}{b + \dfrac{a}{1-a}(1-\beta)}, \qquad (15.8)$$

assuming $b \neq 0$ and $a \neq 0, 1$. If $b = 0$ and $\beta \neq 0$ then, necessarily, $\alpha \neq 0, 1$ and $\beta \neq 1$ and we find

$$\pi = \frac{\beta(1-\alpha)}{\alpha + \beta - \alpha\beta}. \qquad (15.9)$$

Analogous expressions are available in other cases in which there is a single zero in A and B. If there are two zeros in A and B then we must have one of the following two trivial cases:

$$A = B = \begin{pmatrix} 1 & 0 \\ 0 & 1 \end{pmatrix} \tag{15.10}$$

or

$$A = B = \begin{pmatrix} 0 & 1 \\ 1 & 0 \end{pmatrix}. \tag{15.11}$$

Compatibility in such cases is obvious, however there is not a unique compatible distribution. The marginal π in cases (15.10) and (15.11) is completely arbitrary. If (X, Y) have compatible conditionals given by A, B in (15.1) and (15.2), then direct computation yields

$$\begin{aligned}
\text{var}(X) &= (x_2 - x_1)^2 \pi(1 - \pi), & (15.12) \\
\text{var}(Y) &= (y_2 - y_1)^2 \eta(1 - \eta), & (15.13) \\
\text{cov}(X, Y) &= (x_2 - x_1)(y_2 - y_1)\pi(1 - \pi)|B| \\
&= (x_2 - x_1)(y_2 - y_1)\eta(1 - \eta)|A|. & (15.14)
\end{aligned}$$

(Our earlier observation relating diatomic variables to Bernoulli variables will simplify the derivation of (15.12)–(15.14).)

From (15.14) we may observe that

$$\pi(1 - \pi)|B| = \eta(1 - \eta)|A|. \tag{15.15}$$

From (15.12)–(15.14), the correlation between X and Y is found to be

$$\rho(X, Y) = |A|\sqrt{\frac{\eta(1 - \eta)}{\pi(1 - \pi)}} \tag{15.16}$$

$$= |B|\sqrt{\frac{\pi(1 - \pi)}{\eta(1 - \eta)}}. \tag{15.17}$$

Since $\pi, \eta \in (0, 1)$, the expressions inside the radicals in (15.16) and (15.17) are positive real numbers. Thus the sign of the correlation between X and Y is determined by the common sign of $|A|$ and $|B|$. If there are no zeros in A and B, we can give a simple expression for the correlation in terms of the elements of A and B, namely,

$$\rho(X, Y) = |A|\sqrt{\frac{b\beta}{a\alpha}} \tag{15.18}$$

$$= |B|\sqrt{\frac{a\alpha}{b\beta}}. \tag{15.19}$$

Formulas (15.18)–(15.19) continue to be valid when $|A| = 1$, i.e. when A is given by (15.10). In this case $\rho = 1$. If $|A| = -1$, i.e. A as in (15.11), then $\rho = -1$, but formulas (15.18)–(15.19) cannot be used.

From the above discussion it is evident that a joint distribution with diatomic conditionals will be completely determined by π, η, and ρ.

Thus if we are given two arbitrary values $\pi \in (0, 1), \eta \in (0, 1)$ the possible values for ρ are

$$\left\{ \rho = \frac{\theta - \eta\pi}{\sqrt{\pi(1 - \pi)\eta(1 - \eta)}} : 0 < \theta < \min(\pi, \eta) \right\}. \tag{15.20}$$

The value $\rho = 1$ can be obtained from an arbitrary $\pi = \eta$ and the value $\rho = -1$ can be obtained for an arbitrary $\pi = 1 - \eta$ (cases corresponding to matrices (15.10) and (15.11), respectively). The reader is referred to Hurlimann (1993) for a variety of other expressions relating A, B, π, η, and ρ. Note that since the family of Bernoulli distributions can be viewed as an exponential family we can use Theorem 4.1 to characterize the class of all bivariate distributions with conditionals in the given diatomic families with support $\{x_1, x_2\}$ and $\{y_1, y_2\}$, respectively. From that theorem we get

$$\begin{aligned} p_{ij} &= P(X = x_i, Y = y_j) \\ &= \exp[m_{00} + m_{10}i + m_{01}j + m_{11}ij], \end{aligned} \tag{15.21}$$

where m_{00} is chosen so that $\sum_{i=1}^2 \sum_{j=1}^2 p_{ij} = 1$. Of course (15.21) reproduces the standard log-linear representation of the full family of multinomial distributions with $n = 1$ and outcomes $(1, 1), (1, 2), (2, 1)$, and $(2, 2)$.

Hurlimann (1993) observes that certain diatomic random variables can play an extremal role in efforts to bound the stop-loss transform of a random sum $X + Y$, i.e. of $E((X + Y - T)^+)$. Thus:

Theorem 15.1 *If X, Y have a joint distribution with $E(X) = \mu_X$, $E(Y) = \mu_Y$, $var(X) = \sigma_X^2$, and $var(Y) = \sigma_Y^2$, it follows that, for any $T \in \mathbb{R}$,*

$$E((X + Y - T)^+) \le (\sqrt{\sigma^2 + (T - \mu)^2} - (T - \mu))/2, \tag{15.22}$$

in which $\mu \overset{\Delta}{=} \mu_X + \mu_Y$ and $\sigma \overset{\Delta}{=} \sigma_X + \sigma_Y$. Equality in (15.22) is attained by a bivariate diatomic distribution with correlation 1 (i.e., A as in (15.10)) with

$$x_1 = \mu_X - \sigma_X z_0, \quad x_2 = \mu_X + \sigma_X/z_0,$$
$$y_1 = \mu_Y - \sigma_Y z_0, \quad y_2 = \mu_Y + \sigma_Y/z_0,$$

and

$$\pi = 1/(1 + z_0^2),$$

where

$$z_0 = [\sqrt{\sigma^2 + (T - \mu)^2} - (T - \mu)]/\sigma.$$

$\square$

15.3   Failure Rates and Mean Residual Life Functions

In the context of reliability and survival modeling, failure rate functions and mean residual life functions play a prominent role. Gupta (1998) has investigated these and related functions in the case of the bivariate Pareto conditionals distributions (as described in Section 5.2).

Definition 15.1 (Failure rate). For a positive random variable X with density $f_X(x)$ and distribution function $F_X(x)$ we define its failure rate function by

$$r_X(x) = f_X(x)/(1 - F_X(x)), \quad x > 0. \tag{15.23}$$

Definition 15.2 (Mean residual life function). The mean residual life function of a positive random variable X is defined by

$$\mu_X(x) = E(X - x|X > x), \quad x > 0. \tag{15.24}$$

These two functions are related by

$$r_X(x) = [1 + \mu'_X(x)]/\mu_X(x). \tag{15.25}$$

If (X, Y) has the Pareto conditionals distribution with density (5.7), Gupta provides the following expression for the marginal mean remaining life function of X:

$$\mu_X(x) = \left[\frac{1}{\lambda_{10}(\alpha - 1)(\lambda_{00} + \lambda_{10}x)^{\alpha-1}(1 - F_X(x))} - \lambda_{01} \right] \lambda_{11}^{-1}. \tag{15.26}$$

From which by letting $x \to \infty$, he obtains

$$E(X) = \lambda_{11}^{-1} \left(\frac{1}{\lambda_{10}(\alpha - 1)\lambda_{00}^{\alpha-1}} - \lambda_{01} \right). \tag{15.27}$$

He also verifies indirectly that X has a decreasing failure rate function.

The hazard gradient of the Pareto conditionals distribution can also be studied. This is a vector $(h_1(x, y), h_2(x, y))$ where $h_1(x, y)$ is the failure rate function of X given $Y > y$ and where $h_2(x, y)$ is the failure rate function of Y given $X > x$. Without evaluating these functions analytically, Gupta verifies that for each y, $h_1(x, y)$ is a decreasing function of x and for each x, $h_2(x, y)$ is a decreasing function of y.

Conditional failure rates of the Pareto conditionals distribution are well behaved, since the conditionals are Pareto densities. If we define

$$r_{X|Y}(x|y) = \frac{f_{X|Y}(x|y)}{1 - F_{X|Y}(x|y)}, \tag{15.28}$$

we find that for the density (5.2),

$$r_{X|Y}(x|y) = \alpha \left(x + \frac{\lambda_{00} + \lambda_{01}y}{\lambda_{10} + \lambda_{11}y} \right)^{-1}. \tag{15.29}$$

Clearly $r_{X|Y}(x|y) \downarrow$ as $x \uparrow$. As a function of y, rather than of x, $r_{X|Y}(x|y)$ will decrease or increase depending on the sign of the correlation between X and Y (cf. (5.15)). It decreases if $\rho > 0$. The general prevalence of decreasing failure rates associated with the Pareto conditionals distribution continues with the revelation that the random variable $Z = \min(X, Y)$ also has a decreasing failure rate.

Another example of a conditionally specified distribution with computable failure rate functions is the bivariate exponential conditionals distribution with joint pdf,

$$f_{X,Y}(x, y) = \frac{k(c)}{\sigma_1 \sigma_2} \exp[-x/\sigma_1 - y/\sigma_2 - cxy/(\sigma_1\sigma_2)] \, I(x \geq 0)I(y \geq 0), \tag{15.30}$$

where $k(c)$ is defined in (5.22) and (5.23) (Arnold and Strauss (1988a)). The joint survival function is available in closed form. We have

$$\overline{F}_{X,Y}(x, y) = P(X > x, Y > y) = \int_x^\infty \int_y^\infty f_{X,Y}(u, v) \, du \, dv$$

$$= \frac{k(c)e^{-(x/\sigma_1 + y/\sigma_2 + cxy/(\sigma_1\sigma_2))}}{(1 + cx/\sigma_1)(1 + cy/\sigma_2)k \left[\dfrac{c}{(1 + cx/\sigma_1)(1 + cy/\sigma_2)} \right]}. \tag{15.31}$$

Another formulation in terms of the exponential integral function is,

$$\overline{F}_{X,Y}(x, y) = \frac{-\text{Ei}(1/c + x/\sigma_1 + y/\sigma_2 + cxy/(\sigma_1\sigma_2))}{-\text{Ei}(1/c)}. \tag{15.32}$$

The bivariate failure rate function is given by

$$\begin{aligned}
r_{X,Y}(x, y) &= \frac{f_{X,Y}(x, y)}{\overline{F}_{X,Y}(x, y)} \\
&= \frac{(1 + cx/\sigma_1)(1 + cy/\sigma_2)}{\sigma_1 \sigma_2} k \left[\frac{c}{(1 + cx/\sigma_1)(1 + cy/\sigma_2)} \right].
\end{aligned}$$

Using this last expression, it can be shown that the failure rate is increasing in both x and y. This fact is consistent with the observation that X and Y are negatively correlated.

15.4   Hypothesis Testing

Nested within most of our conditionally specified models are usually to be found submodels involving more extensive symmetry, submodels that have simpler distributions and submodels with independent marginals. The bivariate normal conditionals density (3.26) is a case in point. Nested within it are to be found the classical bivariate normal model (when $m_{22} = 0$) and the model with independent marginals (when $\tilde{M} = 0$). For large samples, generalized likelihood ratio tests may be used to determine the appropriateness of such submodels. Indeed, for example, such an approach was illustrated in Arnold and Strauss (1991) within the centered normal conditionals family.

For the case of the bivariate exponential conditionals distribution (Section 4.4), SenGupta (1995) has provided some alternative tests for independence. If, referring to the density in the form (4.14), m_{10} and m_{01} are known, then a UMP test for independence (i.e., $m_{11} = 0$) is available. The rejection region is of the form $\sum_{i=1}^{n} X_i Y_i > c$. When m_{10} and m_{01} are unknown, SenGupta outlines the development of a UMPU test for independence. He also discusses the problems associated with implementing a UMPU test for symmetry (i.e., $m_{10} = m_{01}$). It should be possible to develop analogous results in other conditionally specified bivariate families (e.g., the centered normal conditionals distribution).

15.5 Related Stochastic Processes

We begin by reviewing a conditional characterization problem mentioned in Section 1.2 that turns out to be easy to resolve. Suppose that we are given the family of conditional densities of X given Y and we are given that $X \overset{d}{=} Y$ (i.e., $F_X(x) = F_Y(x), \forall x \in \mathbb{R}$), can we determine from this the joint distribution of (X, Y)? The answer is, in many cases, yes. Consider a Markov chain $X_1, X_2, \ldots$ whose transitions are governed by the given family of conditional densities of X given Y. If this chain is irreducible (which would happen, for example, if $f_{X|Y}(x|y) > 0, \forall x \in S(X), y \in S(Y)$), then a unique long-run distribution exists, say, $F_0(x)$. This must be the common distribution of X and Y and then, armed with a marginal for $Y(F_0)$ and the given conditional distribution of X given Y, the determination of the joint distribution of (X, Y) is straightforward.

Example 15.1 (Bivariate normal distribution). Suppose $S(X) = S(Y) = \mathbb{R}$ and for each $y \in \mathbb{R}$, for some $\alpha \in (-1, 1)$,

$$X|Y = y \sim N(\alpha y + \beta, \sigma^2). \tag{15.33}$$

In addition, assume $X \overset{d}{=} Y$. It follows that (X, Y) has a classical bivariate normal distribution, i.e., that

$$\binom{X}{Y} \sim N^{(2)} \left(\begin{pmatrix} \dfrac{\beta}{1 - \alpha} \\ \dfrac{\beta}{1 - \alpha} \end{pmatrix}, \sigma^2 \begin{pmatrix} \dfrac{1}{1 - \alpha^2} & \dfrac{\alpha}{1 - \alpha^2} \\ \dfrac{\alpha}{1 - \alpha^2} & \dfrac{1}{1 - \alpha^2} \end{pmatrix} \right). \tag{15.34}$$

How was (15.34) obtained? The preamble to the example suggested that we search for a distribution for X, say $F_0(x)$, with density $f_0(x)$ that satisfies

$$f_0(x) = \int_{-\infty}^{\infty} \frac{1}{\sqrt{2\pi}\sigma} e^{-(x-\alpha y-\beta)^2/2\sigma^2} f_0(y)\, dy. \tag{15.35}$$

However, a more constructive approach is possible because of the special properties of the normal distribution. Begin with $X_0 = 0$ and successively generate $X_1, X_2, \ldots$ using the conditional distribution (15.33). For any n we will have

$$X_n \overset{d}{=} \sum_{j=0}^{n-1} \alpha^j Z_j + \beta \sum_{j=0}^{n-1} \alpha^j, \tag{15.36}$$

where $Z_0, Z_1, \ldots$ are i.i.d. $N(0, \sigma^2)$ random variables.

Evidently for each n, X_n is normally distributed with readily computed mean and variance.

Finally, $X_n \overset{d}{\to} N\left(\dfrac{\beta}{1-\alpha}, \dfrac{\sigma^2}{1-\alpha^2}\right)$ and this is the sought-for distribution F_0.

An exactly parallel development is possible if $X|Y = y \sim \text{Cauchy}(\alpha y + \beta, \sigma)$ and $X \overset{d}{=} Y$.

In both the normal and Cauchy case we have the representation

$$X_n = \alpha X_{n-1} + \beta + Z_{n-1}, \tag{15.37}$$

where Z_{n-1}'s are i.i.d. stable random variables.

Except in these cases, solution of (15.35) seems the only possible approach. There is a possibility of generating some other tractable examples if we use some of our available conditionally specified distributions.

Example 15.2 (Normal conditionals distribution). Suppose that

$$S(X) = S(Y) = \mathbb{R}$$

and that for each $y \in \mathbb{R}$ and some $c > 0$

$$X|Y = y \sim N(0, \sigma^4/(\sigma^2 + cy^2)). \tag{15.38}$$

Suppose in addition that $X \overset{d}{=} Y$. We claim that, necessarily, (X, Y) has a centered normal conditionals distribution (as in (3.51)) with $\sigma_1^2 = \sigma_2^2 = \sigma^2$. □

Example 15.2 was easily resolved since we knew that there was a unique bivariate distribution with $X|Y = y$ as in (15.38) and with $X \overset{d}{=} Y$. Recognizing (15.38) from our experience in Chapter 3, we only needed to look for a distribution with all of its conditionals of the centered normal form that had its parameters selected to make $X \overset{d}{=} Y$.

We can embed (15.38) in a stochastic process as follows: for $n = 0, \pm 1, \pm 2, \ldots$,

$$X_n = \sqrt{\sigma^4/[\sigma^2 + c(X_{n-1})^2]}Z_{n-1}, \tag{15.39}$$

where the Z_is are i.i.d. $N(0, 1)$ random variables. It may be noted that this process will be time-reversible.

Analogous considerations lead to a time-reversible process with exponential transition distributions defined by

$$X_n = \left(\frac{\sigma^2}{\sigma + cX_{n-1}}\right)E_{n-1}, \tag{15.40}$$

where $c \geq 0$ and the E_i's are i.i.d. standard exponential variables. Processes of this kind have been studied by Anderson (1990). He also considered processes with exponential transition probabilities involving higher-order dependence.

15.6 Given $E(X|Y = y)$ and $F_X(x)$

Korwar (1974) discussed the possibility of characterizing a joint distribution given one regression function and one marginal. Clearly we would generally be unable to characterize the joint distribution of (X, Y) if we were only given $E(X|Y = y)$ and $F_Y(y)$. Any family of conditional distributions $F_{X|Y}(x|y)$ with means given by $E(X|Y = y)$ could be paired with $F_Y(y)$ and a wide variety of joint distributions could thus be obtained. The only exception occurring when X, Y are both diatomic, in which case $E(X|Y = y)$ and $F_Y(y)$ will determine the distribution of (X, Y). Do the chances look better if we are given $E(X|Y = y)$ and $F_X(x)$? Again, the diatomic case must be dealt with separately.

For, if both X and Y are diatomic then $E(X|Y = y)$ and $F_X(x)$ do determine the joint distribution of (X, Y). The three numbers they furnish are adequate to pin down the three parameters in the joint distribution. Aside from that special case there is no hope of characterizing the joint distribution based on $E(X|Y = y)$ and $F_X(x)$ alone. More must be assumed.

If we state our characterization problem in a parametric setting we can obtain characterization results involving one regression function and one marginal. The following result is due to Kyriakoussis and Papageorgiou (1989) using notation suggested by Wesolowski (1995a):

Theorem 15.2 *For each $\theta \in \Theta$, a possibly infinite interval in $\mathbb{R}$, assume that (X_θ, Y) is a discrete random vector with the property that X_θ has a power series distribution of the form*

$$f_{X_\theta}(x) \propto a(x)\theta^x, \quad x = 0, 1, 2, \ldots,$$

and that for each $y \in \{0, 1, 2, \ldots\}$,

$$E(X_\theta | Y = y) = y + \sum_{j=1}^{\infty} b_j(y)\theta^j.$$

If we assume that the conditional distribution of Y *given* X_θ *does not depend on* θ, *then the conditional distribution of* $Y|X_\theta = x$ *can be determined uniquely for each* x. □

For example, if X_θ is Poisson(θ) and if $E(X_\theta | Y = y) = y + q\theta$ for some $q \in (0, 1)$ then $Y|X_\theta = x \sim$ binomial(x, q).

15.7   Near Compatibility with Given Parametric Families of Distributions

Suppose that we are given the conditional densities $a(x, y)$ and $b(x, y)$, $x \in S(X), y \in S(Y)$, perhaps compatible but not necessarily so. Now consider a given parametric family $\mathcal{E}$ of bivariate joint densities. Arnold and Gokhale (1998c) address the problem of determining that member of $\mathcal{E}$ that is most nearly compatible with the given a and b. They use Kulback-Leibler distance between the conditionals of each member of $\mathcal{E}$ and the conditionals given in a and b as a measure of discrepancy. Thus for example in a Bayesian context, we might seek the bivariate distribution with gamma–normal conditionals that is most nearly compatible with a and b supplied by our informed expert, representing his conditional a priori beliefs about a normal mean and precision. We thus would find the "best" conditionally conjugate prior in this manner. The following toy example illustrates the general ideas involved:

Example 15.3 (Near compatibility). Suppose $S(X) = S(Y) = \{0, 1, 2\}$ and we are given the conditional probability matrices

$$A = \begin{pmatrix} 1/4 & 1/5 & 1/3 \\ 1/2 & 2/5 & 1/3 \\ 1/4 & 2/5 & 1/3 \end{pmatrix} \tag{15.41}$$

and

$$B = \begin{pmatrix} 1/3 & 1/3 & 1/3 \\ 1/2 & 1/4 & 1/4 \\ 1/5 & 2/5 & 2/5 \end{pmatrix} \tag{15.42}$$

(these are not compatible, but that does not matter).

Consider the one-parameter family of joint distributions on $\{0, 1, 2\} \times \{0, 1, 2\}$ defined by

$$P_\theta = \begin{pmatrix} (1-\theta)^4 & 2\theta(1-\theta)^3 & 2\theta^2(1-\theta)^2 \\ 2\theta(1-\theta)^3 & 2\theta^2(1-\theta)^2 & 2\theta^3(1-\theta) \\ 2\theta^2(1-\theta)^2 & 2\theta^3(1-\theta) & \theta^4 \end{pmatrix}, \tag{15.43}$$

where $\theta \in [0,1]$. We wish to determine the member of $\{P_\theta : \theta \in [0,1]\}$ (i.e., we wish to choose θ) that is most nearly compatible with A and B. The conditional distributions corresponding to P_θ, obtained by column and row normalization, respectively, are

$$
A_\theta =
\begin{pmatrix}
\dfrac{(1-\theta)^2}{1+\theta^2} & \dfrac{(1-\theta)^2}{1-\theta+\theta^2} & \dfrac{2(1-\theta)^2}{1+(1-\theta)^2} \\[2mm]
\dfrac{2\theta(1-\theta)}{1+\theta^2} & \dfrac{\theta(1-\theta)}{1-\theta+\theta^2} & \dfrac{2\theta(1-\theta)}{1+(1-\theta)^2} \\[2mm]
\dfrac{2\theta^2}{1+\theta^2} & \dfrac{\theta^2}{1-\theta+\theta^2} & \dfrac{\theta^2}{1+(1-\theta)^2}
\end{pmatrix}
\tag{15.44}
$$

and by symmetry

$$
B_\theta = A_\theta^T.
$$

Using Kulback-Leibler distance as our measure of discrepancy we can write (recalling (2.5))

$$
\begin{aligned}
d((A,B), P_\theta) &= I(A, A_\theta) + I(B, B_\theta) \\
&= \sum_{i=0}^{2} \sum_{j=0}^{2} a_{ij} \log \frac{a_{ij}}{a_{ij}(\theta)} + \sum_{i=0}^{2} \sum_{j=0}^{2} b_{ij} \log \frac{b_{ij}}{b_{ij}(\theta)}.
\end{aligned}
\tag{15.45}
$$

To minimize this objective function we need to solve the following equation:

$$
\sum_{i=0}^{2} \sum_{j=0}^{2} a_{ij} \frac{a'_{ij}(\theta)}{a_{ij}(\theta)} + \sum_{i=0}^{2} \sum_{j=0}^{2} b_{ij} \frac{b'_{ij}(\theta)}{b_{ij}(\theta)} = 0,
\tag{15.46}
$$

where $a_{ij}(\theta)$ and $b_{ij}(\theta)$ denote the elements of A_θ and B_θ. For the particular matrices A and B given in (15.41) and (15.42), the optimal choice of θ is

$$
\tilde{\theta} = 0.510495
\tag{15.47}
$$

and the corresponding matrix in $\{P_\theta : \theta \in [0,1]\}$ most nearly compatible with this A and B is

$$
P_{\tilde\theta} =
\begin{pmatrix}
0.05742 & 0.1198 & 0.1249 \\
0.1198 & 0.1249 & 0.1302 \\
0.1249 & 0.1302 & 0.06792
\end{pmatrix},
\tag{15.48}
$$

which has as conditionals

$$
A_{\tilde\theta} =
\begin{pmatrix}
0.1901 & 0.3194 & 0.3866 \\
0.3965 & 0.3331 & 0.4032 \\
0.4135 & 0.3474 & 0.2102
\end{pmatrix}
\tag{15.49}
$$

and

$$
B_{\tilde\theta} =
\begin{pmatrix}
0.1901 & 0.3965 & 0.4135 \\
0.3194 & 0.3331 & 0.3474 \\
0.3866 & 0.4032 & 0.2102
\end{pmatrix}.
\tag{15.50}
$$

The maximum absolute error deviations between the entries in these most nearly compatible conditional matrices and the given conditional matrices (15.41) and (15.42) are 0.1635 and 0.1898, respectively.   □

15.8   Marginal Maps

To reinforce the idea that conditional densities must play a crucial role in visualizing multivariate densities, we present the following pedagogical exercise. A topographical map of a country showing height above sea level can be, suitably normalized, viewed as the description of a bivariate density.

The marginals of such densities are essentially uninformative. Observe Figures 15.1 and 15.2 in which Y-marginals (from north to south coast) and X-marginals (from west to east coast) are provided for Mexico, Spain, and the United States of America. They are in random order. Can you sort them out?

The conditional densities, even only a few of them, are effectively evocative in describing the terrain and allowing us to recognize the country. Three conditionals (normalized cross sections from west to east coast) are provided, one for each country, in random order in Figure 15.3. Now can you sort them out?

Finally, Figure 15.4 exhibits the corresponding joint densities (i.e., normalized topographical maps) of the three countries, again in the same order as in Figures 15.1–15.3. We can now grade our ability to identify the maps on the basis of marginal or conditional information.

15.9   A Cautionary Tale

It would be remiss in a book, such as this, which has its focus on conditional densities to not admit that there is a famous skeleton in the conditional density closet. Conditional densities are a bit tricky and our intuitions about interpreting them can be wide of the mark. The famous Kolmogorov–Borel example, in which we consider the location of a person at a point uniformly distributed in the surface of the globe, provides an eloquent case in point. If we are told that such a person is in fact located on the prime meridian, what is the distribution of his location on that meridian? Uniform surely! But in fact the answer is: "it depends." It depends on how we define location on the sphere. If, as is customarily done, we locate by latitude and longitude, we find that given our person is on the prime meridian, he is more likely to be near the equator than near the pole. There are ways to measure location on the sphere, that <u>will</u> make his location given he is on the prime meridian, uniform. But as Arnold and Robertson (1998) point out, there are also ways to measure location on the sphere, that will make

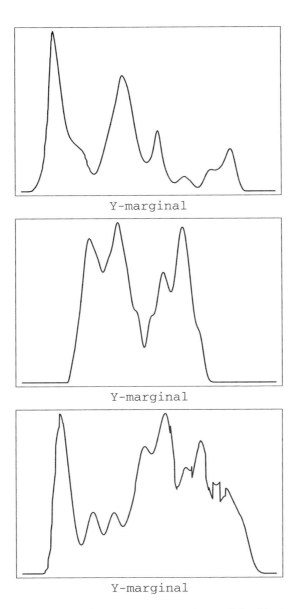

Y-marginal

Y-marginal

Y-marginal

FIGURE 15.1. Y-marginals (from north to south coast) for Mexico, Spain, and the United States of America in random order.

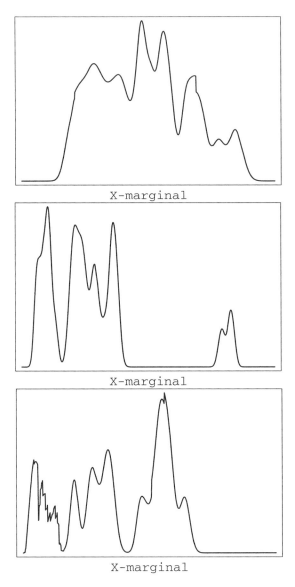

FIGURE 15.2. X-marginals (from west to east coast) for Mexico, Spain, and the United States of America in random order.

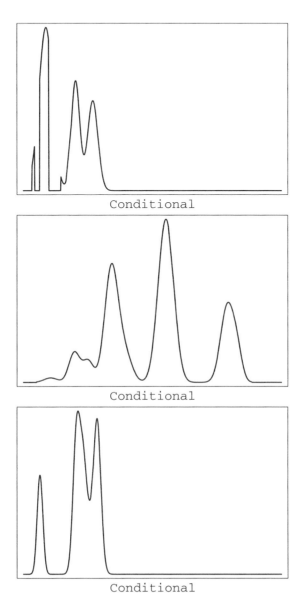

FIGURE 15.3. Conditionals (from west to east coast) for Mexico, Spain, and the United States of America in random order.

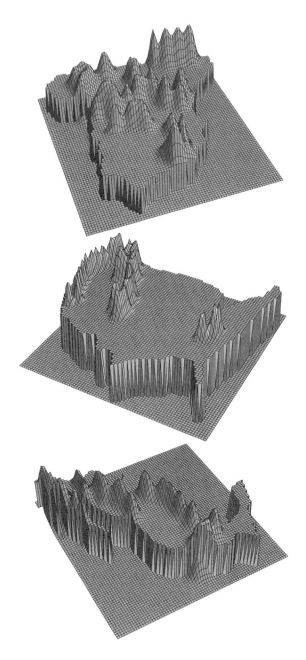

FIGURE 15.4. Joint densities (i.e., normalized topographical maps) of Spain, the United States of America, and Mexico.

his location on the prime meridian be governed by almost any distribution that pleases you! Kolmogorov's message that, for absolutely continuous (X, Y), it doesn't really make sense to talk about the distribution of X given Y equals a particular value y_0, must be remembered, since the event $\{Y = y_0\}$ is the same as the event $\{g(Y) = g(y_0)\}$ for any invertible function g. It <u>does</u> make sense to speak of the family of conditional densities of X given $Y = y$ as y ranges over $S(Y)$.

15.10   Bibliographic Notes

Hurlimann (1993) discussed diatomic conditionals (Section 15.2). Key references for Section 15.3 are Gupta (1998) and Arnold and Strauss (1988a). SenGupta (1995) discusses hypothesis testing in certain conditionally specified models (Section 15.4).

Arnold and Pourahmadi (1988) is useful reference for Section 15.5. Key references for Section 15.6 are Korwar (1974) and Wesolowski (1995a, 1995b). Section 15.7 is based on Arnold and Gokhale (1998c). The maps discussed in Section 15.8 were prepared with the assistance of Carmen Sánchez. Warnings about the dangers of facile interpretations of conditional densities (Section 15.9) date back to Borel. More recent references are Arnold and Robertson (1998), Rao (1993), and Proschan and Presnell (1998).

Exercises

15.1 Suppose X, Y are i.i.d. standard exponential random variables. Verify that $X | X - Y = 0 \sim \Gamma(1, 2)$ while $X | \log X - \log Y = 0 \sim \Gamma(2, 2)$. Can you identify a function g such that $X | g(X) - g(Y) = 0 \sim \Gamma(3, 2)$?

15.2 Suppose that $(X_1, X_2, \ldots, X_k)$ is a k-dimensional random vector such that
$$(X_1, X_2, \ldots, X_{k-1}) \stackrel{d}{=} (X_2, X_3, \ldots, X_k)$$
and that
$$X_n | X_1 = x_1, X_2 = x_2, \ldots, X_{k-1} = x_{k-1} \sim N\left(\alpha + \sum_{j=1}^{k-1} \beta_j x_j, \sigma^2\right).$$

Verify that $(X_1, X_2, \ldots, X_k)$ must have a k-variate normal distribution.

15.3 Consider the following general normal process. Let $\mu : \mathbb{R} \to \mathbb{R}$ and $\sigma : \mathbb{R} \to \mathbb{R}^+$. For each n define $X_n = \mu(X_{n-1}) + \sigma(X_{n-1}) Z_n$ where the Z_n's are i.i.d. $N(0, 1)$ random variables. For what choices of $\mu(\cdot)$ and $\sigma(\cdot)$ does this process have a proper long-run distribution?

15.4 Consider again the incompatible conditional matrices A and B given in (15.40) and (15.41). Find the most nearly compatible joint distribution P_{θ_1,θ_2} which has independent binomial $(2,\theta_1)$ and binomial $(2,\theta_2)$ marginals.

15.5 Suppose that X has a power series distribution i.e., $P(X = x) \propto a(x)\theta^x, x = 0, 1, 2\dots$. Suppose also that $E(X|Y = y) = \sum_{j=0}^{\infty} b_j(y)\theta^j$ with $b_0(y) = y$ for every $y = 0, 1, 2\dots$. Show that the distribution of (X, Y) can be uniquely determined.

(Kyriakoussis and Papageorgiou (1989).)

Appendix A
Simulation

A.1 Introduction

One of the most important tools for deriving the distributional properties
of complicated functions of random variables arising in practice, is the
Monte Carlo method. In fact, on many occasions the analytical treatment
of statistical problems is impossible and we are forced to use simulations
in order to get the desired result.

In this book a broad spectrum of conditionally specified models has been
introduced. The usefulness of these models would be severely limited if simulation of random variables with these distributions were to prove difficult
or impossible. The presence of complicated normalizing constants in most
conditionally specified models might suggest that simulation would be difficult. Despite the fact that we often lack analytic expressions for the densities, it turns out to be quite easy to devise relatively efficient simulation
schemes. As observed by Arnold and Strauss (1988a), a straightforward
rejection scheme will often accomplish this goal. Alternatively, importance
sampling simulation techniques also allow us to forget about the normalizing constant problem. In Sections A.2 and A.3 we will review the fundamental ideas of the rejection and importance sampling methods. In subsequent sections we give more detailed descriptions of simulation strategies
appropriate for some specific conditionally specified models.

For conditionally specified distributions, the simulation strategies outlined in this appendix provide attractive alternatives to the always available
Gibbs sampler simulation method.

A.2   The Rejection Method

The rejection method is based upon the following theorem (see Rubinstein (1981) or Devroye (1986)):

Theorem A.1   *Let $\underline{X}$ be a k-dimensional random vector with density $f_{\underline{X}}(\underline{x})$. Suppose that $f_{\underline{X}}(\underline{x})$ can be represented in the form*

$$f_{\underline{X}}(\underline{x}) = Cg(\underline{x})h(\underline{x}), \qquad (A.1)$$

where $C > 1, 0 < g(\underline{x}) < 1$, and $h(\underline{x})$ is also a probability density function. Let U be a standard uniform $(U[0,1])$ random variable and let $\underline{Y}$ be an independent random variable with pdf $h(y)$. It follows that the conditional distribution of $\underline{Y}$, given that $U \leq g(\underline{Y})$, coincides with the distribution of $\underline{X}$.

In the light of this theorem we may use the following algorithm to simulate pseudorandom variables corresponding to the density $f_{\underline{X}}(\underline{x})$:

Algorithm A.1 (Simulating by the rejection method (Theorem A.1)).

Input. *The density being simulated $f_{\underline{X}}(\underline{x})$ and the density $h(\underline{x})$ used for simulation.*

Output. *A random realization from $f_{\underline{X}}(\underline{x})$.*

Step 1. *Generate one candidate random variate $\underline{X}$ with density $h(\underline{x})$.*

Step 2. *Generate one random variate U uniformly distributed on $[0,1]$ independent of $\underline{X}$.*

Step 3. *Repeat Steps 1 and 2 until $U \leq g(\underline{X})$.*

Step 4. *Output $\underline{X}$.*

Clearly a random number, say N, of iterations is required to generate one pseudovariate $\underline{X}$. This random variable N, the waiting time until the condition in Step 3 is satisfied, has a geometric distribution with parameter p that satisfies

$$p = P(U \leq g(\underline{X})) = \int_{\mathbf{R}^k} g(\underline{x})h(\underline{x})\, d\underline{x} = 1/C. \qquad (A.2)$$

It follows that

$$E(N) = C \qquad (A.3)$$

and

$$\mathrm{var}(N) = C(C-1). \qquad (A.4)$$

When C is large, the efficiency of the above algorithm is low and the rejection rate is high. Consequently, we may have to generate a huge number of random variables from $h(x)$ to obtain a small sample from $f_X(x)$.

To have an efficient algorithm we should endeavor to choose C as small as possible.

In order to implement the rejection method we need:

(i) A convenient representation of the density $f_X(\underline{x})$ in the form (A.1) with a C close to 1.

(ii) A convenient and efficient method to generate pseudorandom variables with the density $h(\underline{x})$ appearing in (A.1).

Note that the precise value of C is not needed to implement the rejection scheme, an important feature when we apply the technique to conditionally specified densities with undetermined normalizing constants. As we shall see, we are often able to choose a density $h(\underline{x})$, to be used in our rejection scheme, which has independent coordinates. This frequently allows relatively simple generation of pseudovariates corresponding to $h(\underline{x})$, since we then only need to simulate one-dimensional random variables with well known and understood distributions. A convenient source of algorithms for generating univariate pseudorandom variables with any of the "usual" distributions is Devroye (1986).

It is in fact often possible to implement the rejection scheme alluded to in Theorem A.1, when both $f_X(\underline{x})$ (the hard to simulate density) and $h(\underline{x})$ (the easy to simulate density) are both only known up to a constant. Suppose that
$$f_X(\underline{x}) \propto g_1(\underline{x})$$
and
$$h(\underline{x}) \propto g_2(\underline{x}).$$
Now, provided we can find a constant k such that
$$g_2(\underline{x}) \geq kg_1(\underline{x}), \quad \forall \underline{x}, \tag{A.5}$$
then we can generate pseudovariates corresponding to the hard density $f_X(\underline{x})$ by generating a value $\underline{X} = \underline{x}$ from the easy density $h(\underline{x})$ and keeping it with probability $kg_1(\underline{x})/g_2(\underline{x})$. The retained $\underline{X}$'s will have density $f_X(\underline{x})$. For efficiency k should be as small as possible.

However, obtaining the value of k is not an easy task. We can avoid this problem for large enough samples. To this end, we modify the rejection method as follows:

Algorithm A.2 (Modified rejection method for use when C in (A.1) or k in (A.5) is not known).

Input. *The density being simulated $f_X(\underline{x})$, the density $h(\underline{x})$ used for simulation, and the sample size n.*

Output. *An approximate random sample of a random size N such that $E(N) = n$ from $f_X(x)$.*

Step 1. *Make $i = 1$, $s_{sum} = 0$ and $s_{max} = 0$.*

Step 2. *Generate one random variate $\underline{X}_i$ with density $h(x)$.*

Step 3. *Calculate $s_i = f_X(\underline{X}_i)/h(\underline{X}_i)$, $s_{max} = \max(s_{max}, s_i)$ and $s_{sum} = s_{sum} + s_i$.*

Step 4. *If $s_{sum}/s_{max} \geq n$ go to Step 5; otherwise make $i = i + 1$ and go to Step 2.*

Step 5. *For $j = 1$ to i do:*

    1. $s_j = s_j/s_{max}$.

    2. Generate a uniform $U(0,1)$ random number U_j.

    3. If $U_j < s_j$ accept $\underline{X}_j$ as one item in the generated sample; otherwise reject it.

Step 6. *Return the generated sample.*

The main shortcoming of this method is that we get a sample of unknown size, but close to n. Note that the algorithm is designed to provide $E(N) = n$.

A.3   The Importance Sampling Method

The acceptance-rejection algorithm, however, can be made more efficient by the following modification. Write $f_X(x)$ as

$$f_X(x) = \frac{f_X(x)}{h(x)} h(x) = s(x)h(x), \tag{A.6}$$

where

$$s(x) = \frac{f_X(x)}{h(x)} \tag{A.7}$$

is a *score function.* Thus, the score of the event x is the ratio of the population distribution, $f_X(x)$, to the simulation distribution, $h(x)$.

From (A.1) and (A.7), we see that $s(x) = c\,g(x)$, that is, the score is proportional to $g(x)$. Therefore, instead of rejecting a number x generated from $h(x)$, we assign it a probability proportional to $s(x)$ or $g(x)$. Then at the end we normalize the scores (by dividing each score by the sum of all scores) and use the normalized scores to estimate the probability of any event of interest. This leads to a much higher efficiency of the simulation process.

Note however that in this approach we, in fact, approximate the distribution of the random variable instead of obtaining a simulated sample from its distribution.

The above discussion suggests a general framework for simulation methods. Let $X = \{X_1, \ldots, X_n\}$ be a set of variables with joint probability density function $f_X(\underline{x})$.

After a simulated sample of n realizations from $h(\underline{x})$, $\underline{x}^j = \{x_1^j, \ldots, x_n^j\}$, $j = 1, \ldots, n$, is obtained, the distribution of the random variable $\underline{X}$ is approximated by that of the discrete random variable whose support is the set associated with the sample and whose probability mass function is given by the set of normalized scores $\{s_i / \sum_{j=1}^n s_j : i = 1, 2, \ldots, n\}$.

The above procedure is described in the following general algorithm:

Algorithm A.3 (General simulation framework).

- **Input.** *The population distribution* $f_X(\underline{x})$, *the simulation probability density* $h(\underline{x})$, *and the sample size* n.

- **Output.** *An approximation of the population distribution by that of a discrete random variable with its probability mass function.*

1. *For $j = 1$ to n:*

    - *Generate $\underline{x}^j$ using $h(\underline{x})$.*

    - *Calculate $s(\underline{x}^j) = \dfrac{f_X(\underline{x}^j)}{h(\underline{x}^j)}$.*

2. *Return the support $S(\underline{X}^*) = \{\underline{x}^j : j = 1, 2, \ldots, n\}$ and the probability mass function $P(\underline{X}^* = \underline{x}^j) = s(\underline{x}^j) / \sum_{j=1}^n s(\underline{x}^j)$.*

The selection of the simulation distribution influences the quality of the approximation considerably. Sampling schemes leading to similar scores for all realizations are associated with high quality and those leading to substantially unequal scores have low quality.

The accuracy of the approximation obtained using Algorithm A.3 depends on the following factors:

- The population distribution $f_X(\underline{x})$.

- The simulation distribution $h(\underline{x})$ chosen to obtain the sample.

- The method used to generate realizations from $h(\underline{x})$.

- The sample size n.

To facilitate the simulation process, selection of a Bayesian network model for $h(\underline{x})$ is a good choice. Careful selection of the method to generate realizations from $h(x)$ is also very important. Many of the existing methods are variants of the above method; they usually differ only in one or both of the last two components.

Given a population distribution $f_X(\underline{x})$, each of the above three algorithms generates a sample of size n (exact or approximate) from $f_X(\underline{x})$ or an approximation of the population distribution. They differ only in how the sample is generated (simulation distribution) and in the choice of the scoring function. As an illustration we give below one simple example.

Example A.1 (Uniform sampling method). Suppose we are interested in approximating the distribution of an n-dimensional random variable $(X_1, X_2, \ldots, X_k)$. In this method the simulation distribution of the variable X_i, is uniform, that is,

$$h(x_i) = \frac{1}{S}, \tag{A.8}$$

where $S = \int_{S(X_i)} dx_i$.

Once a realization $\underline{x}$ is generated, the associated score becomes

$$s(\underline{x}) = f_X(\underline{x}), \tag{A.9}$$

where we have ignored the factor $1/S$ since it is identical for all realizations.

This method can be applied to the set of variables $\{X_1, X_2, \ldots, X_k\}$ in any order because for each i, $h(x_i)$ does not depend on the value of any other variable.

More details and examples of the importance sampling method can be seen in Castillo, Gutiérrez, and Hadi (1997), Salmerón (1998a, 1998b), or Hernández, Moral, and Salmerón (1998).

A.3.1 Systematic Sampling Method

In this subsection we mention the systematic sampling techniques that have been shown to be much more efficient than the stochastic ones. Recently, Bouckaert (1994) and Bouckaert, Castillo, and Gutiérrez (1996) introduced a new method for generating the realizations forming a sample in a systematic way. Unlike the algorithms introduced in the previous sections, which are stochastic in nature, this method proceeds in a deterministic way.

The original idea comes from stratified sampling. It is well known that if we divide the sample space in several regions and allocate the sample in an optimal number of subsamples, each taken from the corresponding region, the resulting sample leads to estimates with smaller variance. The limiting case consists of dividing the sample space in n regions such that only one sample is taken from each region, and even the sample values can be deterministically selected. For a detailed description of this method and some examples see Castillo, Gutiérrez, and Hadi (1997), Salmerón (1998a, 1998b), or Hernández, Moral, and Salmerón (1998).

A.4   Application to Models with Conditionals in Exponential Families

Simulation of two-dimensional random variables corresponding to conditionally specified models can be approached in three ways:

(a) by direct simulation of the bivariate random variables using the joint density;

(b) by simulating one marginal and then simulating a corresponding covariate using the appropriate conditional density; and

(c) by using the importance sampling method. This is useful only when we aim to approximate the expectation of a sample statistic and not to actually simulate a sample.

Consider the general class of conditionals in exponential families (CEF) distributions, with joint density given by (4.5), i.e.,

$$f_{X,Y}(x,y) = r_1(x)r_2(y) \exp\{\underline{q}^{(1)}(\underline{x})' M \underline{q}^{(2)}(y)\}. \qquad (A.10)$$

In order to utilize the rejection scheme based on Theorem A.1, it will be sufficient to determine an upper bound for the factor responsible for dependence in $f_{X,Y}(x,y)$. Thus we define

$$\Delta = \sup_{x,y} \left[\exp[\tilde{\underline{q}}^{(1)}(x)' \tilde{M} \tilde{\underline{q}}^{(2)}(y)] \right], \qquad (A.11)$$

where $\tilde{M}$ is as defined in (4.6) and $\tilde{\underline{q}}^{(1)}(x)$ (respectively $\tilde{\underline{q}}^{(2)}(y)$) is $\underline{q}^{(1)}(x)$ (respectively $\underline{q}^{(2)}(y)$) with its first coordinate deleted. We may then define

$$g(x,y) = \Delta^{-1} \exp[\tilde{\underline{q}}^{(1)}(x)' \tilde{M} \tilde{\underline{q}}^{(2)}(y)] \qquad (A.12)$$

and

$$h(x,y) \propto r_1(x)r_2(y) \exp \left[\sum_{i=1}^{\ell_1} m_{i0}q_{1i}(x) + \sum_{j=1}^{\ell_2} m_{0j}q_{2j}(y) \right] \qquad (A.13)$$

(which corresponds to the easily simulated independent marginals model) and use our rejection scheme to generate pseudo-observations from (A.10).

The second approach available to us involves simulation of observations from the marginal $f_X(x)$ of (A.10) and then simulating the corresponing values of the Y variable using the known (exponential family) conditional distribution of Y given X. We may illustrate this approach in the exponential conditionals case, i.e., when $f_{X,Y}(x,y)$ is given by equation (4.14). In this case, the marginal density of X is given by

$$f_X(x; m_{10}, m_{01}, m_{11}) = \theta \left(-\frac{m_{11}}{m_{10}m_{01}} \right) \frac{m_{01}}{m_{01} - m_{11}x} m_{10} e^{-m_{10}x}, \qquad x > 0.$$

$$(A.14)$$

To simulate observations from (A.14) we may use our rejection scheme (Theorem A.1) with the following choices for $h(x)$ and $g(x)$:

$$h(x) = m_{10}e^{-m_{10}x}, \quad x > 0, \tag{A.15}$$

and

$$g(x) = m_{01}/(m_{01} - m_{11}x) \tag{A.16}$$

(recall $m_{11} < 0$ so that $g(x) \leq 1$). Note that (A.15) corresponds to an exponential distribution which is particularly easy to simulate. Having generated a pseudovalue X^* for X using this scheme, we may generate an appropriate value Y^* to pair with it, by recalling that the conditional distribution of Y given $X = X^*$ is exponential with mean $(m_{01} - m_{11}X^*)^{-1}$. So we need to generate independent exponential variates until we accept one for X^* and then need to generate just one more exponential variate to get the corresponding value of Y^*. In this fashion we obtain (X^*, Y^*), a pseudorandom variable corresponding to the density (4.14).

The third approach is to use the importance sampling technique. In this case we can proceed as above but forgetting about the problem of the normalizing constant, i.e., we do not need to calculate the Δ bound in (A.11). We use

$$s(x, y) = \exp[\tilde{q}^{(1)}(x)' \tilde{M} \tilde{q}^{(2)}(y)/h(x, y)]. \tag{A.17}$$

However, at the end we must normalize the weights (or scores) $s(x, y)$, replacing them by

$$s^*(x, y) = \frac{s(x, y)}{\displaystyle\sum_{(x,y)} s(x, y)}.$$

The end product of this exercise is an approximation to the distribution of (X, Y), rather than a simulated sample.

A.5   Other Conditionally Specified Models

The same approach, involving rejection, can usually be used to generate pseudorandom variables corresponding to the distributions catalogued in Chapter 5. We will illustrate by considering the Pareto conditionals density (5.7). In this case our joint density is of the form

$$f_{X,Y}(x, y) = k(\lambda_{00}, \lambda_{10}, \lambda_{01}, \lambda_{11})(\lambda_{00} + \lambda_{10}x + \lambda_{01}y + \lambda_{11}xy)^{-(\alpha+1)}, \quad x, y > 0. \tag{A.18}$$

In this case we may use the rejection scheme of Theorem A.1 or the importance sampling technique with the following choices for $h(x, y)$ and

$g(x, y)$:

$$h(x, y) \propto (\lambda_{00} + \lambda_{10}x + \lambda_{01}y)^{-(\alpha+1)} \tag{A.19}$$

$$g(x, y) = \left(\frac{\lambda_{00} + \lambda_{10}x + \lambda_{01}y}{\lambda_{11} + \lambda_{10}x + \lambda_{01}y + \lambda_{11}xy} \right)^{(\alpha+1)}. \tag{A.20}$$

Note that $h(x, y)$ corresponds to the Mardia bivariate Pareto distribution which is easily simulated (see, e.g., Devroye (1986), p. 602).

Alternatively we can first generate a variate corresponding to the X marginal of (A.18) and then generate the corresponding Y value using the Pareto conditional density of Y given X. In this case the X marginal takes the form

$$f_X(x; \lambda_{00}, \lambda_{10}, \lambda_{01}, \lambda_{11}) \propto [(\lambda_{01} + \lambda_{11}x)(\lambda_{00} + \lambda_{10}x)^{\alpha}]^{-1}, \quad x > 0. \tag{A.21}$$

We may simulate variates with the density (A.21), using our rejection scheme by setting

$$h(x) \propto (\lambda_{00} + \lambda_{10}x)^{-\alpha} \tag{A.22}$$

and

$$g(x) = \frac{\lambda_{01}(\lambda_{00} + \lambda_{10}x)^{\alpha}}{(\lambda_{01} + \lambda_{11}x)(\lambda_{00} + \lambda_{10}x)^{\alpha}}. \tag{A.23}$$

The density (A.22) is an easily simulated Pareto density. Thus we simulate independent Pareto's until one is accepted as X^*, and then simulate one more Pareto variate to get a corresponding value of Y^*.

A.6   A Direct Approach Not Involving Rejection

For our conditionally specified models, we typically have no trouble generating a value of Y to pair with a value of X corresponding to the marginal density $f_X(x)$. So, effectively, the only challenging problem is generating pseudovariates corresponding to a marginal density

$$f_X(x; \underline{\theta}) \propto g(x, \underline{\theta}), \tag{A.24}$$

where g is known and relatively simple in form. For given values of the parameters $\underline{\theta}$, we can determine the normalizing constant in (A.24) by numerical integration. We may then apply the inversion method. That is we simulate a uniform $U(0, 1)$ variate, say U, and then, by numerical integration, determine a value X^* such that

$$\int_{-\infty}^{X^*} f_X(t, \underline{\theta}) \, dt = U. \tag{A.25}$$

Due to the monotone character of the integral, this can be efficiently done by the bisection method. With this generated value of X^* in hand, we then generate the corresponding value Y^* using the conditional density of Y given X.

A.7   Bibliographic Notes

Rubinstein (1981) and Devroye (1986) are excellent references for simulation techniques including rejection techniques. The observation that we only need to know kernels of densities to implement a rejection scheme does not seem to be explicitly mentioned in most simulation texts but it is undoubtedly not new to simulation experts. A convenient reference for discussion of importance sampling is Tanner (1996) or Castillo, Gutiérrez and Hadi (1997).

Exercises

A.1 In Bayesian analyses it is common to use posterior means as the working parameters. If the posterior distribution cannot be easily integrated, two alternative approaches are:

(a) Simulate a sample from the posterior and calculate the sample mean.

(b) Replace the posterior distribution by an approximation.

Design a simulation method and an importance sampling method to solve this problem in the case of the posterior density of the form (13.27).

A.2 Adapt Algorithms A.1, A.2, and A.3 to simulate a normal random variable using the Gumbel distribution with cdf,

$$F(x) = \exp\left[-\exp\left(-\frac{x-a}{b}\right)\right], \quad -\infty < x < \infty,$$

as the simulation distribution.

A.3 Suggest a simulation method for:

(a) The Student-t distribution in (5.84).

(b) The Pearson type VI conditionals distribution in (5.17).

(c) The generalized Pareto conditionals distribution in (5.31).

(d) The Cauchy conditionals distribution in (5.47).

A.4 Obtain, by simulation, 1000 replications of a $\Gamma(2, 0.3)$ variable using, as the simulation distribution:

(a) A $\Gamma(1, 0.3)$ distribution.

(b) A $\Gamma(1, 1)$ distribution.

(c) The distribution with cdf,

$$F(x) = 1 - e^{-x^2}, \quad x \geq 0.$$

Graph the exact and the three simulated distributions above.

Compare the empirical cdf of the samples with the true distribution, and discuss the results.

A.5 Use the importance sampling method described in Section A.4 for the normal conditionals distribution in (3.26) (1000 replications).

Draw the true density and the histogram of the simulated sample, and compare them.

Repeat the process for 10000 and 100000 replications.

Appendix B
Notation Used in This Book

In this appendix we list the main notations used in this book. We have attempted to keep the notations consistent throughout the book as much as possible.

Notations

$a_{ij}, p_{1|2}$ .... Conditional probability of $X = x_i$ given $Y = y_j$

a_{ijk} .... Conditional probability of $X = x_i$ given
$\qquad\qquad Y = y_j$, $Z = z_k$

$a(x, y)$ .... Conditional density of X given Y

A .... Matrix whose (i, j)th element is the conditional
$\qquad\qquad$ probability of $X = x_i$ given $Y = y_j$

b_{ijk} .... Conditional probability of $Y = y_i$ given
$\qquad\qquad X = x_j$, $Z = z_k$

$b(x, y)$ .... Conditional density of Y given X

$ba(x|z)$ .... $\int_{S(Y)} a(x, y)b(z, y) \; d\mu_2(y)$

$\text{binomial}(n, p)$ ... Binomial random variable with parameters
$\qquad\qquad n$ and p

B .... Matrix whose (i, j)th element is the conditional
$\qquad\qquad$ probability of $Y = y_j$ given $X = x_i$

$B(p, q)$ .... The beta function

$B2(p, q, \sigma)$ .... The beta distribution of the second kind

c_{ijk} .... Conditional probability of $Z = z_i$ given
$\qquad\qquad X = x_j$, $Y = y_z$

cdf .... Cumulative distribution function

cosh .... Hyperbolic cosine

coth .... Hyperbolic cotangent

$\text{cov}(X, Y)$ .... Covariance between X and Y

CEF .... Conditionals in exponential families

CS .... Conditional specification models

C_{ij} .... Subvector of $\underline{X}_{(i)}$

$C(\mu, \sigma)$ .... Cauchy distribution with location and
scale parameters

$\pi(C)$ .... Cone generated by the columns of C

d_{ij} .... The (i, j) element in the D matrix

D .... Cross-product ratio matrix of a 2×2 matrix

exp .... Exponential function

$E(X)$ .... Expectation of the random variable X

$E(X|Y = y), \psi(y)$ Conditional expectation of X given $Y = y$

$-\text{Ei}(x)$ .... Classical exponential integral function

$\text{Exp}(\lambda)$ .... Exponential random variable with mean $1/\lambda$

$\text{Exp}(\alpha, \lambda)$ .... Exponential distribution with location
parameter α and scale parameter $1/\lambda$

f_{ij} .... Element (i, j) of the cross-product ratio matrix

$f_X(x)$ .... Marginal pdf of X

$f_{X|Y}(x|y)$ .... Conditional density of X given Y

$f_{Y|X}(y|x)$ .... Conditional density of Y given X

$f(x|Y > y)$ ..... pdf of the random variable X given $Y > y$

$F(a, b)$ .... Complete elliptic integral of the first kind

$F(a, b; c, d)$ ...... Hypergeometric function

F .... The cross-product ratio matrix

$\bar{F}(x; \underline{\theta})$ .... Parametric family of survival functions

$F_{X,Y}(x, y)$ .... Bivariate cumulative distribution function

$F_{X|Y}(x|y)$ .... Conditional cdf of X given Y

$\bar{F}(x, y)$ .... Bivariate survival function

$\bar{F}(x_1, \ldots, x_k)$ .... Multivariate survival function

$\mathcal{G}(p)$ .... Geometric distribution

$\mathcal{GP}(\sigma, \delta, \alpha)$ ...... Generalized Pareto random variable

$\mathcal{GP}^*(\mu, \sigma, \delta, \alpha)$ ... Generalized Pareto random variable
with location parameter μ

$h(x|Y > y)$ ...... Conditional hazard function

$I(a, x)$ .... Incomplete gamma function

$I(A)$ .... Indicator function, which is equal to 1 if A holds,
and 0 otherwise

IG .... Inverse Gaussian distribution

$I(Q, P)$ .... Kullback-Leibler information distance between
two matrices Q and P

$i_{\theta_i \theta_j}$ .... The (i, j)th element of the Fisher
information matrix

J .... Jacobian
$k(c)$ .... Normalizing constant
$k_{r,s}(c)$ .... Normalizing constant in gamma conditionals
                    Model II
$\ell(\underline{\theta})$ .... The log-likelihood function
$\log$ .... Logarithm function
$L(\underline{\theta})$ .... The likelihood function
$L(V)$ .... Linear space generated by the columns of V
$M_f(t)$ .... The Laplace transform of f
$\mathrm{ML}(\underline{\theta})$ .... The marginal likelihood function
$M_{X,Y}(s,t)$ .... The joint moment generating function
$\mathrm{normal}(\mu,\sigma^2)$ .... Normal distribution with mean
                    μ and variance σ^2
$\mathbb{N}$ .... Set of natural numbers
$N(\mu,\sigma^2)$ .... Normal distribution with mean μ
                    and variance σ^2
$N_k(\underline{\mu},\Sigma)$ .... k-dimensional normal distribution with
                    mean vector $\underline{\mu}$ and covariance matrix Σ
N^A, N .... Incidence set of A
pdf .... Probability density function
$p_i.$ .... Marginal probability of X
$p_{.j}$ .... Marginal probability of Y
p_{ij} .... Joint probability of (X,Y)
$p_{ij}^{(n)}$ .... nth iteration of an iterative scheme
$\mathrm{PdH}(k,\alpha)$ .... Pickands–de Haan random variable
$\mathrm{PL}(\underline{\theta})$ .... The pseudolikelihood function
$P(\sigma,\alpha)$ .... Pareto random variable with pdf proportional
                    to $(1+x/\sigma)^{-(\alpha+1)}$
$\mathrm{PS}(\theta)$ .... Power series distribution
$\mathrm{Poisson}(\lambda)$ .... Poisson random variable with mean λ
$P(X=x_i)$ ...... Probability of $X=x_i$
$P(X>x), \bar{F}(x)$. The survival function
$P(X>x, Y>y)$ Bivariate survival function
$P(X>x|Y>y)$. Conditional survival function
$P(X=x_i|Y=y_j)$ Conditional probability
                    of $X=x_i$ given $Y=y_j$
$P(Y=y_j)$ .... Probability of $Y=y_j$
$P(Y=y_j|X=x_i)$ Conditional probability of $Y=y_j$ given $X=x_i$
Q .... Quadratic measure of discrepancy
$r_X(x)$ .... Failure rate of X
$r_{X|Y}(x|y)$ .... Conditional failure rate of X given Y
$\mathbb{R}$ .... Set of real numbers
R_{XY} .... The sample correlation of (X,Y)
RR2 .... Reversed rule of order 2

$\text{sign}(a)$ . . . . Sign of a

$\sinh$ . . . . Hyperbolic sine

SE . . . . Simultaneous equations models

$S(X)$ . . . . Support of the random variable X

S_X^2 . . . . The sample variance of X

S_{XY} . . . . The sample covariance of (X, Y)

S_ψ . . . . Contraction Mapping

$\text{SBC}(\underline{\alpha}, A)$ . . . . Scaled beta conditionals distributions

TP2 . . . . Total positivity of order 2

U . . . . Standard uniform random variable

$U(a, b, z)$ . . . . Confluent hypergeometric function

$\text{var}(X|Y = y)$ . . . . Conditional variance of X given $Y = y$

$V(X)$ . . . . Variance of the random variable X

$W(c)$ . . . . Weibull random variable such that $W^{1/c}$
        is unit exponential

$\underline{w}_i$ . . . . One of the generator vectors of a cone

w_{ij} . . . . Slack variable

$\underline{x}$ . . . . k-dimensional vector

$\underline{x}_{(i)}$ . . . . $(k-1)$-dimensional vector obtained
        from $\underline{x}$ by deleting x_i

$\underline{x}_{(i,j)}$ . . . . The vector $\underline{x}$ with its ith and jth
        coordinates deleted

X, Y . . . . Univariate random variables

$\underline{X}_{(i)\ell}$ . . . . Subvector of $\underline{X}_{(i)}$ with ℓ coordinates

(X, Y) . . . . Two-dimensional random variable

$X|Y$ . . . . Conditional random variable X given Y

$\underline{X}$ . . . . k-dimensional random vector

$\underline{X}_{(i)}$ . . . . $(k-1)$-dimensional random vector
        obtained from $\underline{X}$ by deleting X_i

$\underline{X}_{(i,j)}$ . . . . The random vector $\underline{X}$ with its ith
        and jth coordinates deleted

$\underline{X} = (\dot{\underline{X}}, \ddot{\underline{X}})$ . . . . . A partitioning of $\underline{X}$ into two subsets with k_1
        and $k - k_1$ coordinates

$\overline{X}$ . . . . The sample mean of X

Z . . . . Normal random variable with mean 0 and
        variance 1

$(\varepsilon_1, \varepsilon_2)$ . . . . Error terms in bivariate SE models

η_j . . . . Probability of $Y = y_j$

$\Gamma(a)$ . . . . Gamma function

$\Gamma(\alpha, \sigma)$ . . . . Gamma random variable with pdf
        proportional to $x^{\alpha-1}e^{-\sigma x}$

Λ . . . . Positive definite matrix

μ . . . . Location parameter

$\mu_X(x)$ . . . . Mean residual life of X

$\mu(y)$ . . . . Conditional expectation of X given $Y = y$

$\Omega(\pi(C))$......... The dual or polar cone of $\pi(C)$

$\prod_{i=1}^{n} x_i$ .... $x_1 \times \ldots \times x_n$

$\phi_z(x)$............ The Wöhler field

$\Phi(x)$ .... Standard normal distribution function

$\phi_{X,Y}(t_1, t_2)$...... Bivariate characteristic function

$\Psi(x)$ .... The digamma function

$\Psi_{X,Y}(t_1, t_2)$ ..... Bivariate Laplace transform

$\pi(W)$ .... Cone generated by W

$\rho(X,Y)$ .... Coefficient of linear correlation between
                           X and Y

σ .... Scale parameter

$\sigma^2(y)$........... Conditional variance of X given $Y = y$

Σ .... A nonnegative definite matrix

$\sum_{i=1}^{n} x_i$ .... $x_1 + \ldots + x_n$

τ_i .... Probability of $X = x_i$

$\overset{a.s.}{\rightarrow}$ .... Almost sure convergence

$\approx$ .... Approximately equal to

$\overset{\Delta}{=}$ .... Equal by definition

$\sim$ .... Distributed as

$\emptyset$ .... Empty set

$\exists$ .... Exists at least one

$\propto$ .... Proportional to

$\cap$ .... Set intersection

$\cup$ .... Set union

References

Abrahams, J. and Thomas, J. (1984), A Note on the Characterization of Bivariate Densities by Conditional Densities. *Communications in Statistics, Theory and Methods*, 13:395–400.

Abramowitz, M. and Stegun, I. (1964), *Handbook of Mathematical Functions*. National Bureau of Standards Applied Mathematical Sciences. U.S. Government Publishing Office, Washington, DC.

Aczél, J. (1966), *Lectures on Functional Equations and their Applications*. Academic Press, New York.

Ahsanullah, M. (1985), Some Characterizations of the Bivariate Normal Distribution. *Metrika*, 32:215–218.

Ahsanullah, M. and Wesolowski, J. (1993), Bivariate Normality via Gaussian Conditional Structure. *Preprint*.

Amemiya, T. (1975), Qualitative Response Models. *Annals of Economic and Social Measurement*, 4:363–372.

Anderson, D. N. (1990), *Some Time Series Models with Non-Additive Structure*. PhD thesis, Department of Statistics, University of California, Riverside, CA.

Anderson, D. N. and Arnold, B. C. (1991), Centered Distributions with Cauchy Conditionals. *Communications in Statistics, Theory and Methods*, 20:2881–2889.

Angus, J. E. (1989), Application of Extreme Value Theory to the BEC Family of Distributions. *Communications in Statistics, Theory and Methods*, 18:4413–4419.

Arnold, B. C. (1983), *Pareto Distributions*. International Cooperative Publishing House, Fairland, MD.

Arnold, B. C. (1987), Bivariate Distributions with Pareto Conditionals. *Statistics and Probability Letters*, 5:263–266.

Arnold, B. C. (1988a), Pseudo-likelihood Estimation for the Pareto Conditionals Distribution. Technical report 170, Department of Statistics, University of California, Riverside, CA.

Arnold, B. C. (1990), A Flexible Family of Multivariate Pareto Distributions. *Journal of Statistical Planning and Inference*, 24:249–258.

Arnold, B. C. (1991), Dependence in Conditionally Specified Distributions. In IMS Lecture Notes, Topics in Statistical dependence, Monograph 16, 13–18.

Arnold, B. C. (1995), Conditional Survival Models. In Balakrishnan, N., editor, *Recent Advances in Life-Testing and Reliability, a Volume in Honor of Alonzo Clifford Cohen Jr.*, pages 589–601. CRC Press, Boca Raton, FL.

Arnold, B. C. (1996), Marginally and Conditionally Specified Multivariate Survival Models. In Ghosh, S., Schucany, W., and Smith, W., editors, *Statistics of Quality*, pages 233–252. Marcel Dekker, New York.

Arnold, B. C., Athreya, K. B., and Sethuraman, J. (1998), Notes on the Specification of Bivariate Distributions by Marginal and/or Conditional Distributions. Technical report, Department of Statistics, University of California, Riverside, CA.

Arnold, B. C., Castillo, E., and Sarabia, J. M. (1992), *Conditionally Specified Distributions*. Lecture Notes in Statistics, Vol. 73, Springer-Verlag, New York.

Arnold, B. C., Castillo, E., and Sarabia, J. M. (1993a), Conditionally Specified Models: Structure and Inference. In Cuadras, C. M. and Rao, C. R., editors, *Multivariate Analysis: Future Directions 2*, pages 441–450. North-Holland, Amsterdam, Series in Statistics and Probability, Vol. 7.

Arnold, B. C., Castillo, E., and Sarabia, J. M. (1993b), Conjugate Exponential Family Priors for Exponential Family Likelihoods. *Statistics*, 25:71–77.

Arnold, B. C., Castillo, E., and Sarabia, J. M. (1993c), Multivariate Distributions with Generalized Pareto Conditionals. *Statistics and Probability Letters*, 17:361–368.

Arnold, B. C., Castillo, E., and Sarabia, J. M. (1993d), A Variation on the Conditional Specification Theme. *Bulletin of the International Statistical Institute*, 49:51–52.

Arnold, B. C., Castillo, E., and Sarabia, J. M. (1994a), A Conditional Characterization of the Multivariate Normal Distribution. *Statistics and Probability Letters*, 19:313–315.

Arnold, B. C., Castillo, E., and Sarabia, J. M. (1994b), Multivariate Normality via Conditional Specification. *Statistics and Probability Letters*, 20:353–354.

Arnold, B. C., Castillo, E., and Sarabia, J. M. (1995), General Conditional Specification Models. *Communications in Statistics. Theory and Methods*, 24:1–11.

Arnold, B. C., Castillo, E., and Sarabia, J. M. (1995a), Distribution with conditionals in the Pickands-deHaan generalized Pareto family. *Journal of the Indian Association for Productivity, Quality and Reliability (IAPQR Trans.)*, 20, 27–35, 1995.

Arnold, B. C., Castillo, E., and Sarabia, J. M. (1996a), Priors with Convenient Posteriors. *Statistics*, 28:347–354.

Arnold, B. C., Castillo, E., and Sarabia, J. M. (1996b), Specification of Distributions by Combinations of Marginal and Conditional Distributions. *Statistics and Probability Letters*, 26:153–157.

Arnold, B. C., Castillo, E., and Sarabia, J. M. (1997), Comparisons of Means Using Conditionally Conjugate Priors. *Journal of the Indian Society of Agricultural Statistics*, 49:319–344.

Arnold, B. C., Castillo, E., and Sarabia, J. M. (1998a), Bayesian Analysis for Classical Distributions Using Conditionally Specified Priors. *Sankhya, Ser. B*, 60, 228–245.

Arnold, B. C., Castillo, E., and Sarabia, J. M. (1998b), A Comparison of Conditionally Specified and Simultaneous Equation Models. Technical report. Department of Statistics. University of California, Riverside, CA.

Arnold, B. C., Castillo, E., and Sarabia, J. M. (1998c), Some Alternative Bivariate Gumbel Models. *Environmetrics* 9, 599-616.

Arnold, B. C., Castillo, E., and Sarabia, J. M. (1998d), The Use of Conditionally Priors in the Study of Ratios of Gamma Scale Parameters. *Computational Statistics and Data Analysis*, 27:125–139.

Arnold, B. C., Castillo, E., and Sarabia, J. M. (1999a), Exact and Near Compatibility of Discrete Conditional Distributions. *To appear.*

Arnold, B. C., Castillo, E., and Sarabia, J. M. (1999b), Multiple Modes in Densities with Normal Conditionals. *To appear.*

Arnold, B. C. and Gokhale, D. V. (1994), On Uniform Marginal Representations of Contingency Tables. *Statistics and Probability Letters*, 21:311–316.

Arnold, B. C. and Gokhale, D. V. (1998a), Distributions Most Nearly Compatible with Given Families of Conditional Distributions. The Finite Discrete Case. *Test*, 7, 377-390.

Arnold, B. C. and Gokhale, D. V. (1998b), Remarks on Incompatible Conditional Distributions. Technical report #260, Department of Statistics, University of California, Riverside, CA.

Arnold, B. C. and Gokhale, D. V. (1998c), Distributions in Given Parametric Families Most Nearly Compatible with Given Conditional Information. Technical report. Department of Statistics. University of California, Riverside, CA.

Arnold, B. C. and Kim, Y. H. (1996), Conditional Proportional Hazard Models. In Jewell, N. P., Kimber, A. C., Lee, M. L. T., and Whitmore, G. A., editors, *Lifetime Data: Models in Reliability and Survival Analysis*, pages 21–28. Kluwer Academic, Dordrecht, Netherlands.

Arnold, B. C. and Pourahmadi, M. (1988), Conditional Characterizations of Multivariate Distributions. *Metrika*, 35:99–108.

Arnold, B. C. and Press, S. J. (1989a), Bayesian Estimation and Prediction for Pareto Data. *Journal of the American Statistical Association*, 84:1079–1084.

Arnold, B. C. and Press, S. J. (1989b), Compatible Conditional Distributions. *Journal of the American Statistical Association*, 84:152–156.

Arnold, B. C. and Press, S. J. (1990), Pseudo-Bayesian Estimation. Technical report #186, Department of Statistics, University of California, Riverside, CA.

Arnold, B. C. and Robertson, C. A. (1998), The Conditional Distribution of X given $X = Y$ Can Be Almost Anything. Technical report #257, Department of Statistics, University of California, Riverside, CA.

Arnold, B. C. and Strauss, D. (1988a), Bivariate Distributions with Exponential Conditionals. *Journal of the American Statistical Association*, 83:522–527.

Arnold, B. and Strauss, D. (1988b), Pseudolikelihood Estimation. *Sankhya, Ser. B*, 53:233–243.

Arnold, B. C. and Strauss, D. (1991), Bivariate Distributions with Conditionals in Prescribed Exponential Families. *Journal of the Royal Statistical Society, Ser. B*, 53:365–375.

Arnold, B. C. and Thoni, J. (1997), Analysis of Contingency Tables Using Conditionally Conjugate Priors. Technical report #248, Department of Statistics, University of California, Riverside, CA.

Arnold, B. C. and Wesolowski, J. (1996), Multivariate Distributions with Gaussian Conditional Structure. *Stochastic Processes and Functional Analysis*, pages 45–59.

Barlow, R. E. and Proschan, F. (1981), *Statistical Theory of Reliability and Life Testing: Probability Models*. Silver Springs, MD.

Berger, J. O. (1985), *Statistical Decision Theory and Data Analysis, 2nd Ed.* Springer-Verlag, New York.

Besag, J. E. (1974), Spatial Interaction and the Statistical Analysis of Lattice Systems. *Journal of the Royal Statistical Society, Ser. B*, 36:192–236.

Bhattacharyya, A. (1943), On Some Sets of Sufficient Conditions Leading to the Normal Bivariate Distribution. *Sankhya*, 6:399–406.

Bickel, P. J. and Mallows, C. L. (1988), A Note on Unbiased Bayes Estimates. *The American Statistician*, 42:132–134.

Bischoff, W. and Fieger, W. (1991), Characterization of the Multivariate Normal Distribution by Conditional Normal Distributions. *Metrika*, 38:239–248.

Bischoff, W. (1993), On the Greatest Class of Conjugate Priors and Sensitivity of Multivariate Normal Posterior Distributions. *Journal of Multivariate Analysis*, 44:69–81.

Bischoff, W. (1996a), Characterizing Multivariate Normal Distributions by Some of its Conditionals. *Statistics and Probability Letters*, 26:105–111.

Bischoff, W. (1996b), On Distributions Whose Conditional Distributions Are Normal. A Vector Space Approach. *Mathematical Methods of Statistics*, 5:443–463.

Bouckaert, R. R. (1994), A Stratified Simulation Scheme for Inference in Bayesian Belief Networks. In *Proceedings of the Tenth Conference on Uncertainty in Artificial Intelligence*, pages 110–117, Portland (Oregon). Morgan Kaufmann, San Mateo, CA.

Bouckaert, R. R., Castillo, E., and Gutiérrez, J. M. (1996), A Modified Simulation Scheme for Inference in Bayesian Networks. *International Journal of Approximate Reasoning*, 14:55–80.

Brucker, J. (1979), A Note on the Bivariate Normal Distribution. *Communications in Statistics, Theory and Methods*, 8:175–177.

Bryc, W. and Plucinska, A. (1985), A Characterization of Infinite Gaussian Sequences by Conditional Moments. *Sankhya*, 47:166–173.

Cacoullos, T. and Papageorgiou, H. (1995), Characterizations of Discrete Distributions by a Conditional Distribution and a Regression Function. *Annals of the Institute of Statistical Mathematics*, 35:95–103.

Castillo, E. (1988), *Extreme Value Theory in Engineering*. Academic Press, New York.

Castillo, E., Cobo, A., Jubete, F., and Pruneda, R. E. (1998), *Orthogonal Sets and Polar Methods in Linear Algebra: Applications to Matrix Calculations, Systems of Equations and Inequalities, and Linear Programming*. Wiley, New York.

Castillo, E., Fernández-Canteli, A., Esslinger, V., and Thürliman, B. (1985), Statistical Model for Fatigue Analysis of Wires, Strands and Cables. *IABSE Proceedings P-82/85*, pages 1–40.

Castillo, E. and Galambos, J. (1985), Modelling and Estimation of Bivariate Distributions with Conditionals Based on Their Marginals. In Patil, G., editor, *Conference on Weighted Distributions*. Penn. State University.

Castillo, E. and Galambos, J. (1987a), Bivariate Distributions with Normal Conditionals. In *Proceedings of the International Association of Science and Technology for Development*, pages 59–62. Acta Press, Anaheim, CA.

Castillo, E. and Galambos, J. (1987b), Lifetime Regression Models Based on a Functional Equation of Physical Nature. *Journal of Applied Probability*, 24:160–169.

Castillo, E. and Galambos, J. (1989), Conditional Distributions and the Bivariate Normal Distribution. *Metrika*, 36:209–214.

Castillo, E. and Galambos, J. (1990), Bivariate Distributions with Weibull Conditionals. *Analysis Mathematica*, 16:3–9.

Castillo, E., Galambos, J., and Sarabia, J. M. (1990), Caracterización de Modelos Bivariantes con Distribuciones Condicionadas tipo Gamma. *Estadística Española*, 31:439–450.

Castillo, E., Gutiérrez, J. M., and Hadi, A. S. (1997), *Expert Systems and Probabilistic Network Models*. Springer-Verlag, New York.

Castillo, E. and Hadi, A. S. (1995), Modelling Lifetime Data with Applications to Fatigue Models. *Journal of the Americal Statistical Association*, 90:1041–1054.

Castillo, E. and Ruiz-Cobo, R. (1992), *Functional Equations in Science and Engineering*. Marcel Dekker, New York.

Castillo, E. and Sarabia, J. M. (1990a), Bivariate Distributions with Second Kind Beta Conditionals. *Communications in Statistics, Theory and Methods*, 19:3433–3445.

Castillo, E. and Sarabia, J. M. (1990b), Caracterización de distribuciones bivariantes con condicionadas tipo beta. In *Proceedings of the Jornadas matemáticas hispano-lusas*, pages 121–216, Évora (Portugal). Comisión Organizadora de las XV Jornadas Hispano Lusas de Matemáticas. Departamento de Matemáticas de la Universidad de Évora.

Castillo, E. and Sarabia, J. M. (1991), Bivariate Distributions with Type Burr Conditionals. In de Cantabria, U., editor, *Aportaciones Matemáticas. Libro en memoria a Víctor Onieva*, pages 105–115. Universidad de Cantabria, Santander.

Cox, D. R. and Oakes, D. (1984), *Analysis of Survival Data*. Chapman and Hall, London.

Darroch, J. and Ratcliff, D. (1972), Generalized iterative scaling for log-linear models. *Annals of Mathematical Statistics*, 43:1470–1480.

DeGroot, M. H. (1970), *Optimal Statistical Decisions*. McGraw-Hill, New York.

Devroye, L. (1986), *Non-uniform Random Variate Generation*. Springer-Verlag, New York.

Durling, F. C., Owen, D. B., and Drane, J. W. (1970), A New Bivariate Burr Distribution (abstract). *Annals of Mathematical Statistics*, 41:1135.

Dyer, D. (1981), Structural Probability Bounds for the Strong Pareto Law. *Canadian Journal of Statistics*, 9:71–77.

Eaton, M. L. (1983), *Multivariate Statistics*. Wiley, New York.

Fisher, R. A. (1936), The use of Multiple Measurements in Taxonomic Problems. *Annals of Eugenics*, 7:179–188.

Fraser, D. and Streit, F. (1980), A Further Note on the Bivariate Normal Distribution. *Communications in Statistics, Theory and Methods*, 9:1097–1099.

Galambos, J. (1978), *The Asymptotic Theory of Extreme Order Statistics*. Wiley, New York.

Galambos, J. (1987), *The Asymptotic Theory of Extreme Order Statistics*. Robert E. Krieger, Malabar, FL.

Galambos, J. and Kotz, S. (1978), *Characterizations of Probability Distributions*. Lecture Notes in Mathematics 675. Springer-Verlag, Heidelberg.

Gelman, A. and Meng, X. L. (1991), A Note on Bivariate Distributions that are Conditionally Normal. *The American Statistician*, 45:125–126.

Gelman, A. and Speed, T. P. (1993), Characterizing a Joint Probability Distribution by Conditionals. *Journal of the Royal Statistical Society, Ser. B*, 55:185–188.

González-Vega, L. (1998), A Combinatorial Algorithm Solving Some Quantifier Elimination Problems. In B. F. Caviness and J. R. Johnson, editors. *Quantifier Elimination and Cylindrical Algenraic Decomposition*, Texts and Monographs in Symbolic Computation, pages 365–375, Springer-Verlag, New York.

Gourieroux, C. and Montfort, A. (1979), On the Characterization of a Joint Probability Distribution by Conditional Distributions. *Journal of Econometrics*, 10:115–118.

Gumbel, E. J. (1960), Bivariate Exponential Distributions. *Journal of the American Statistical Association*, 55:698–707.

Gumbel, E. J. and Mustafi, C. K. (1967), Some Analytical Properties of Bivariate Extremal Distributions. *Journal of the American Statistical Association*, 62:569–588.

Gupta, R. C. (1998), Reliability Studies of Bivariate Distributions with Pareto Conditionals. Technical report 868, University of Maine, Department of Mathematics, Orono, ME.

Haavelmo, T. (1943), The Statistical Implications of a System of Simultaneous Equations. *Econometrica*, 11:1–12.

Hamedani, G. G. (1992), Bivariate and Multivariate Normal Characterizations: A Brief Survey. *Communications in Statistics, Theory and Methods*, 21:2665–2688.

Hernández, L. D., Moral, S., and Salmerón, A. (1998), A Monte Carlo Algorithm for Probabilistic Propagation Based on Importance Sampling and Stratified Simulation Techniques. *International Journal of Approximate Reasoning*, 18:53–92.

Hobert, J. P. and Casella, G. (1996), The Effect of Improper on Gibbs Sampling in Hierarchical Linear Mixed Models. *Journal of the American Statistical Association*, 91:1461–1473.

Hurlimann, W. (1993), Bivariate Distributions with Diatomic Conditionals and Stop-Loss Transforms of Random Sums. *Statistics and Probability Letters*, 17:329–335.

Hutchinson, T. P. and Lai, C. D. (1990), *Continuous Bivariate Distributions Emphasising Applications*. Rumsby Scientific, Adelaide.

Inaba, T. and Shirahata, S. (1986), Measures of Dependence in Normal Models and Exponential Models by Information Gain. *Biometrika*, 73:345–352.

James, I. R. (1975), Multivariate Distributions which Have Beta Conditional Distributions. *Journal of the American Statistical Association*, 70:681–684.

Joe, H. (1997), *Multivariate Models and Dependence Concepts*. Chapman and Hall, London.

Johnson, N. L., Kotz, S. and Balakrishnan, N. (1994), *Continuous Univariate Distributions*, Vol. 1, 2nd ed. Wiley, New York.

Johnson, N. L., Kotz, S. and Balakrishnan, N. (1995), *Continuous Univariate Distributions*, Vol. 2, 2nd ed. Wiley, New York.

Johnson, N. L., Kotz, S., and Balakrishnan, N. (1997), *Discrete Multivariate Distributions*. Wiley, New York.

Johnson, N. L., Kotz, S., and Kemp, A. W. (1992), *Univariate Discrete Distributions*, 2nd ed. Wiley, New York.

Kagan, A. M., Linnik, Y. U., and Rao, C. R. (1973), *Characterizations Problems in Mathematical Statistics*. Wiley, New York.

Kagan, A. and Wesolowski, J. (1996), Normality via Conditional Normality of Linear Forms. *Statistics and Probability Letters*, 29:229–232.

Korwar, R. M. (1974), On Characterizing Lagragian Poisson and Quasi-Binomial Distributions. *Communications in Statistics, Theory and Methods*, 6:1409–1415.

Kyriakoussis, A. and Papageorgiu, H. (1989), On Characterization of Power Series Distributions by a Marginal Distribution and a Regression Function. *Annals of the Institute of Statistical Mathematics*, 41:671–676.

Lajko, K. (1980), On the Functional Equation $f(x)g(y) = h(ax + by)k(cx + dy)$. *Periodica Mathematica Hungarica*, 11:187–195.

Levi-Civita, T. (1913), Sulle funzioni che ammettono una formula d'addizione del tipo $f(x + y) = \sum_{i=1}^{n} X_i(x)Y_i(y)$. *Atti della Accademia Nazionale dei Lincei, Rendiconti*, 5:181–183.

Lindsey, J. K. (1974), Comparison of Probability Distributions. *Journal of the Royal Statistical Society, Ser. B*, 36:38–47.

Lindsey, J. K. and Mersch, G. (1992), Fitting and Comparing Probability Distributions with Log-Linear Models. *Computational Statistics and Data Analysis*, 13:373–384.

Liu, J. (1996), Discussion of "Statistical inference and Monte Carlo algorithms." *Test*, 5:305–310.

Lwin, T. (1972), Estimating the Tail of the Paretian Law. *Skand. Aktuarietidskrift.*, 55:170–178.

Mardia, K. V. (1962), Multivariate Pareto Distributions. *Annals of Mathematical Statistics*, 33:1008–1015.

Mardia, K. V. (1970), *Families of Bivariate Distributions*. Griffin, London.

Moschopoulos, P. and Staniswalis, J. G. (1994), Estimation Given Conditionals from an Exponential Family. *The American Statistician*, 48:271–275.

Mosteller, F. (1968), Association and Estimation in Contingency Tables. *Journal of the American Statistical Association*, 63:1–28.

Nair, K. R. M. and Nair, N. V. (1988), On Characterizing the Bivariate Exponential and Geometric Distributions. *Annals of the Institute of Statistical Mathematics*, 40:267–271.

Narumi, S. (1923), On the General Forms of Bivariate Frequency Distributions which Are Mathematically Possible when Regression and Variation Are Subjected to Limiting Conditions. I, II. *Biometrika*, 15:77–88, 209–21.

Nerlove, M. and Press, S. J. (1986), Multivariate Log-Linear Probability Models in Econometrics. In Mariano, R., editor, *Advances in Statistical Analysis and Statistical Computing*, pages 117–171. JAI Press, Greenwich, CT.

Nguyen, T. T., Rempala, G., and Wesolowski, J. (1996), Non-Gaussian Measures with Gaussian Structure. *Probability and Mathematical Statistics*, 16:287–298.

Obrechkoff, N. (1963), *Theory of Probability*. Nauka i Izkustvo, Sofia.

Ord, J. K. (1972), *Families of Frequency Distributions*. Griffin, London.

Papageorgiou, H. (1983), On Characterizing some Bivariate Discrete Distributions. *Australian Journal of Statistics*, 25:136–144.

Patil, G. P. (1965), On a Characterization of Multivariate Distribution by a Set of Its Conditional Distributions. In *Handbook of the 35th International Statistical Institute Conference in Belgrade*. International Statistical Institute.

Pickands, J. III (1975), Statistical Inference Using Extreme Order Statistics, *Annals of Statistics*, 3:119-131.

Press, S. J. (1982), *Applied Multivariate Analysis: Using Bayesian and Frequentist Methods of Inference*. Krieger, Melbourne, FL.

Proschan, M. A. and Presnell, B. (1998) Expect the Unexpected from Conditional Expectation. *The American Statistician* 52, 248–252.

Ramachandran, B. and Lau, K. S. (1991), *Functional Equations in Probability Theory*. Academic Press, New York.

Rao, C. R. and Shanbhag, D. N. (1994), *Choquet–Deny-type Functional Equations with Applications to Stochastic Models*. Wiley, New York.

Rao, M. M. (1993), *Conditional Measures and Applications*. Marcel Dekker, New York.

Resnick, S. (1987), *Extreme Values, Regular Variation, and Point Processes*. Springer-Verlag, New York.

Rice, J. A. (1975), *Mathematical Statistics and Data Analysis*. Duxbury, Belmont, CA.

Robert, C. P. (1991), Generalized Inverse Normal Distributions. *Statistics and Probability Letters*, 11, 37–41.

Rubinstein, R. Y. (1981), *Simulation and the Monte Carlo Method*. Wiley, New York.

Salmerón, A. (1998a), Algoritmos de Propagación II. Métodos de Monte Carlo. In Gámez, J. A. and Puerta, J. M., editors, *Sistemas Expertos Probabilísticos*, pages 65–88. Ediciones de la Universidad de Castilla-La Mancha, Cuenca.

Salmerón, A. (1998b), *Precomputación en Grafos de Dependencias Mediante Algoritmos Aproximados*. PhD thesis, Departamento de Matemáticas, Universidad de Granada.

Sarabia, J. M. (1994), Distribuciones Multivariantes con Distribuciones Condicionadas *t* de Student. *Estadística Española*, 36:389–402.

Sarabia, J. M. (1995), The Centered Normal Conditionals Distribution. *Communications in Statistics, Theory and Methods*, 24:2889–2900.

Sarabia, J. M. and Castillo, E. (1991), Extensiones Bivariantes de la Familia Logaritmo-Exponencial. In de Evora, U., editor, *XV Jornadas Luso-Espanholas de Matematica. Vol. IV: Probabilidades, Estatistica e Investigacao Operational*, pages 127–132, Evora, Portugal. Departamento de Matematica.

SenGupta, A. (1995), Optimal Tests in Multivariate Exponential Distribution. In Balakrishnan, N. and Basu, A. P., editors, *The Exponential Distribution. Theory, Methods and Applications*, pages 351–376. Gordon and Breach, New York.

Seshadri, V. and Patil, G. P. (1964), A Characterization of a Bivariate Distribution by the Marginal and the Conditional Distributions of the Same Component. *Annals of the Institute of Statistical Mathematics*, 15:215–221.

Shaked, M. and Shanthikumar, J.G. (1987), Multivariate Hazard Rates and Stochastic Ordering. *Advances in Applied Probability*, 19:123–137.

Silverman, B. W. (1986), *Density Estimation and Data Analysis*. Chapman and Hall, Bristol.

Simiu, E. and Filliben, B. (1975), Structure Analysis of Extreme Winds. Technical report 868, National Bureau of Standards, Washington, DC.

Stephanos, C. (1904), Sur une Categorie d'Equations Fonctionnelles. *Rend. Circ. Mat. Palermo*, 18:360–362.

Stoyanov, J. M. (1987), *Counterexamples in Probability*. Wiley, New York.

Suto, O. (1914), Studies on Some Functional Equations. *Tohuku Mathematics Journal*, 6:1–15.

Takahasi, K. (1965), Note on the Multivariate Burr's Distribution. *Annals of the Institute of Statistical Mathematics*, 17.

Tanner, M. A. (1996), *Tools for Statistical Inference. Methods for the Exploration of Posterior Distributions and Likelihood Functions*. Springer-Verlag, New York.

Tiago de Oliveira, J. (1962), Structure Theory of Bivariate Extremes: Extensions. *Estudos de Math. Est. Econom.*, 7.

Vardi, Y. and Lee, P. (1993), From Image Deblurring to Optimal Investments: Maximum Likelihood Solutions for Positive Linear Inverse Problems. *Journal of the Royal Statistical Society, Ser. B*, 55:569–612.

Wesolowski, J. (1995a), Bivariate Discrete Measures Via a Power Series Conditional Distribution and a Regression Function. *Journal of Multivariate Analysis*, 55:219–229.

Wesolowski, J. (1995b), Bivariate Distributions Via a Pareto Conditional Distribution and a Regression Function. *Annals of the Institute of Statistical Mathematics*, 47:177–183.

Author Index

Subject Index

Springer Series in Statistics

(continued from p. ii)

Springer Series in Statistics